T0298684

STRUCTURAL MECHANICS IN REACTOR TECHNOLOGY

TRANSACTIONS OF THE 9TH INTERNATIONAL CONFERENCE ON
STRUCTURAL MECHANICS IN REACTOR TECHNOLOGY / LAUSANNE
17-21 AUGUST 1987

Structural Mechanics in Reactor Technology

VOLUME F
LWR Pressure Components

Editor
FOLKER H.WITTMANN

Coordinators
DESIDERIUS G.H.LATZKO
BRIAN TOMKINS

CRC Press
Taylor & Francis Group
Boca Raton London New York

CRC Press is an imprint of the
Taylor & Francis Group, an **informa** business

A BALKEMA BOOK

Preface

The area to be covered by Division F can be briefly summarized as follows:

Problems of vibration and fatigue loading under normal conditions and integrity under overload conditions, fluid structural interactions. PWR and BWR blowdown analysis. Stress analysis of components. Piping dynamics including flow induced vibrations and seismic loading. Behaviour of reactor core structures under seismic loading. Analysis and testing of pipe supports and whip restraints. Effect of environment on materials and structures.

In addition, thermal shock and failure aspects pertinent to piping will be dealt with in a number of joint sessions with Division G.

More than fifty papers have been submitted for publication and are included in this volume of Transactions of SMIRT-9. Contributions to a joint session DF are published in Vol. D of the Transactions of SMIRT-9. Those papers which did not arrive in time are included in Vol. A of the Transactions.

Divisions F and G have jointly invited two Principal Lecturers: D.Munz and H.Okamura. They respectively presented papers on Development of Leak-before-Break Methodology, and An Expert-Interaction Algorithm of Structural Integrity Evaluation and its Application to PTS.

Both lectures will be published in a Special Volume of the Transactions of SMIRT-9.

Together with Division D a Panel Session has been organized, i.e. DF 10, Impact of Material Problems on Structural Experience.

Lausanne, April 1987   F.H.Wittmann

Table of contents

Environmentally assisted cracking I: Data evaluation

Blowdown and associated effects: Transient analysis I

Stability of self actuating safety valves in liquid service

P.Coppolani & J.M.Henry
Pump and Valves Department, Framatome, Paris la Défense, France
P.Caumette, J.L.Huet & M.Lott
Commissariat à l'Energie Atomique, St-Paul-lez-Durance, France

1 INTRODUCTION

PWR auxiliary systems and pressurizer mechanical safety valves operating experience shows that liquid discharge often involves high flutter or chatter of the disc inducing valve damages and high amplitude pressure sollicitations. As the failure of a safety valve often leads, on a nuclear plant, to the release of radioactive fluid and to costly repairs, a systematic study has been undertaken by the Commissariat à l'Energie Atomique (C.E.A) and FRAMATOME to understand and find a solution to this problem.

2 SPRING ACTUATED SAFETY VALVE PROBLEM

2.1. Historical background

Some cases of mechanical safety valves instabilities have been observed on nuclear plants and reproduced afterwards on test loops :
- high frequency, high amplitude oscillations of the pressure in the inlet piping and of the stem of pressurizer valves set with a loop seal (AUBLE, 1982).
- valve flutter with binding of the stem blocking the valve in the open position on a heat removal system valve.

Symbols

A	pipe inside area	P_t	set pressure
C	valve damping coefficient		
C_D	valve discharge coefficient	Pa	atmospheric absolute pressure
		Q	volumetric flowrate
Co	sound velocity	S	nozzle area
D	nozzle diameter	z	pipe abscissa
F_H	hydraulic force	$\emptyset$	pipe inside diameter
		λ	friction factor
K	spring stiffness	ρ	specific mass
		β	Hydraulic stiffness
l	valve lift		
lm	maximum valve lift		
Lo	pipe length		
M	valve moving mass		
P	relative pressure		

- Excessive pressure induced vibration causing rupture by fatigue on the lines connected to a reactor volumetric control system valve.

2.2. Valve instability mechanisms

Instabilities occur mainly for valves not fitted directly to the capacity to be protected and generally at low flowrate. Indeed for installations maintenance and safety reasons this cannot always be done. Instabilities occur also when the length to diameter ratio of the capacity to be protected is great.

Safety valve manufacturers usually quote the line flow resistance as the main cause of instability. In fact, in most cases, oscillations occured at very low flows for which the line pressure losses are negligible and this explanation is not sufficient.

We can consider two different types of instabilities :
- static and low frequencies instabilities for which line resistance, inertance and inlet conditions (constant pressure tank, more or less steep head flow pump curve) have some influence.
- dynamic instability which is mainly influenced by the acoustic parameters of the line and the mechanical parameters of the valve.

The first type of instability can be explained by the operation of the valve which results from a balance between the hydraulic force due to pressure and momentum of the fluid on the disc and between the force of the spring. The hydraulic force, as evidenced on experiments, is proportional to pressure differential at the valve. The hydraulic force versus lift curve, the rate of which is the hydraulic stiffness, generally looks like an S curve. Hydraulic stiffness higher than spring stiffness induces statically unstable lift position for constant pressure feeding condition (fig. 1) and combined with line pressure losses and inertia can give low frequency oscillations. However with pump feeding condition this position can become stable and give the surprising result of pressure rising as valve closes (fig. 2).

This type of static or low frequency instability does not induce high sollicitations and therefore, has not been a major concern of our study.

The second type of instability results from an interaction between disc lift and pressure propagation through the line : as the valve opens, stagnation pressure upstream of the valve drops due to sudden expansion giving rise to an expansion wave travelling towards the high pressure vessel through the upstream pipe. As the expansion wave reaches the pipe end connected to the vessel the wave gets reflected as a compression wave travelling back towards the valve. This compression wave opens again the valve closed by the hydraulic force decrease due to pressure drop. This oscillation phenomenon can be amplified for a given combination of pipe acoustic and valve mechanical parameters. As this type of instability induces strong high frequency pressure sollicitations and stem oscillations our study has been focussed on it.

3 NUMERICAL SIMULATION

3.1. Transient computation

3.1.1 Model description
Previous works (SINGH, 1982 ; CATALANI, 1983) led FRAMATOME to use a monodimensional transient model of the line with resolution of the equation of flow by the method of characteristics

The boundary conditions are assigned pressure at the inlet of the line connected to the tank and pressure/flow function at the valve computed from the disc lift by equation 2. Motion of the lift submitted to a hydraulic force F_H is modelled by equation 3.

(1)
$$\begin{cases} \dfrac{1}{A}\dfrac{dQ}{dt} + \dfrac{\varepsilon}{\rho Co}\dfrac{dP}{dt} + \dfrac{\lambda Q\,|Q|}{2\emptyset A2} = 0 \\[2mm] \dfrac{dz}{dt} = \varepsilon\,Co \\[2mm] \varepsilon = \pm 1 \end{cases}$$

(2)
$$Q = Cd\,\pi\,D1\sqrt{\dfrac{2P}{\rho}}$$

(3)
$$M\dfrac{d^2 1}{dt^2} + Cd\dfrac{dl}{dt} + Kl = F_H - P_t \cdot S$$

We used 3 different models of hydraulic force :

1. linear curve with statically stable behaviour (low hydraulic stiffness) (eq. 4)

(4)
$$\begin{cases} F_H = P.S\,(1+\beta 1) \\[2mm] \beta < \dfrac{K}{P_t \cdot S} \end{cases}$$

2. linear curve with statically unstable behaviour (high hydraulic stiffness) (eq. 5)

(5)
$$\begin{cases} F_H = P.S\,(1+\beta 1) \\[2mm] \beta > \dfrac{K}{P_t \cdot S} \end{cases}$$

3. S curve or "3-linear" (eq. 6)

(6)
$$\begin{cases} \beta < \dfrac{K}{P_t \cdot S} \quad \text{for } 0 \leqslant 1 \leqslant 1_1 \\[2mm] \beta > \dfrac{K}{P_t \cdot S} \quad \text{for } 1_1 < 1 \leqslant 1_2 \\[2mm] \beta < \dfrac{K}{P_t \cdot S} \quad \text{for } 1 > 1_2 \end{cases}$$

The calculation results used in this study are the transient valve lift, upstream valve pressure and valve flow.

3.1.2 Parametric studies
Parametric studies were realized to investigate the influence of the parameters we assumed to be the most significant apart from stem friction (stem friction influence can be seen from valve damping sensibility but, in fact, it should be modelled by Coulomb damping) :
   - pipe length,
   - tank pressure time rate and final transient pressure,

- valve hydraulic force,
- valve damping.

For easy understanding of the phenomena we did not reproduce in these studies an experimental test configuration (line constituted with only one pipe diameter).

Parametric studies were made from a basic case shown on table 1 with a linear statically stable hydraulic force model. These studies showed that the valve/pipe system is dynamically unstable particularly at low opening of the valve.

The safety valve must be set very close to the capacity to be protected to eliminate any risk of vibration (see table 2)*

When this is not possible damping must be introduced in the mechanical system of the valve. One way damping at the closure of the valve is sufficient to stabilize the system (see table 3) and allows to keep the valve's characteristics. Pressure time rate parametric studies showed that the valve must remain a very short time at low opening to reduce instability. This means, first, that quick full opening must be prefered, second, that low discharge flow must be avoided. However the modelization of valve with higher hydraulic stiffness and large blowdown did not give stable results in all cases. Therefore valve poppet modifications, such as liquid trims setting, are not useful to eliminate valve/pipe instabilities in all valve/pipe combinations.

3.1.3. Comparison with test results

Some elements make this comparison difficult :
- different time scale between experiment and computation,
- unknown experimental stem friction computed with unadequate model (linear damping),
- upstream transient boundary conditions uneasy to control experimentally.

Taking into account these limitations we can compare numerical results with steady-state vibration experimental results for comparative mean lift and peak to peak lift amplitude.

For this comparison we use the actual hydraulic force and an appropriate damping coefficient to simulate stem friction (peak to peak lift adequation). Numerical valve mean lift is a little higher than the experimental one (3.5 mm compared to 2.0). Pressure peak amplitudes are comparable and frequency responses are very similar.

With the limitations stated above comparison with test results is quite fair and our model can be considered as a good tool for analysis and prediction of valve/pipe behaviour.

3.2. Modal computation

In a complementary work CEA made a direct stability analysis with a linearized model. We look for a time variation in $e^{\alpha t}$ of a small initial perturbation. A positive real part of α will correspond to an unstable mode.

---

* Higher valves of maximum pressure can be seen on this table for some lengths. This is obtained when coupling between valve and flow through the line is maximum. This can explain the experimentally observed instability reduction with longer length.

We shall use the following reduced parameters :

$$p = \frac{P+Pa}{Pa} \quad q = \frac{\sqrt{\rho' Q}}{\sqrt{2}\ Pa\,c_D\ \pi D_{1m}} \qquad Kf = \frac{\lambda Lo}{\phi} \qquad\qquad a = \frac{Klm}{PtS} \quad c = \beta\,lm$$

$$\tau_m^2 = \frac{Mlm}{SPa} \qquad x = \frac{1}{lm} \qquad \sigma = \frac{c_D\ \pi D_{1m}}{A} \qquad \tau = Lo\sqrt{\rho/Pa}$$

$\sim$ refers to parameter perturbation

3.2.1. Rigid column analysis

The set of original and linearized equations are reproduced on table 4, considering the line as a rigid column of water and lead to the system (7).

$$\begin{bmatrix} \tau_m^2\alpha^2 + a1 & a2 & 0 \\ b_1 & b_2 & -2q \\ 0 & 1 & \alpha\tau + 2\sigma^2 q(1+Kf) \end{bmatrix}\begin{bmatrix} \tilde{y} \\ \tilde{p} \\ \tilde{q} \end{bmatrix} = \begin{bmatrix} 0 \\ 0 \\ 0 \end{bmatrix} \qquad (7)$$

The α's are solutions of the third degree equation $\Delta(\alpha) = 0$. One (real) solution corresponds to the time for the flow to be established in the pipe, the other two to the valve vibration. Figure 6 shows α, by its real and imaginary parts, as a function of pipe length Lo through the reduced parameters Kf. The other parameters are :

$p = 10$ $c = 1.$
$\sigma = 0.1$ $\tau = 0.1$ s
$a = 0.9$ $\tau_m^2 = 10^{-4}$ s^2
$x = 0.5$

The figure shows statical instability for Kf < 5

3.2.2 Acoustics

Solving the wave equations in the pipe one is led to define a potential function $\varphi(x, t)$ with :

$$(8) \qquad \tilde{u} = \frac{\partial\varphi}{\partial x} \qquad \tilde{p} = -\rho\left(\frac{\partial\varphi}{\partial t} + \frac{\lambda u}{\phi}\varphi + u\frac{\partial\varphi}{\partial x}\right)$$

where u is the fluid velocity

A solution $\varphi = \psi(z)e^{\alpha t}$ of the resulting equation (9) is sought

$$(9) \qquad \frac{\partial^2\varphi}{\partial t^2} + 2u\frac{\partial^2\varphi}{\partial t\partial x} + (u^2 - c_o^2)\frac{\partial^2\varphi}{\partial x^2} + \frac{\lambda u}{\phi}\left(\frac{\partial\varphi}{\partial t} + u\frac{\partial\varphi}{\partial x}\right) = 0$$

A linear differential equation with constant coefficients is obtained, the characteristic equation of which is eq. 10.

$$(10) \qquad n^2(c_o^2 - u^2) - u\left(2\alpha + \frac{\lambda u}{\phi}\right)n - \alpha\left(\alpha + \frac{\lambda u}{\phi}\right) = 0$$

One finally gets the same system (7) but for the third equation, which now writes :

$$(11) \qquad c2\ \tilde{p} + c3\ \tilde{q} = 0$$

With C2 = $\pi_1 l_0 e^{\pi_1 l_0} - \pi_2 l_0 e^{\pi_2 l_0}$

C3 = $2q\sigma^2 C_2 + \tau\left(\alpha + \frac{2\sigma^2}{\tau} k_p q\right)\left(e^{\pi_1 l_0} - e^{\pi_2 l_0}\right)$

ri solutions of eq. 10

Numerical application for the case mentioned shows that valve vibration, which was attenuated in the rigid column study, is now amplified.

Beside this unstable mode a series of decaying modes are observed (fig. 7) which correspond to the pipe eigen modes. Figure 8 shows valve frequency variation with valve moving mass through the reduced parameter τm^2. Broken line represents the rigid column calculation, solid line the acoustical one. When valve frequency approaches a line frequency it becomes unstable. Figure 9 shows how successive unstable modes appear when valve moving mass is reduced.

4 EXPERIMENTS

4.1. Test rig

Experiments were carried out by the "Commissariat à l'Energie Atomique" (C.E.A) on a test loop named CLAUDIA in Cadarache (C.E.A center near Aix-en-Provence).
The main characteristics of this loop (fig. 10) are the following ones :
- pump rated flow : 140 m3/h
- pump rated head : 480 m
- tank volume : 10 m3

Water tank pressure is controlled by nitrogen blanket compression.
The instrumentation used for the tests is the following one :
- tank and line pressure transducers (Po, P1, P2)
        (bandwidth 0 to 4 KHz)
- ultrasonic flowmeter (Q)
- stem position by proximity probe (l1)
- load cell for hydraulic force discharge tests.

4.2. Valve discharge tests

These tests consist in a pressurization of the loop until set pressure is reached and attempts to keep flow when the valve is opened.
Different tests were carried out under the following conditions :
- two different safety valves : 4 "M6" and 3" K4"
- two inlet lay out configurations :
    . 21 meter   1st pipe length 8.5 m diameter 6"
                   2nd pipe length 12.5 m diameter 4"
    . 10 meter   1st pipe length 8.5 m diameter 6"
                   2nd pipe length 1.5 m diameter 4"
- set pressure : 1.5 to 2.5 MPa
- temperature : ambient
- pressure time rate and flow : variable and controlled by pump discharge valve operation
- use of dampers on the top of the valves.
Without damper both valves vibrate at the opening for both lines lay out configurations. However these vibrations are not systematic with the longer line and the pressure peak amplitudes are less than on the shorter line where they can reach 10 MPa.

8

The reason for instability reduction with the longer line is flow/valve decoupling as it can be seen from numerical analysis.
The reason for non systematic vibration might be non reproducible stem friction.
Spectral analysis of the vibrations shows identical high frequencies on valve lift and pressure (fig. 11). These frequencies seem to vary with valve mean lift as it can be seen on figure 12 where 130 Hz frequency at low mean opening becomes 69 Hz as lift is increased.
With the dampers which realize damping only on the closure way of the valve, whatever the test conditions, high frequency oscillations were eliminated. Low frequency (0.1 Hz) lift oscillations which do not induce high amplitude pressure oscillations were sometimes observed. Damping modify these low frequencies as it can be seen on figure 13 where an opening in the flow paths of the damping device (damping reduction) increases the lift frequency.

4.3. Hydraulic force measurements

Tests were realized to determine the hydraulic force on the valve 3" K4".
These tests consisted in variable pressure discharge through the valve with the stem fixed at a given lift. A load cell measured the difference between hydraulic and spring forces. The hydraulic force to pressure ratio variation with lift is given on fig. 14.
This characteristics is "2-linear" :

$$F_H = P(S + \gamma 1)$$

with $\begin{cases} \gamma = 13.5 \text{ N/bar/mm for } 0 \leqslant 1 \leqslant 5 \text{ mm} \\ \gamma = 9.2 \text{ N/bar/mm for } 1 > 5 \text{ mm} \end{cases}$

This characteristic is very close to spring characteristic when pressure is about set pressure. These tests allowed also to confirm the discharge coefficient of the valve (Cd = 0.64) which can be obtained from the linear lift-flow curve on figure 15.

5 CONCLUSIONS

Dynamic models of self actuating safety valves operating in liquid service have been developped to investigate the stability of the pipe/valve system. Though direct comparison with test results has to be improved general trends obtained with the different test configurations have been confirmed by numerical simulations.
Stability predicted from our models can be considered as conservative since the stem friction is not taken into account.
The best way to improve stability of the safety valves as it can be seen from test results and numerical simulations seems to be damping of the valves. This needs development of damping devices fitted to the valves.

REFERENCES

- AUBLE T.E. Full scale pressurized water reactor safety valve test results ASME Winter Annual Meeting nov. 15-19, 1982

- SINGH A. On the stability of a coupled safety valve piping system ASME Winter Annual Meeting nov. 15-19, 1982
- SINGH A. A model for predicting the performance of spring loaded safety valve ASME Winter Annual Meeting nov. 15-19, 1982.

- CATALANI L. Dynamic stability analysis of spring loaded safety valves
  Elements for improved valve performance through assistance devices.
  7th SMIRT International Conference Aug. 22-26, 1983.

```
                        TOTAL TRANSIENT TIME : 0.5 s
                        Fluid : Ambient temperature water

        Pipe                                        Valve

Length            4.5 m                  Mass                 3.223 Kg
Inside diameter   146 mm                 Spring stiffness 19,6 Kg/mm
Friction factor   0.1
                                         damping coefficient 0 N/(m/s)
Tank boundary condition
                                         Nozzle area          13,2 cm2
Initial pressure    15.5 bar             set pressure         15.5 bar
Pressure time rate   3 bars              maximum lift         10.5 mm
Final pressure      16 bar
```

Table 1 : Transient flow model parametric studies - Basic case

Pipe length m	1.1	1.7	2.3	4.5	6.8	8.9	11.3	15.8	19.8*
Peak to peak lift oscillation mm	0	0	3.5	← full lift →					
Maximum pressure bar	16	16	33	54	48	38	54	47	40
Numerical frequency Hz	-	200	142	100	78	63	100	78	65
Theoretical closed pipe frequency Hz	314	203	150	77	51	39	31	22	17
Note	2.2 mm stable opening	weak transient Oscillatory pressures	Oscillation resorption at 0.3 s	← Steady state vibration →					

Table 2 : Transient flow model parametric studies - pipe length variation

* Computation results shown on fig. 3

PIPE LENGTH : 4.5 m	2 way damping			closure way damping		
Damping Coefficient N/(m/s)	0	1575	4723	1575	4723	15750
Peak to peak lift oscillation mm	full	4.5	0.02	5.1	1.3	0.3
Maximum pressure bar	54	36	16.5	29.3	21	19
Numerical frequency Hz	100	88	87	82	80	88
Theoritical closed pipe frequency Hz	←————— 77 —————→					
Notes	Steady state oscillations		Very weak lift oscillations around 1 mm mean lift	Steady state oscillations	Transient oscillations	

Table 3 : Transient flow model parametric studies damping coefficient variation

	ORIGINAL	LINEARIZED
VALVE MOVEMENT	$\tau_m{}^2 \ddot{x} = \dfrac{P_1 - \sigma^2 x^2}{1 - \sigma^2 x^2}(1+cx) - p_o(1+ax)$	$\tilde{x}(\tau_m{}^2 a^2 + a_1) + \tilde{P}_1 a_2 = 0^{(*)}$
FLOW AT VALVE	$q^2 = \dfrac{x^2}{1-\sigma^2 x^2}(P_1 - 1)$	$\tilde{x}\, b_1 + P_1\, b_2 - 2q\tilde{q} = 0^{(**)}$
WATER MOVEMENT	$P_e - P_1 = \tau\dfrac{dq}{dt} + k_f \sigma^2 q^2$	$\tilde{P}_1 - \tilde{P}_e + (\tau a + 2k_f \sigma^2 q)\tilde{q} = 0$
JUNCTION AT TANK	$P_r = P_e + \sigma^2 q^2$	$\tilde{P}_e + 2\sigma^2 q\tilde{q} = 0$

$$(*) \quad a_1 = P_o a - \frac{1}{(1-\sigma^2 x^2)^2}\left[c\sigma^4 x^4 + c\sigma^2 x^2(P_1 - 3) + 2\sigma^2 x(P_1 - 1) + cP_1 \right]$$

$$a_2 = -\frac{1+cx}{1-\sigma^2 x^2}$$

$$(**) \quad b_1 = 2(P_1 - 1)\frac{x}{(1-\sigma^2 x^2)^2} \qquad b_2 = \frac{x^2}{1-\sigma^2 x^2}$$

TABLE 4 : MODAL COMPUTATIONS : LINEARIZED EQUATIONS

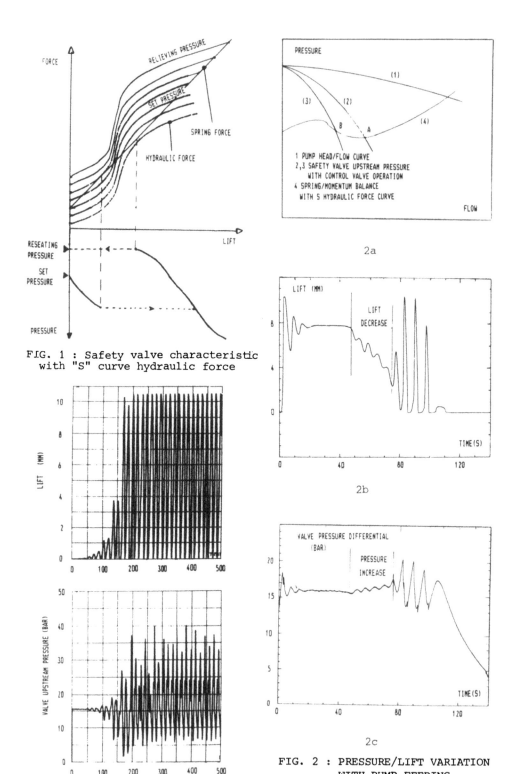

FIG. 1 : Safety valve characteristic
with "S" curve hydraulic force

PRESSURE

(1)
(3) (2)
B A
(4)

1 PUMP HEAD/FLOW CURVE
2,3 SAFETY VALVE UPSTREAM PRESSURE
WITH CONTROL VALVE OPERATION
4 SPRING/MOMENTUM BALANCE
WITH S HYDRAULIC FORCE CURVE

FLOW

2a

LIFT (MM)

LIFT
DECREASE

TIME(S)

2b

VALVE PRESSURE DIFFERENTIAL
(BAR)

PRESSURE
INCREASE

TIME(S)

2c

FIG. 2 : PRESSURE/LIFT VARIATION
WITH PUMP FEEDING

FIG. 3 : TYPICAL COMPUTATION RESULTS

12

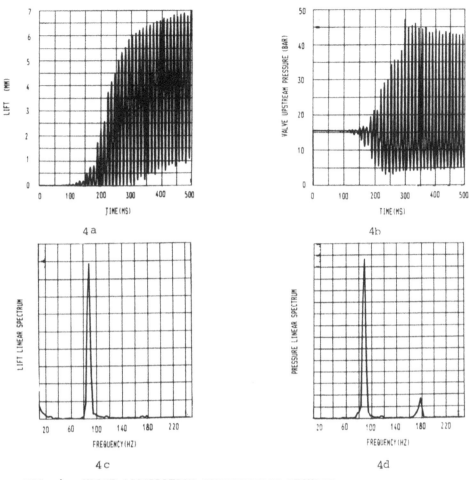

FIG. 4 : VALVE OSCILLATION COMPTUTATION RESULTS

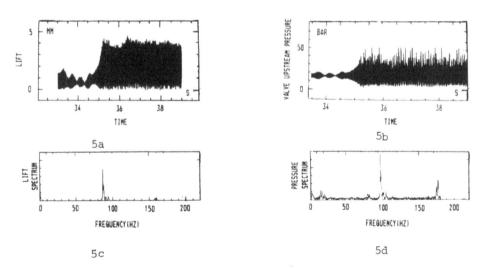

FIG.5 : VALVE OSCILLATION EXPERIMENTAL RESULTS

13

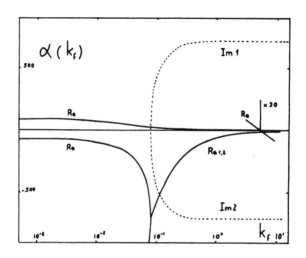

FIGURE 6 : VARIATION OF THE ROOTS IN FUNCTION
OF LINE LENGTH

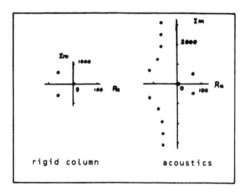

FIGURE 7 : POSITION OF THE ROOTS IN THE
COMPLEX PLANE

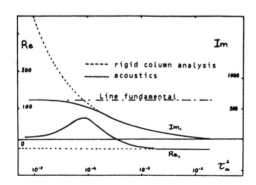

FIGURE 8 :

VARIATION OF THE ROOTS IN
FUNCTION OF VALVE MASS

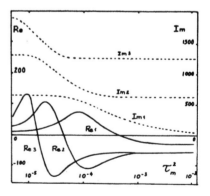

FIGURE 9 :

ACOUSTICAL STUDIES
SUCCESSIVE UNSTABLE MODES

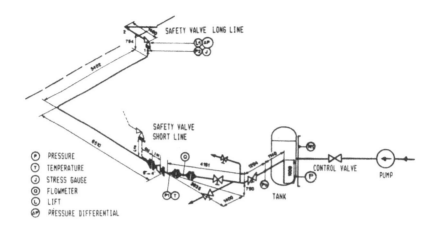

FIGURE 10 : CLAUDIA TEST LOOP

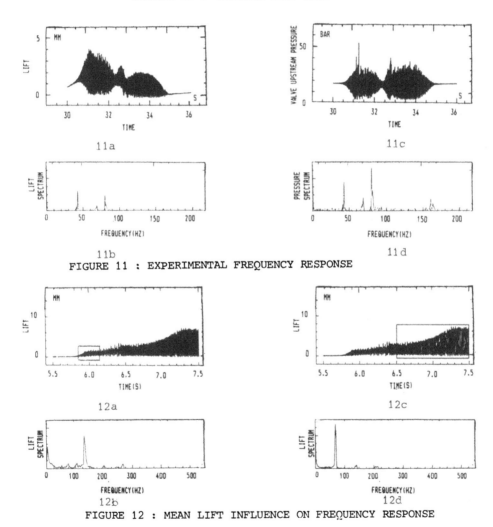

11a

11c

11b

11d

FIGURE 11 : EXPERIMENTAL FREQUENCY RESPONSE

12a

12c

12b

12d

FIGURE 12 : MEAN LIFT INFLUENCE ON FREQUENCY RESPONSE

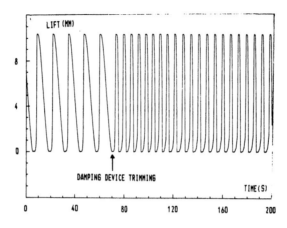

FIGURE 13 : DAMPING INFLUENCE ON LOW
FREQUENCY OSCILLATIONS

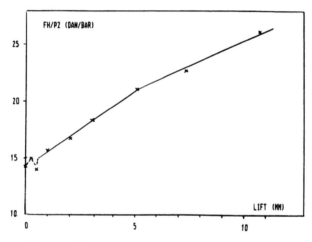

FIGURE 14 : HYDRAULIC FORCE TEST RESULTS

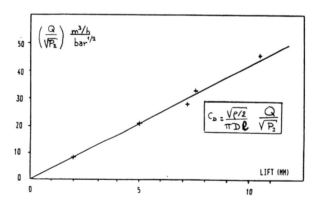

$$C_D = \frac{\sqrt{P/2}}{\pi D \ell} \frac{Q}{\sqrt{P_2}}$$

FIGURE 15 : VALVE FLOW CHARACTERISTICS

16

Analysis and modelling of a swing check valve dynamic behaviour during blowdown experiments

P.Dumazert
Framatome, Paris, France

ABSTRACT

We exhibit in the present paper the phenomena occuring during the closure of a swing check valve and propose a modelling of its dynamic behaviour. Experimental and calculated results match quite well.

1 - INTRODUCTION

The aim of the swing check valve used in PWR lines is to prevent the complete emptying of the line and the capacity which bounds it when the
flow is reversed. This occurs particularly when the line breaks. Unex-
pected large fluctuations may result from the stop of the reversed flow
by the check valve. It appears essential to estimate as precisely as
possible the intensity and the duration of the pressure surges to size
the line in order to avoid a new break between the valve and the capacity.
Experiments on the test facility CLAUDIA (CEA Cadarache, FRANCE) have
been conducted to analyze the closing behaviour of a swing check valve.
This paper deals with these experiments. After applying ourself to point
out the main phenomena, we expose a comparison between test data and
calculation results. The study emphasizes the fluid-disc interaction.

2 - EXPERIMENTAL

In order to understand the phenomena met during the closure of a swing
check valve, we choose and compare, among all the experiments, three tests.

2.1 Experimental apparatus and initial conditions

The figure 1 shows a sketch of the test line.
The expansion results from the burst of the membrane closing the line.
The pressure transducers (points A and B, figure 1) allow us to get in-
formations about the transient. The disc course is known by a displacement
sensor.
More details on the experimental apparatus can be found in (1).
The valve is shown in figure 2.

From one test set as reference (Test 1 : 3 MPa, 20 C, complete opening),
two parameters have been explored :
- reduced opening (test 2 : 3 MPa, 20 C, mid-opening)
- temperature (test 3 : 3 MPa, 190 C, complete opening)
Figures 3 to 5 refer to the first test, figures 6 to 8 to the second and
figures 9 and 10 to the third. Unfortunately, the displacement sensor
failed during the third test.

2.2 Tests analysis

A detailed analysis and a comparison between the three experiments reveal four main phenomena in the dynamic behaviour of the valve.

- head loss effect
- wave effect
- induced flow rate effect
- disc bounce effect

2.2.1 Head loss effect

The head loss coefficient in reversed flow K is measured in permanent flow.
Its curve versus θ (1) has the following form : nearly constant if θ is
greater than fifteen degrees and decreasing in $1/\theta^2$ if not. By comparing
PA and PB for each test, it appears that the head loss effect become significant only during the last degrees of closure. This allows us to use
the "permanent curve" of K in transient flow. As shown by figures 9
and 10, when no wave travels through the valve during the major part
of the transient, the head loss effect is indeed the closing force.

2.2.2 Wave effect and induced flow rate effect (fig. 3,4,5 and 6,7,8)

The initial angular velocity of the disc is more important in the second test (2,7 rd/s) than in the first (0 rd/s). The initial expansion
wave is obviously the same and a greater head loss coefficient in the
second test cannot fully explain such a discrepancy.
In the second test, the initial valve contracted area is smaller than
the pipe section. The expansion, partly reflected by the contracted
area, is higher up-stream than down-stream of the valve and the disc is
accelerated (figure 8). It pushes then in front of it some fluid toward
the break. This added flow rate in the pipe induces the small pressure
surge showing by the figure 6 around 6 ms. This surge reduces the velocity of the disc. When it vanishes, the disc will be accelerated again
and a new pressure surge appears.
The valve disc motion is responsible for a fluid transfer whereby some
fluid has to take the place left by the disc displacement. To cross the
valve, the pipe flow rate splits into the flow rate beneath the disc
and the induced flow rate.
During the braking, the induced flow rate is slowed down leading to a
small compression wave which goes toward the capacity. It superimposes
on the initial expansion the slope of which is smoother down-stream of
the valve. (figure 7)
In the surrounding of the valve, the fluid vena geometry is rapidly modified by the disc angular position and, as for a flexible pipe, the
wave velocity falls down : the reflected compression wave cannot reach
the valve before its closure. In the same way, later in the transient,
this flexible pipe behaviour damps the two effects when a wave impinges

the disc (figures 3,4 and 5).

2.2.3 Disc bounce effect   (fig. 6, 7, 8, 9, 10)

In the last instants before closure, the section of the gap between the
disc and its seat rapidly decreases. The fluid is quickened and the pressure tends to the local saturation pressure. If it is reached (temperature
or no compression wave) the flow beneath the disc becomes choked. To
close, the disc must flatten out the fluid vena. This causes a pressure
increase that opposes the pressure drop.
This forces the disc to stop or to bounce according to the local fluid
compressibility and condensation. If it bounces, the inversion of the
induced flow rate generates down-stream a pressure surge.

In the third test, the flow is choked at 124ms (figure 9). Down-stream
the pressure increases slightly from 125 to 136ms (figure 10). Then the
final pressure surge occurs. Since no bounce pressure surge appears in
figure 10, the disc in the third test stops when the flow is choked.
The delay between the instant of bounce (58ms, figure 8) and the instant
where the surge is in A (62ms, figure 7) backs up the assertion of a
flexible pipe behaviour of the fluid vena. The total closure surge
superimposes on this surge.
The experimental results are summarized in table 1.

3 - MODELLING OF THE DISC BEHAVIOUR

From the test analysis, a model of swing check valve has been developed
to take the three first effects into account. The fourth effect is more
difficult to model and is not presented here. No attempt was made to
model the fluid vena behaviour in the surrouding of the valve. This model is implanted in the ATHIS code (2) which solves for complex piping
systems and for water, vapour or two phase mixture the one dimensional
equations of fluid motion by the method of characteristics.
The model is illustrated by the sketch in figure 12.

3.1 Cross Section

The check valve is represented by a short pipe. Its cross section, the
area let free by the valve disc in the pipe, is a fonction of θ. This
section, constant along the pipe, is obtained   from drawings.

3.2 Induced flow rate and head loss

We suppose that the flow induced by the disc motion is accurately represented by : $\dot{m} = \rho \pi R^2 d\dot{\theta}$
Where ρ is the arithmetical mean of the local fluid density at
each part of the junction :

junction 1 : $\rho = (\rho a + \rho b)/2$
junction 2 : $\rho = (\rho c + \rho d)/2$

At each end of the valve, the two characteristics, the continuity and
the energy equations are solved with the following assumptions :

-The pipe flow rate is the sum of the flow rate through the contracted area of the valve and the induced flow rate :

junction 1 : $\rho_a\, v_a\, A = \rho_b\, v_b\, A_v + \dot{m}$
junction 2 : $\rho_d\, v_d\, A = \rho_c\, v_c\, A_v + \dot{m}$

- The head loss is located at the up-stream end of the valve and calculated with the flow velocity beneath the valve door (3). This allows us to match the experimental determination of the head loss coefficient in reverse flow K (permanent flow).
A mesh point is used to join the ends of the valve by the mean of the usual equations.

3.3 Disc dynamics (figure 12)

If we assume that :
- the centre of pressure and the centre of gravity are identical during the motion,
- the hydrodynamic forces resultant on the disc can be approximate by S (P_a - P_d) where S is the part of the disc area actually fritting in the flow (figure 2) and determined from drawings, the disc movement equation can be written as :
$I\ddot{\theta} = - Mg\, d \sin \theta + S (P_a - P_d)$
I is the total inertia of the valve. It can be set as (4) :

$I = I_o + I_{add}$

I_o is the disc inertia
I_{add} is the added inertia which represents the inertia of the fluid dragged by the disc and depends on the fluid temperature and the length of the pipe representing the valve.

4 - COMPARISON BETWEEN TEST DATA AND CALCULATIONS

The model was run for three tests presented above. The figures 3 to 10 show the results of the calculations which are summarized in table 1.
For the first test, the computed disc course and closure time agree very well with the data. In spite of a mismatch on the wave propagation and a little saturation of the pressure transducer PB, computed intensity and duration of the waterhammer are in good agreement with the measured results.
The absence of the fourth effect in the model and the lack of the fluid area flexible pipe behaviour modelisation is somewhat detrimental to a good match between calculations and experiments for the second and third tests.
The figures 6 to 8 show tha the fluid-disc interaction assumed for the calculations is less strong than in the tests :
- waves generated by the initial disc acceleration exist in the calculations but they are less important than in the test and hardly visible on the figure 7.
- the initial expansion wave slope is not smoothed off by the disc motion and the reflected wave reaches the valve before closure. So the flow beneath the disc cannot be choked.
Nevertheless, outside the bounce zone, the disc course is well computed. It appears clearly that the wave effect is essential when the disc is at rest and that the head loss effect is the closing force when the disc moves.

Since the disc closes earlier in the calculations, the flow rate in the pipe is necessarily lower and the waterhammer less important. The gap between calculated and measured values is less than 10 % for the pressure surge.

In the third test, condensation occurs in P A at 118ms but the initial expansion wave is stronger in the experiment (figure 9). The valve closes at 125ms and the pressure surge reaches point B at 126ms. The calculated waterhammer is less than 10 % smaller than the actual one.

5 - CONCLUSION

The experiments carried out on the test facility CLAUDIA at CEA Cadarache (FRANCE) reveal four main effects during the closure of a swing check valve : the head loss effect, the wave effect, the induced flow rate effect and the bounce effect. A model has been developed to take the three first effects into account. The lack of the fourth effect, more difficult to compute, is obviously detrimental to a good agreement between the test data and the calculated results, but the mismatch remains reasonable : the calculated waterhammer is less than 10 % smaller than the actual one. When the fourth effect does not affect the transient, the model is very near the actual behaviour of the valve disc and the waterhammer very well reproduced.

SYMBOLS

A : pipe cross section	K : head loss coefficient
A_v: valve contracted area	M : disc weight
d : distance between the axis of rotation and the centre of gravity	P : pressure
	R : disc radius
g : gravity acceleration	v : fluid velocity
I : total valve inertia	ρ : fluid density
	θ : disc angle

REFERENCES

HUET, J.L., GARCIA, J.L., COPPOLANI, P. & ZIEGLER, B. 1987 - Experimental and analytical studies on waterhammer generated by the closing of check valves. This Conference - Division F
MONHARDT, D. & ROUSSET, P. Programme de calcul ATHIS, Manuel d'utilisation. Rapport technique FRAMATOME FRATEC 211.
PROVOOST, G.A. 1983. A critical analysis to determine dynamic characterics of non-return valves. 4th Int. Conf. on Pressure Surges
THORLEY, A.R.D. 1983. Dynamic response of check valves. 4th Int. Conf. on Pressure Surges.

Table 1

TEST	CLOSURE TIME (M S)		PRESSURE SURGE IN B (bar)	
	measured	computed	measured	computed
1	121	123	250	257
2	105;58*	57	155 - 160	144
3	-	125	175 - 180	162

* first bounce

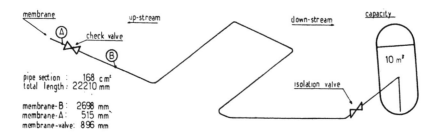

Figure 1

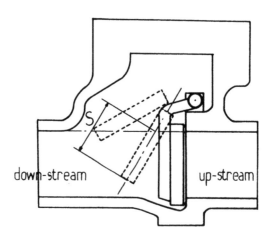

Figure 2

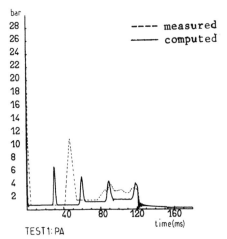

---- measured
—— computed

TEST1: PA

Figure 3

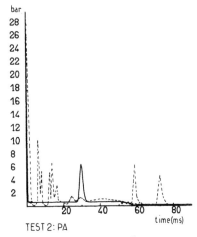

TEST 2: PA

Figure 6

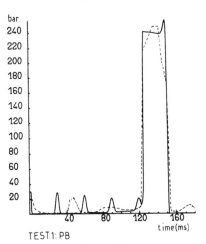

TEST1: PB

Figure 4

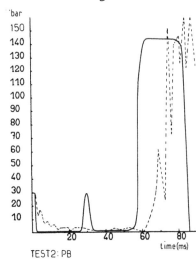

TEST2: PB

Figure 7

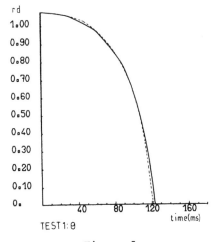

TEST1: θ

Figure 5

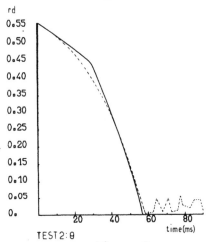

TEST2: θ

Figure 8

23

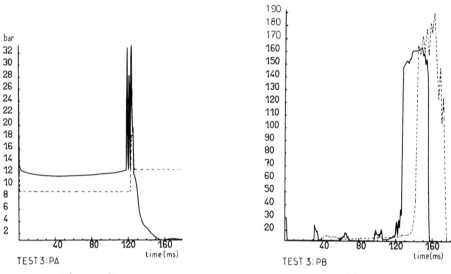

TEST 3: PA

Figure 9

TEST 3: PB

Figure 10

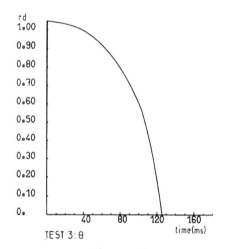

TEST 3: θ

Figure 11

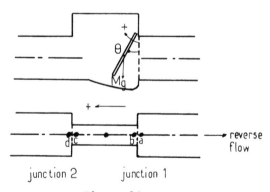

junction 2   junction 1

Figure 12

Experimental and analytical studies on waterhammer generated by the closing of check valves

J.L.Huet & J.L.Garcia
CEA-CEN Cadarache, IRDI/DRE/STRE, St-Paul-lez-Durance, France
P.Coppolani
Framatome, Paris la Défense, France
B.Ziegler
EDF Les Renardières, Ecuelles, France

1 INTRODUCTION

A double-guillotine rupture on a water line upstream from a check valve generates a severe transient between the check valve and the pressure vessel on the downstream side. Successively following phenomena occur :
- decrease then reversal of the flow,
- closing of the check valve with impact of the plug on its seat,
- waterhammer propagating in the pipe downstream from the check valve |1,2,3|.
The COMMISSARIAT A L'ENERGIE ATOMIQUE (C.E.A.) FRAMATOME and ELECTRICITE DE FRANCE (E.D.F.) have undertaken a joint program in order to :
- investigate the behavior of the check valve in the event of a sudden closure,
- evaluate the pressure and flow transient in the line.
The program includes :
- full scale tests in two loops, CLAUDIA (C.E.A.) and ECLAIR (E.D.F.),
- analytical studies in order to qualify the calculation codes
This paper describes the experimental program and presents the analysis results for a benchmark test.

2 TEST DESCRIPTION

2.1 Test facilities

The main characteristics of the two test facilities are given in the following table :

	Maximal pressure (MPa)	Maximal temperature (° C)	Volume (m³)
CLAUDIA	3.5	240	10
ECLAIR	12	25	7

2.2 Test sections

The test sections consist of a U-shaped pipe connected to CLAUDIA or ECLAIR vessel via a fixed pipe supporting the shuttoff valve.
The U-shaped pipe is built-in in two points, weight supports are also used. The test valves are installed at the end of the line (figure 1).
The major test parameters measured include the following :
. pressure at various points in the line and in the reservoir,
. flow rate in the line,
. valve plug displacement,
. pipe skin strain at the built-in supports.

2.3 Experimental procedure

The test section simulates the line downstream from a check valve. The line upstream is limited to a 2 m long pipe closed by a membrane system. The test is performed with no initial water flow in the line, the check valve being held open artificially. When the desired test conditions are obtained the membranes burst and the water flow stream causes the valve to close. This procedure is conservative compared with the actual phenomenon as the flow reversal period is not simulated.

3 TEST PROGRAM

3.1 Test conditions

A large number of tests have been conducted using lift type or swing check valves in 2", 3", 4" and 6" lines. Major test parameters included the following :
. initial valve opening (fully open or partially open),
. water pressure (3 - 12 MPa),
. temperature (room temperature, 130 and 190 °C)
The tests provided a substantial data base covering the valve closing time, waterhammer amplitude and maximum fluid flow velocity.

3.2 Major experimental results

Room temperature tests with an initial pressure p_0 of 3 MPa and the valve fully open resulted in overpressure values of between 6 and 8 p_0, with valve closing times of 65 and 120 ms, respectively. With an initial pressure p_0 of 12 MPa the overpressure reached from 3 to 6 p_0 for closing times of 13 and 52 ms, respectively.
High temperature tests showed that the waterhammer amplitude was reduced from 10 to 30 % depending on the test temperature.
Typical test recordings are shown in figures 2 to 5 for the benchmark configuration used to qualify the computer codes, and for a high temperature test.

4. ANALYSIS

A test involving a 6" swing check valve (figure 6) at a pressure of 3 MPa was analyzed by each of the participating organizations using their own calculation codes.
A complementary test was performed in order to determine the pressure loss coefficient versus valve opening, this result (figure 6) was entered into calculations.

4.1 Calculation codes

4.1.1 ATHIS (FRAMATOME) |4|

The mass, momentum and energy conservation equations are expressed for a homogeneous fluid in transient one-dimensional flow conditions, and are solved by the characteristics method.

4.1.2 PLEXUS (C.E.A.) |5|

PLEXUS is a finite-element code using an explicite algorithm to compute fast transients. It may also be used to analyze fluid-structure interaction by means of an Euler-Lagrange formulation.

4.1.3 GRACOB (E.D.F.) |6|

The mass and momentum conservation equations are solved by the characteristics method.

4.2 Calculation results

The three codes were used to compute the benchmark test configuration with identical hypotheses, notably,
  - the same pressure loss coefficient versus valve opening curve determined experimentally,
  - no allowance for added inertia due to the water.
The valve model used is shown in figure 7. The results are given below and compared with the experimental results.

Calculation and test results

	Maximal pressure (MPa)	Closing time (ms)	Maximal flow rate (m^3/s)
Test	26.0	119	0.27
ATHIS	23.5	99	0.21
PLEXUS	21.0	98	0.22
GRACOB	23.0	100	0.21

It can be seen that the results are in relatively close agreement for the three codes and close to the experimental values.

The maximum deviation between the calculated and experimental pressure peak is 25 %.
Better accuracy could be obtained by refining the valve model and notably by taking account of the added inertia due to the water, and of the relative flow rate to calculate the valve pressure drop. Further calculations were done with ATHIS |4| to confirm these points.

V CONCLUSION

The experimental series provided an extensive data base which was used to qualify the analysis codes and to identify the significant parameters affecting the check valve behavior and the generated waterhammer :
  - opening of the valve and pressure drop law,
  - geometry of the pipe (length, inside diameter),
  - test conditions (pressure, temperature)
The high amplitudes of pressure peaks lead presently the partners to initiate a new joint program on damped check valves.

REFERENCES

Wylie Streeter. Fluid transients. Max Graw Hill Inc New-York (1978)
Provoost (BHRA 1980). The dynamic behavior of non return valves
Provoost (BHRA 1983). A critical analysis to determine dynamic characteristics of non return valve
Dumazert SMIRT 9 LAUSANNE. Analysis and modelling of a swing check valve dynamic behavior during blowdown experiment
Lepareux and al SMIRT 8 BRUXELLES Paper F1 2/1. PLEXUS. A general computer program for the fast dynamic analysis.The case of pipe circuits
Jolas Alvkarleby September 1985. The behavior of a non return valve in a reverse flow

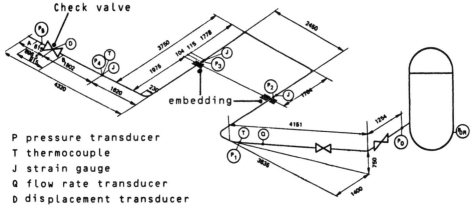

Check valve

P pressure transducer
T thermocouple
J strain gauge
Q flow rate transducer
D displacement transducer

embedding

Figure 1 : Test section (CLAUDIA facility)

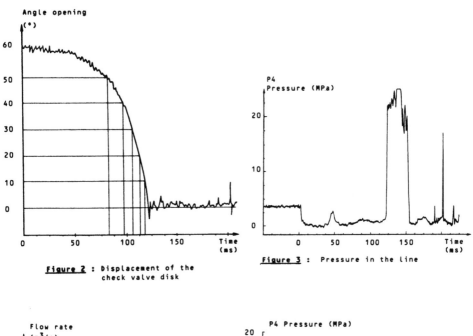

Angle opening
(°)

Figure 2 : Displacement of the
check valve disk

P4
Pressure (MPa)

Figure 3 : Pressure in the line

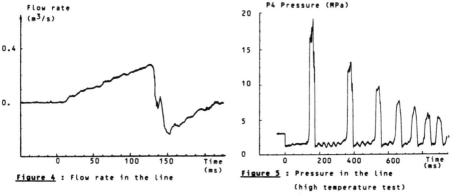

Flow rate
(m³/s)

Figure 4 : Flow rate in the line

P4 Pressure (MPa)

Figure 5 : Pressure in the line
(high temperature test)

29

K pressure drop coefficient

$$\Delta P = \frac{1}{2} K \rho v^2$$

Opening angle (°)	K
60	0.38
50	0.74
30	1.49
15	2.75
10	3.86
6	6.03
4	24.1

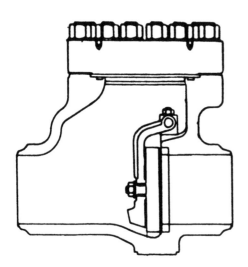

Figure 6 : 6" swing check valve

Motion equation of the disk : $I \dfrac{d^2\alpha}{dt^2} = M$

I   moment of inertia of the disk

α rotation angle of the disk

M   moment of external forces applied to the disk

$M = M_{weight} + M_{pressure} + M_{friction}$

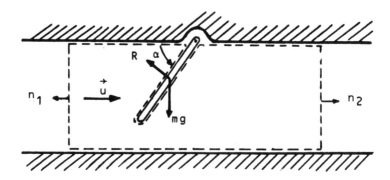

Figure 7 : Check valve model

Uncoupled and coupled analysis of a large HDR pipe

W.Ch.Müller

Gesellschaft für Reaktorsicherheit, Garching, FR Germany

1.   INTRODUCTION

Large scale piping experiments have been carried out recently in Germany at the HDR test facility. The aim of the experiments is to investigate the closing behaviour of a feedwater check valve and the fluid dynamics in its respective pipeline as a result of a simulated break in the pipeline and ensuing rapid closure of the check valve. In connection with these fluid-dynamic investigations the structural dynamic behaviour of the piping system, when subjected to fluid forces, was measured.

GRS is engaged in the analytical evaluation of these tests and is developing advanced methods to predict the measured data for the valve dynamics, the fluiddynamics and the structural dynamic behaviour of the pipeline.

GRS has carried out pre- and posttest calculations of these experiments using the fluiddynamic code DAPSY, which also simulates the dynamic behaviour of a feedwater check valve, and the structural code ADINA.

In this paper it is demonstrated how both codes perform in uncoupled and coupled calculations. The state-of-the-art of transient piping analysis is the uncoupled calculation where first the fluiddynamic behaviour is calculated. The calculated pressure are used to compile load fuctions which are used as input for the structural dynamic analyses. No feedback from the structural behaviour is taken into account.

In contrast to this in a coupled analysis the fluid and structural behaviour are calculated simultaneously. At each time step pressures and displacements are exchanged between the fluid and the structural code. This technique is more precisely and - as will be shown in this paper - gives different results.

2.   DESCRIPTION OF THE EXPERIMENT

The test which is analyzed in this paper is planned at the HDR test facility, which is an old test reactor nowadays used for nuclear safety experiments. It allows to run full scale tests under nuclear reactor conditions. The test pipe is shown in fig. 1, the check valve in fig. 2. The test under consideration is the test T21.4 /1/ which is a pressure transient test designed to lead to plastic response of the pipeline. In the test the pipe system is to be filled with subcooled water (p = 90 bar, T = 220°) and a pipe break is simulated. The decompression wave from the break causes the check valve to close. Since the damping mechanism has been removéd the valve will close rapidly which will result in check valve slam and a series of heavy pressure pulses. Due to these pressure waves the piping structure will undergo strong vibrations.

3. THE FLUID - AND STRUCTURAL DYNAMIC CODES

The code DAPSY /2/ is a one dimensional fluiddynamic code solving the EULER equation for nonequilibrium two phase flow. DAPSY uses a four equation model for the four fundamential variables

- p   pressure
- h   enthalpy
- v   velocity
- α   volumetric steam quality.

The four equations are

- conservation of the water mass
- conservation of the steam mass
- conservation of momentum
- conservation of energy

DAPSY's numerical solution technique is explicit integration and the method of characteristics on a fixed grid.

DAPSY has been developed to predict the pressure waves in nuclear piping system due to a pipe break. The effects of two phase flow are modelled by homogeneous thermodynamical nonequilibrium.

DAPSY has implemented a dynamic valve model to simulate the closing behaviour of a typical self-operating feedwater check valve as used in German nuclear reactors (figure 1).

It is a well-known fact that the elasticity of the pipe wall has a substantial effect on the velocity at which the pressure waves travel in the pipe. For a straight infinite pipe an analytical formula can be derived to account for this effect which has been implemented in DAPSY as a first approximation.

For the structural analysis the computer code ADINA has been used. Most current computer codes for piping analysis use straight or curved beams to model the pipelines. Special features of piping systems like the flexibility of the elbows due to ovalization or stress intensification at elbows or tees are accounted for by the use of special factors.

In contrast to this the ADINA pipe element is more sophisticated /3/. It consists of curved isoparametric beam elements with additional degrees of freedom for the ovalizations. The effect of flanges and internal pressure on the ovalization is taken into account directly. This makes the element superior to the standard pipe element.

Each ADINA pipe element has four nodes and each node has up to 12 degrees of freedom:

- "beam" displacements: 3 translations and 3 rotations

- up to 6 Fourier coefficients for ovalization: up to 3 coefficients for ovalization due to in-plane bending and up to 3 coefficients for ovalization due to out-of-plane bending.

32

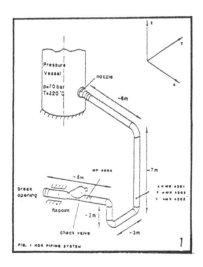

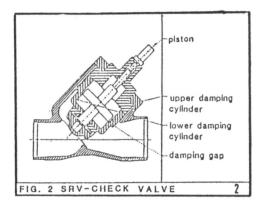

FIG. 2 SRV-CHECK VALVE 2

FIG. 1 HDR PIPING SYSTEM 1

4. THE COUPLED ANALYSIS CODE DAPS

In a typical piping analysis the fluiddynamic analysis and the calculation of the piping response are carried out separately. In this way the effects of fluid-structure-interation (FSI) are neglected though FSI may contribute substantially as will be pointed out later. A second disadvantage of a decoupled analysis is that the analyst has to prepare and compile the load function to be used as input in the structural analysis from the fluid dynamic results "by hand".

For a more advanced analysis coupled methods are being developed by GRS. As a first step the coupled code DAPS has been programmed.

DAPS consists of DAPSY as the fluiddynamic part and uses eigenforms and eigenvalues obtained from ADINA for the structural analysis. The structural response is calculed by explicit time integration in the modal domain. Only the lowest important eigenforms are selected and the pressure loads are transformed to modal loads. By retransformation the nodal displacements are calculated.

In DAPS the fluid and the structure are integrated in a staggered manner and in each time step pressure and volume change of the pipe elements are exchanged. The volume change has been calculated from the dynamic axial elongation and quasistatic radial extension of the pipe element. The static radial extension is obtained by assuming, that radial extension shows no dynamic behaviour i.e. the pressure acts in a quasi-static way.

On the other hand the pressure acts on the pipe element as a nodal reaction force. This technique gives better results at lower computing costs than a direct integration.

5. COMPARISON OF COUPLED AND UNCOUPLED RESULTS

The prediction of both the uncoupled and coupled analysis gives similar fluiddynamic results. The pressures, temperatures, mass flows and the valve dynamics correspond well to the theoretical exspected values and the main fluiddynamic features of the transient are characteristically represented by time function of the pressure upstream of the valve MP 9008. (fig. 3).

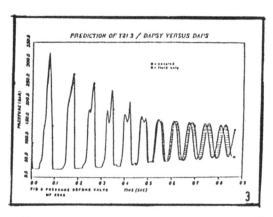

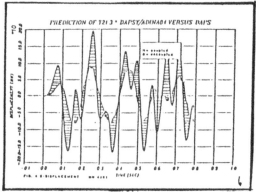

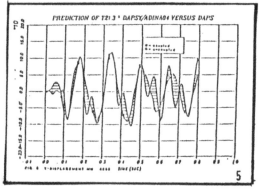

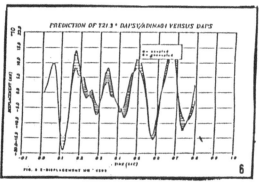

The fig. 3 shows that due to break the pressure drops to saturation values and a pressure wave is travelling from the break to the pressure vessel and back. The mass flow through the valve causes it to close rapidly in 60 msec. This rapid closure stops the flow column and results in a severe pressure pulse of about 320 bar. As a consequence a series of water hammers with decreasing pressure peaks occurs. The first five of these are in the two phase region, but after 500 msec no vaporation occurs between pressure tubes and the occillation stay in the single phase domain (subcooled water).

In the coupled calculation the effect of wall flexibility has been taken into account. The comparison between uncoupled and coupled calculated pressures shows that in the two phase domain the effect of the coupling on the fluiddynamic behaviour is very small. This is true because the compressibility of the two-phase-fluid is the dominant effect. In the single phase domain the effect of the coupling on the velocity of sound can be clearly seen fom fig. 3. This results in a different loading time history for the structure.

The structural response of the piping system to the fluiddynamic loading is a low frequency vibration of the first eigenmodes. As representative data of the structural response of a point 42 (fig. 1) (x-displacement MN4201, the y-displacement MN4202 and the z-displacement MN4203) is given in fig. 4 to fig. 6.

The figures shows that the coupled calculation gives different results than the uncoupled calculation and some typical FSI effects can be recognized.

The presumed tendency "coupled calculation - reduced displacements and strains" does not hold generally. For the x-displacement MN4201 (fig. 4) the coupling eliminates the higher frequency and thereby reduces the amplitudes. For the z-displacement MN4203 (fig. 6) the coupling increases the amplitudes, e.g. the first minimum is increased from - 145 mm to 190 mm.

The fig. 5 and 6 shows that in the course of time the differences grow. This can be expected since the coupling influences the eigenfrequencies. The change in frequency holds for both fluiddynamic and structural dynamic. It is interesting to see that the coupling has a substantial effect on the structural response even in first 500 msec in which the differences in the fluiddynamic are very small.

There are two central effects which influence the structural response

1) The main loads on the structure are the pressure differences between the elbows. That means that the errors in the load function may be large even if the errors in the pressures are small because the differences of the functions can be incorrect.

2) The structural response essentially corresponds to the Duhamel integral. In the integration the high frequencies of the load function are smoothed out.

6.   CONCLUSIONS

Coupled and uncoupled piping transient analyses show similar results for the fluiddynamic data. The differences are less than 10% and as long as the fluid is in the two phase domain they can almost be neglected.

The main differences are in the structural response. There is no clear tendency that a coupled calculation will result in lower amplitudes of the structural response, but it can be seen from the results that there is a typical difference between coupled and uncoupled analysis which increases with time. This increase is mainly due to the fact that in a coupled analysis the speed of sound of the fluid and the eigenmodes of the piping system structure are lower than in the uncoupled analysis.

REFERENCES

/1/   H. Hunger, D. Schrammel
      Strukturverhalten einer Rohrleitung mit Speisewasserrückschlag-
      ventil bei Kühlmittelverluststörfall
      Technischer Fachbericht Nr. 58-86
      PHDR 58-86, September 1986

/2/   T. Grillenberger, W.Ch. Müller, D. Müller-Ecker, S. Meier,
      H. Bartalsky, U. Grzesik
      Dapsy, Ein Programmsystem zur Druckwellenausbreitung im Primär-
      system
      Programmhandbuch-Version DAPSY B/82
      GRS-A-892/I-IV (Dezember 1983)

/3/   K.J. Bathe, C.A. Almeida
      A simple and effective pipe elbow element-linear analysis
      J. Appl. Mech., March 1980, Vol. 47, pp. 93-100

Analytical validation of a numerical solution for a simple fluid-structure interaction case

D.Monhardt

Framatome, Paris, France

1. Introduction

In order to obtain hydraulic loads on lines with check valves during blowdown, the dynamics of the valve disc has to be coupled with the fluid transient. This has been done notably by Dumazert (1987) who compared the numerical results of the ATHIS code to experiment. In order to gain more insight in the modeling aspects of this problem and its numerical solution, a derived fluid-structure interaction (FSI) problem is computed' with this code and compared to the. analytical solution.

2. FSI problem

The problem to be investigated is to compute transient pressures in a fluid filled pipe provided with a flexible bottom and subjected to harmonic pressure condition at the other end. Nomenclature is defined with reference to figure 1. The pipe has length L and area A ; fluid density is ρ and its sound velocity c. The pipe is fitted with a frictionless piston of mass m and stiffness k. A sinusoïdal pressure with amplitude po is imposed at x = 0.

3. Analytical solution

By assuming small amplitudes of disturbance no wall friction and linearizing mass and impulse conservation equations of the fluid, we obtain the linear waterhammer equations

$$\frac{\partial P}{\partial t} + c^2 \frac{\partial Q}{\partial x} = 0$$

$$\frac{\partial Q}{\partial t} + \frac{\partial P}{\partial x} = 0$$

Where Q=ρV and V is the fluid velocity
End conditions are :

- at x = 0 $p(0,t) = P_o \sin\omega t$
- at x = L $\dot{y} = V(L,t)$ where y is the piston displacement

The transient pressure p(x,t) has been shown (Stefik, 1982) to be given by : p(x,t) =

$$P_o\cos(\frac{\omega x}{c}) \sin \omega t + P_o\sin\omega t \frac{(C_1-\omega^2 C_2) \sin(\frac{\omega L}{c}) + c\omega \cos (\frac{\omega L}{c})}{(C_1-\omega^2 C_2) \cos(\frac{\omega L}{c}) - c\omega \sin (\frac{\omega L}{c})} \quad (1)$$

$$+ 2\sum_{n=1}^{\infty}\left[\sin(\frac{Znx}{c})\sin Znt\frac{\omega P_o}{\omega^2-Zn^2}\frac{(C_1-Zn^2C_2)\sin(\frac{ZnL}{c}) + CZn \cos (\frac{ZnL}{c})}{(C_1-Zn^2C_2)\frac{L}{2}\sin(\frac{ZnL}{c})+(L+2C_2)Zncos(\frac{ZnL}{c})+csin\frac{ZnL}{c}}\right]$$

where gravity effects have been neglected and where Zn are solutions of equation :

$$tg (ZnL/c)= \frac{k\,l^2/mc^2 - (ZnL/c)^2}{(\rho\,AL/m)(ZnL/c)} \quad (2)$$

and $C_1 = k/\rho A$ $C_2 = m/\rho A$

4.Numerical solution

The fluid transient analysis code with fluid-structure interaction code ATHIS (Analyse de Transitoires Hydrodynamiques avec Interaction de Structures) has been used for numerical solution. This code solves the one dimensionnal conservation equations of a homogeneous fluid in a network of pipes. Fluid structure coupling is accounted for either by variation of pipe cross section areas deduced from a dynamic model of the pipe wall or by a matching condition between pipe end fluid and piston velocities. (Monhardt,Rousset 1986).

4.1. Fluid equations :

Neglecting friction, gravity and heat transfer effects, and assuming a constant area for the pipes, the conservation equations for the fluid become.

$$\frac{\partial\rho}{\partial t} + \frac{\partial Q}{\partial x} = 0 \qquad\qquad \text{Conservation of mass}$$

$$\frac{\partial Q}{\partial t} + \frac{\partial VQ}{\partial x} + \frac{\partial P}{\partial x} = 0 \qquad\qquad \text{Conservation of momentum}$$

$$\frac{\partial\rho H}{\partial t} + \frac{\partial QH}{\partial x} - \frac{\partial P}{\partial t} = 0 \qquad\qquad \text{Thermal energy equation}$$

Where $H = h+\rho V^2/2$ and h is the specific enthalpy of fluid, given by state equation $h = h(p,\rho)$
Linear combination of the momentum and thermal energy equations yields the mechanical energy equation ; assuming then small disturbances and since there is no thermal effect, there remain two equations together with the definition $c^2 = (\partial P/\partial\rho)s$ wich are then transformed by the Method of Characteristics into following ordinary differential equations:

$$\frac{dQ}{dt} + \frac{1}{c}\frac{dP}{dt} = 0 \quad \text{for} \quad \frac{dx}{dt} = V + c$$

$$\frac{dQ}{dt} - \frac{1}{c}\frac{dP}{dt} = 0 \quad \text{for} \quad \frac{dx}{dt} = V - c$$

(3)

The left hand side equations are then integrated along the characteristic
lines defined by the right hand side equations by explicit, Courant-
Isaacson-Rees (CIR) first order method (Ranganath & Clifton, 1972) to
yield pressure and mass velocity time history at the points of an eule-
rian grid defined by meshing each pipe of the network model (figure 2).

4.2. Piston equations :

Piston motion is computed by solving the dynamic equation
$$m\ddot{y} + ky = P(L,t)A$$
At each time step the hydraulic force is computed, from which acceleration, velocity and displacement are deduced by explicit integration.

4.3. Fluid-piston coupling

Piston velocity is imposed as an end condition to the fluid at x = L
by the relation
$$Q(x,L) = \rho\dot{y} \quad \text{at each time step}$$

5. Results

5.1. Fixed piston case :

In order to assess the mesh size needed for accuracy of solution, a
first parametric study was performed by closing the pipe with a fixed
piston and using several mesh sizes. A good coïncidence between numerical and analytic solutions has been obtained as shown in figures 3a
and 3b with a mesh size to wave length of perturbation ratio $\Delta x/\lambda$ of
1/84, and assuming $\omega = 1000$ rd/s, L = 2.033m, A = 5.45 $10^{-2}m^2$,
$\rho = 10^3 kg/m^3$, c=1512 m/s $\quad p_o = 10^5 Pa$ and $\sigma = 0.98$ (Notice that since
they are too close, analytic and numerical curves are not superposed
but rather shown apart.)
This low $\Delta\dot{x}/\lambda$ ratio can be explained by calculating the radius of
convergence of the numerical method that is used in the code.
Returning to the equation set (3), we may perform a von Neumann type
of analysis ; referring to figure 2 for definitions, we have following
relations :

$$\frac{1}{c}(P_j^{n+1} - P_R) + (Q_j^{n+1} - Q_R) = 0$$

$$\frac{1}{c}(P_j^{n+1} - P_S) - (Q_j^{n+1} - Q_S) = 0$$

$$P_R = P_j^n + \sigma(P_{j-1}^n - P_j^n)$$

$$P_S = P_j^n + \sigma(P_{j+1}^n - P_j^n)$$

and
$$Q_R = Q_j^n + \sigma (Q_{j-1}^n - Q_j^n)$$
$$Q_S = Q_j^n + \sigma (Q_{j+1}^n - Q_j^n)$$
where $\sigma = c\dfrac{\Delta t}{\Delta x}$ is the Courant number

Defining : $U = \binom{P}{Q}$

We deduce
$$U_j^{n+1} = AU_j^n + BU_{j-1}^n + CU_{j+1}^n$$
with
$$A = (1 - \sigma) I$$
$$B = \frac{\sigma}{2} \begin{pmatrix} 1 & -C \\ -1/c & 1 \end{pmatrix}, \quad C = \frac{\sigma}{2} \begin{pmatrix} 1 & -C \\ -1/C & 1 \end{pmatrix}$$

At time $t = n\Delta t$ an harmonic solution U^n computed on the segment $[0, L]$ can be decomposed in a Fourier series ; therefore at point $x = j\Delta x$ of the mesh, we have :

$$U_j^n = \sum_k V_k^n e^{ik(j\Delta x)} \quad \text{where } i^2 = -1 \text{ and } k \text{ is the wave number } 2\pi/\lambda$$

Hence
$$G = A + Be^{-ik\Delta x} + Ce^{ik\Delta x}$$

Diagonalizing the matix G and computing its eigenvalues λ_1, λ_2, we deduce the radius of convergence of the method $r = |\lambda_1| = |\lambda_2|$ which is given by :
$$r^2 = 1 + 2\sigma(1 - \sigma)(\cos k\Delta x - 1)$$
The radius r has been plotted in figure 4 for different values of the Courant number. It can be seen that the first order CIR method is convergent for $\sigma \leqslant 1$. It is non-dissipative for $\sigma = 1$; however, a margin has to be taken against troncature errors, so that in practice, $\sigma < 1$ has to be used. From figure 4 it is apparent that a least-dissipative solution requires the use of a ratio $\Delta x/\lambda$ that is quite small, the more so for smaller Courant numbers.

5.2. Flexible piston case :

Applying the previous procedure, the flexible piston numerical solution has been investigated for different decreasing values of $\Delta x/\lambda$. A good coïncidence is obtained for $\Delta x/\lambda \sim 1/8.4$, assuming following values :
$\omega = 1000$ rd/s, $L = 20.33$m, $A = 1.87210^{-2}$ m^2, $\rho = 10^3$ kg/m^3,
$c = 1512$ m/s, $k = 4.1410^6$ N/m, $m = 116$kg, $p_o = 1$ bar and $\sigma = 0.98$.
Corresponding analytical and numerical solutions are shown in figures 5a and 5b. The result indicates in this case not only that the Courant condition $\sigma < 1$ is sufficient for convergence of the coupled problem, but also that the computation scheme is less dissipative than for the uncoupled case ; this would mean that the radius of convergence of the coupled solution would decrease more slowly with $\Delta x/\lambda$ than the corresponding radius for the uncoupled case.
Indeed, analytic solution shows that when comparing the dominant frequencies for both coupled and uncoupled case obtained by solving equation (2) these frequencies are quite close, and somewhat higher for the coupled case (817,1000 and 1050 rd/s vs. 839,1000 and 1070 rd/s)

A similar result is obtained when instead of increasing Δx compared to the rigid wall case the same value of $\Delta x/\lambda \sim 1/84$ is used, but with a smaller Courant number, $\sigma = 0.91$. Corresponding analytical and numerical results are shown in figures 6a and 6b for $\omega = 300$ rd/s, L = 6.78 m, A = $1.87210^{-2} m^2$, $\rho = 10^3$ kg/m3, c = 1512 m/s, k = 1.24210^7 N/m, m = 38.67 kg and po = 1 bar.

5.3. Non-linear piston

Finally two non-linear cases of piston caracteristics are compared to a linear case in figure 7 and 8. Data are : $\omega = 100$ rd/s, L = 20.33 m, A = $1.87210^{-2}m^2$, k = 4.14106^6 N/m, m = 116 kg, $\rho = 10^3$ kg/m^3, c = 1510m/s, po = 0.69 bar, $\sigma = 0.91$, $\Delta x/\lambda = 1.1910^{-2}$.
The first case corresponds to a geometric non-linearity, for which linear stiffness acts when a gap of 5.10^{-4}m is closed.
The second case corresponds to an ideal, elasto-plastic material, with a yield point at 5.10^{-4}m. Comparing with the linear case, we conclude that non-linearity introduces high frequency components and a lower amplitude in the pressure response for analyzed cases.

SUMMARY AND CONCLUSIONS

The numerical solution of the ATHIS code has been compared sucessfully with analytical solution of a wave propagation-structure interaction problem. Convergence and dissipative properties of the solution have been analyzed ; it was found that in the considered cases, the coupling does not introduce any more restrictive condition on the time step than needed for convergence of the hydraulic problem : rather, dissipative properties are enhanced by the coupling.The code is then used to analyse two simple non-linear problems which shows that non-linearity introduces high frequencies and yield a decrease in pressure response for considered cases. It should finally be noticed that the ATHIS code is not restricted to acoustic conditions nor piston type interaction, but may be used to represent any homogeneous fluid resistive flow with heat transfer and phase change together with more sophisticated wall dynamics, using modal displacement method together with Newmark-β time integration.

Aknowledgements

The author would like to aknowledge the help of Mr G. Fert, who obtained most of the figures shown in this paper.

REFERENCES

P. Dumazert (1987) Analysis and modelling of swing check valve dynamic behaviour during blowdown experiments. This conference, Division F.
L.J. Stefik, Jr. (1982) Fluid/Structure interaction in cylindrical containers. ASME Publication PVP Vol.64.
D. Monhardt, P. Rousset (1986) Programme de calcul ATHIS. Manuel d'utilisation. Rapport technique FRAMATOME FRATEC 211
S. Ranganath and R.J. Clifton (1972). A second order accurate difference method for systems of hyperbolic partial differential equations. Comp. Meth. in Applied Mechanics and Eng. 1, p 173-138.

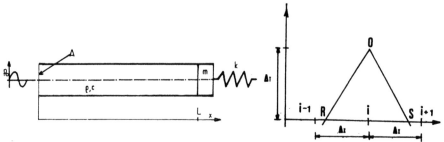

Fig. 1 : 1-D FSI model

Fig. 2 : grid and
characteristics

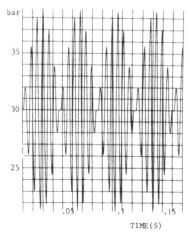

Fig. 3a : pressure at piston (analytic)

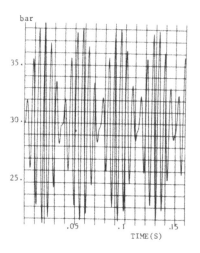

Fig. 3b : pressure at piston (ATHIS)

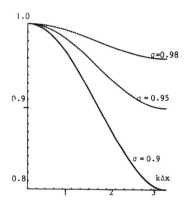

Fig. 4 : spectral radius

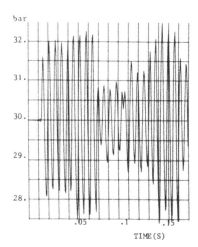

Fig 5a : pressure at piston (analytic

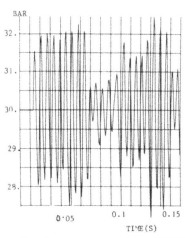

Fig. 5b : pressure at piston (ATHIS)

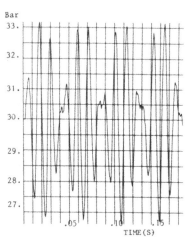

Fig. 6a : pressure at piston
(analytic)

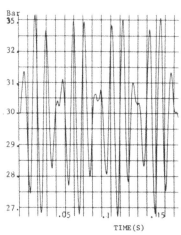

Fig 6b : pressure at piston (ATHIS)

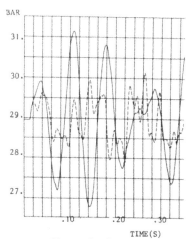

———— linear elastic
----- non linear with gap
Fig.7 : pressure at piston (ATHIS)

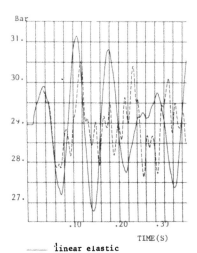

———— linear elastic
----- elastoplastic
Fig.8 : pressure at piston (ATHIS)

43

Some considerations on water-hammer and cavitation following the sudden closure of a check-valve

P.Caumette

CEA-CEN Cadarache, DRE/STRE, St-Paul-lez-Durance, France

M.Lott

CEA-CEN Cadarache, DRE/SCOS, St-Paul-lez-Durance, France

INTRODUCTION

The function of a no-return check-valve is to obturate a pipe when the current in it has a tendancy to revert : in normal operation the valve is maintained in open position by the flow and it closes, in case of flow-rate inversion, under the action of its own weight and of the hydrodynamic forces. At the precise time of closure, the stream most of the time is suddenly interrupted, which determines a water-hammer and the separation of a water column followed by other water-hammers.

We analyze the movement of a swing check-valve before closure and the subsequent evolution of the fluid state.

This study takes its origin in an experimental program described by Huet et al. in a companion paper to this Conference.

1 VALVE MOVEMENT

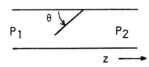

When a liquid flow is concerned, the effect of the valve on the flow is entirely determined by a relation between P_1, P_2 and the velocity u.

From the energy and the momentum balances over the valve one obtains two relations between P_1, P_2, u and θ , and therefore the effect of the valve on the flow, and the valve movement.

1.1 Energy balance

As the valve is energetically involved in the flow, the usual head loss equation corresponding to a fixed obstacle ($\Delta p = K \frac{\rho u^2}{2}$) has to be replaced by the global energy balance :

$$M (h_2 - h_1) + \frac{d E_c}{dt} = 0 .$$

M is the mass flow-rate, h_1 and h_2 specific enthalpies and E_c total valve energy.

$$E_c = \frac{1}{2} I \dot\theta^2 - mgr\sin\theta \qquad I \quad \text{moment of inertia of the moving part}$$

m mass of the moving part

r distance of c.o.m. to axis

$$h_2 - h_1 = \frac{P_2 - P_1}{\rho} + K u |u| .$$

We obtain

$$M \left(\frac{P_2 - P_1}{\rho} + K u |u| \right) + \frac{d E_c}{dt} = 0 . \qquad (1)$$

Head loss coefficient K, function of θ, can be directly measured with clamped valve.

Now it is quite certain that, when the valve accompanies the fluid movement, the head loss is modified so that it is reasonable to replace u in (1) by a relative velocity of the fluid with respect to the valve and thus to write

$$M \left[\frac{P_2 - P_1}{\rho} + K \, \text{sgn} \, (u) \, (u - I \sin\theta \, \dot\theta)^2 \right] + \frac{d E_c}{dt} = 0 ; \qquad (2)$$

I is a length of the order of the pipe radius.

1.2 Valve movement

If C (θ) is the torque applied by the fluid on the valve, one gets the following movement equation :

$$I \ddot\theta = mgr\cos\theta + C .$$

Just as for K, C can be directly measured with clamped valve and naturally put into the form

$$C \, (\theta) = c \, (\theta) \, M_c \, u^2 , \text{ where } M_c \text{ has the dimension of a mass and}$$
c has no dimension.

Like the head loss coefficient, C will be related to the relative velocity so that we shall write

$$I \ddot\theta = mgr\cos\theta + M_c \, c \, (\theta) \, (u - I \sin\theta \, \dot\theta) |u - I \sin\theta\dot\theta| . \qquad (3)$$

Equations (2) and (3) give the effect of the valve on the fluid, and the valve movement.

2 WATER-HAMMER AND OUTSET OF CAVITATION

2.1 Water-hammer

To be specific, we shall assume the valve suddenly interrupts a stationary flow in a pipe with constant cross section and without friction at distance L of a feeding tank with constant pressure P_r.

It is most convenient, to describe the state of the fluid in the pipe, to resort to the method of characteristics.

At each point in the pipe the state (u, p) is determined by the two characteristics C^+ and C^- passing at this point.

The C^+, of slope $u + C_0 \backsim C_0$, carries the invariant $J = u + \dfrac{P}{\rho c}$;

The C^-, " " $u - C_0 \backsim - C_0$, " " " $K = u - \dfrac{P}{\rho c}$.

Initial state is (U_0, P_0). Limiting conditions are, at the reservoir $p = c^t = P_r$, at the valve $u = 0$.

The succession of events is shown on the opposite (z, t) diagram.

From the closure time $(t = 0)$ to the time $\dfrac{2L}{C_0}$,a water-hammer develops in the pipe. The continuation of events, from the time $\dfrac{2L}{C_0}$ on, depends on the value of the pressure at the valve between times $\dfrac{2L}{C_0}$ and $\dfrac{4L}{C_0}$. If that pressure, the value of which is $-(P_0 + \rho_0 C_0 U_0 + 2 P_r$, is greater than P_{sat}, saturation pressure, the evolution proceeds by successive water-hammers of decreasing amplitude, with a period of $\dfrac{4L}{C_0}$. If on the contrary the pressure is lower than P_{sat}, cavitation occurs. It is the common case after a high intensity water-hammer.

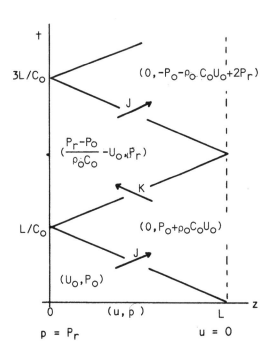

For instance, let us consider an experiment reported by Huet et al. performed with water at 150 °C (P_{sat} = 4.7 bar) and a reservoir pressure of 25 bar.

A water-hammer $(P_0 + \rho C_0 U_0)$ of 150 bar was observed. The subsequent depression would have been $-(P_0 + \rho_0 C_0 U_0) + 2 P_r = - 100$ bar, largely below P_{sat} : there was indeed cavitation.

47

2.2 Self-similar expansion

At time $\frac{2L}{C_o}$ immediately preceding cavitation, the flow is uniform in the pipe. As concerns the immediate vicinity of the valve, everything happens as though the pipe was infinite on the reservoir side.

We assume we shall observe, for the expansion of our liquid, the type of self-similar expansion we would obtain for a gas.

Since the invariant J of the characteristics C^+ is constant throughout the domain, the characteristics C^- are straight lines.

The (z, t) space is divided into a series of simple waves, separated by C^-'s.

We note U_2 the velocity of initial constant state :

$$U_2 = -U_o + \frac{P_r - P_o}{\rho_o C_o}.$$ Since J is a constant, u and p are related by

$$u - U_2 + \int_{P_r}^{P} \frac{dp}{\rho c} = 0.$$

Zone I   initial constant state

$p = P_r$ slope of the right limiting C^- : $U_2 - C_o$

$u = U2$

Zone II   expansion until P_{sat}

$P_r > p > P_{sat}$ slope of limiting C^- : $U_{sat} - C_o$
$U_2 < u < U_{sat}$

$$U_{sat} = U_2 + \frac{P_r - P_{sat}}{\rho_o C_o}$$

Zone III   constant state : saturated liquid

$p = P_{sat}$ slope of limiting C^- : $U_{sat} - C_{do}$
$u = U_{sat}$

C_{do} is the sound velocity in the homogeneous equilibrium model at zero vapor rate.

Zone IV   two-phase expansion till zero velocity

$P_{sat} > p > P_c$ slope of limiting C^- : $- C_d (P_c)$
$U_{sat} < u < 0$

P_c is such that $U_2 = \int_{P_r}^{P_c} \frac{dp}{\rho c}$,

an equation which was solved graphically, using the tables of water properties.

<u>Zone V</u>   constant state adhering to the valve

$p = P_c$ the zone is limited to the right by $z = L$ (motionless
$u = 0$ valve) : it is <u>not</u> a characteristic.

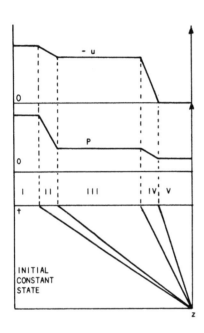

The figure represents, with no attempt to respect the scaling, the fan constituted by the five regions in the (z, t) plane after the time $\frac{2L}{C_0}$ and the corresponding pressure and velocity distributions.

Numerically, one gets for the case mentioned above (after a water-hammer of 150 bar) :

Zone I   p = 25 bar

                u = - 9,4 m/s C_0 = 1450 m/s

Zone II   25 > p > 4,76 bar

                - 9,4 < u < - 7,9 m/s

Zone III   p = 4,76 bar

                u = - 7,9 m/s C_{do} = 5 m/s

Zone IV   4,76 > p > 4,5 bar

                - 7,9 < u < 0 m/s C_d = 11 m/s

Zone V   p = 4,5 bar

                u = 0

In zone V some 0,5 % vapor has formed.

3 MOVEMENT OF THE WATER AFTER COLUMN SEPARATION

For the continuation of the process, one may generally assume that everything goes as though a water column was separating from the valve, leaving behind a cavitating zone at saturation pressure.

3.1 In a __first phase__, up to the time t_1 of standstill due to the coun-
ter-pressure, the column is moving back in the pipe. With no friction,
position y of the front satisfies the following differential equation :

$$P_r - P_{sat} = \Delta P = \rho\, y\, \frac{du}{dt} \qquad\qquad y\,(0) = L$$

$$u = \frac{dy}{dt} \qquad\qquad u\,(0) = U_i = U_{sat}.$$

Considering u as a function of y, we get $\frac{du}{dt} = \frac{du}{dy}\, u$ and   then,   by   a
quadrature,

$$u = \frac{dy}{dt} = -\left[\, U_i^2 + \frac{2\Delta P}{\rho}\, Log\, \frac{y}{L}\,\right]^{1/2}. \qquad\qquad (4)$$

It follows the minimum length L_m corresponding to u = 0, t = t_1 :

$$\frac{L_m}{L} = exp\,\left(-\,\frac{U_i^2\,\rho}{2\Delta P}\,\right).$$

Putting it back into (4), one gets

$$\frac{dy}{dt} = -\left[\,\frac{2\Delta P}{\rho}\, Log\, \frac{y}{L_m}\,\right]^{1/2}$$

which gives, setting $z = \sqrt{\dfrac{2\Delta P}{\rho}}\ \dfrac{t_1}{L_m}$, $x = \dfrac{L}{L_m}$,

$$z = \int_1^x \frac{dx'}{\sqrt{Log\, x'}} = 2\int_0^{\sqrt{Log\, x}} exp\,(u^2)\, du.$$

3.2 In a __second phase__ the column moves forward again, up to the moment
t_2 where it reaches the closed valve. Its movement is given by the
following differential equation :

$$\Delta P = \rho\, y\, \frac{du}{dt} + \rho\, \frac{u^2}{2} \qquad\qquad u\,(t_1) = 0$$

$$u = \frac{dy}{dt} \qquad\qquad y\,(t_1) = L_m.$$

As above, u (y) is obtained by a quadrature :

$$u = \left[\,\frac{2\,\Delta P}{\rho}\,(1 - \frac{L_m}{y})\,\right]^{1/2} ,$$

hence the velocity U_f at impact on the valve :

$$U_f = \left[\,\frac{2\,\Delta P}{\rho}\,(1 - exp\,(-\frac{U_i^2\,\rho}{2\,\Delta P})\,)\,\right]^{1/2}.$$

A second integration gives the movement law and the time t_2 of impact:

$$t_2 - t_1 = (\frac{\rho L_m^2}{2 \Delta P})^{1/2} \left[(\frac{L}{L_m} (\frac{L}{L_m} - 1))^{1/2} + \cosh^{-1}(\frac{L}{L_m})^{1/2} \right].$$

At return time, conditions are again fulfilled for a new water-hammer. Estimating the invariant J as $J = U_f + \frac{P_{sat}}{\rho_o C_o}$, the amplitude of the second water-hammer is obtained by $P_f = P_{sat} + \rho_o C_o U_f$.

CONCLUSION

This paper presents some propositions for a better understanding of water-hammer and induced cavitation.

They still need to be confirmed by a thorough analysis of available experimental data.

REFERENCE

Huet, Garcia et al. Experimental and analytical studies on water-hammer generated by the closing of a check-valve. This Conference.

Structural response of calandria tube to a spontaneous rupture of the pressure tube

P.S.Kundurpi & A.P.Muzumdar

Nuclear Studies and Safety Department, Ontario Hydro, Toronto, Canada

ABSTRACT

The primary function of calandria tubes in a CANDU* pressurized heavy water reactor is to minimize the heat loss from the heat transport system to the cool moderator by providing an insulating gas annulus. Circulation of an inert gas in the annulus also provides an effective leak detection capability. During certain postulated accident scenarios in which the pressure tube ruptures, the surrounding CT is subjected to transient pressure loading due to discharge of the coolant into the annulus. Experimental investigation of the calandria tube response under such a scenario has indicated that the interaction of the two tubes is a complex phenomenon. Some of the main observed structural deformations during the tests are reported in this paper. Theoretical models developed to simulate these observed responses are also reported.

1 INTRODUCTION

In the CANDU reactor the pressure tubes in the reactor core contain the fuel bundles and also form part of the heat transport system boundary. Each pressure tube (PT) is separated from the cool moderator by a calandria tube (CT) and a gas annulus between the two tubes during normal operation. Separation between the two tubes is maintained by means of a set of spacers (either two or four) in the form of coiled springs called the garter springs. These pressure/calandria tube assemblies, called fuel channels (up to 480 in number), pass through the stainless steel calandria vessel in which the heavy water moderator is circulated. A typical arrangement of the fuel channel is shown in Figure 1.

The pressure tubes which are designed as Class 1 pressure vessels are normally expected to show "Leak-before-break" characteristics based on their ductile material behaviour. However, in the licensing analysis a broad range of instantaneous PT breaks are considered. In these scenarios the calandria tube is likely to withstand failure of the pressure tube[1]. In order to understand the structural interaction

* CANada Deuterium Uranium

53

of the two tubes, experiments and modelling have been performed and the results of this work are reported.

2 SEQUENCE OF PT/CT INTERACTION FOLLOWING PT RUPTURE

A review of the experiments[2,3] shows that there are a number of modes in which the ruptured PT and the surrounding CT interact. Some of these interactions occur very early (within 5 ms) after the break while some other modes of CT deformations occur later (up to 500 ms) after rupture.

In the early phase the discharge of the fluid causes a large asymmetric pressure loading on the calandria tube which acts for a short duration and results in impulse loading. The PT is subjected to reaction thrust force. This results in the movement of the tubes in opposite directions culminating in an impact at a location diametrically opposite the break location. Following the impact the CT will be subjected to inextensional deformation resulting in ovaling of the CT. This mode of deformation can be detected by the negative strain on the sides of CT. The asymmetric pressure loading on the CT also results in an impact loading in the transverse direction which can lead to permanent bending of the tube in the direction of the discharging jet. During this phase of pipe movement other minor impacts on the CT due to the ejection of fuel elements and the impact of the ruptured lips of the pressure tube also occur.

After the initial phase the CT will be subjected to uniform hoop stress due to internal pressure. The magnitude of the pressure can exceed the system pressure due to waterhammer type of loading. If the stress level in the CT exceeds the elastic stress limit, the waterhammer pressure is considerably reduced due to the plastic strain in the CT[2]. As the Zr-2 CT material is anisotropic plastic straining of the tube in the hoop direction induces axial compressive strain. Since the CT is fixed at the ends, the axial strain appears as a tensile force in the tube.

This paper describes the modelling of the above phenomena, viz permanent bending, inextensional deformation and the axial compression of the CT. The predicted results from these models are compared with the relevant experimental values.

3 RESPONSE IN THE INITIAL PHASE

The calandria tube response in the initial stage is dominated by the progressive pressure loading on the inner surface of the tube, in which the pressure is non-uniform in both the axial and circumferential direction. The net effect is a transient unbalanced lateral load which results in an impulse loading on the CT due to its short duration. When the impulse delivered is large, the dynamic motion can result in plastic deformation (in bending). To conservatively estimate the bending deflections, the CT is assumed to respond as a beam fixed at its ends. The kinetic energy deposited is assumed to be transformed

into strain energy of the beam loaded at the centre without any
dissipation. The strain energy in the beam will comprise of the
elastic component (which results in subsequent vibration of CT) and the
plastic component. The latter is obtained by assuming the beam to
deflect under a constant load F_y (the yield load), beyond the elastic
limit.

3.1 Estimation of Permanent Bending

The impulse (I) delivered to the CT is obtained by integrating the
pressure transient over area and time. Knowing the impulse delivered
the kinetic energy deposited in the tube is estimated by the relation

$$KE = \frac{I^2}{2m} \tag{1}$$

The plastic displacement in the tube is estimated by the relation

$$d_p = \frac{KE - Ve}{F_y} \tag{2}$$

3.2 Estimation of Inextensional Deformation

The strain gauges placed at the sides of the calandria tube in the
experiment[2] indicated a short duration compressive strain before
indicating the expected tensile strain due to pressurization of the
annulus. A typical strain gauge response is shown in Figure 2. The
magnitude of the compressive strain can be obtained on the basis of the
inextensional deformation of a thin cylindrical shell[4]. The net
force acting on the calandria tube before and after it contacts the
ruptured pressure tube is illustrated in Figure 3. After contact, the
CT exhibits a temporary compressive strain on the outer surface of the
CT at point A.

As the calandria tube is supported at the garter spring locations,
the inextensional deformation is limited to the portion of the CT
between the garter springs. The distribution of the bending moment
around the circumference is given by[4]:

$$M_\theta = \frac{F_1 r}{\Pi L} \sum_{n=2,4,6} \frac{\cos n \theta}{(n^2 - 1)} \tag{3}$$

The bending moment at point A ($\theta = \frac{\Pi}{2}$) is

$$M_{\theta A} = \frac{F_1 r}{\Pi L} \sum_{2,4,6} \frac{(-1)^{n/2}}{(n^2 - 1)} = 0.285 \frac{F_1 r}{\Pi L} \tag{4}$$

The corresponding maximum strain on the surface of the CT due to the
bending stress at point A is:

$$\varepsilon_{\theta A} = \frac{1}{E} \frac{6 M_{\theta A}}{h^2} = 0.54 \frac{F_1 r}{E h^2 L} \tag{5}$$

55

4 SUBSEQUENT RESPONSE

The calandria tube response subsequent to the transient discharge loading phase is governed by the annulus pressurization phase. The magnitude of the pressure rise is dependent on the fluid discharge rate, fluid compressibility and the volume changes resulting from CT strain. Any overpressurization of the annulus is similar to the waterhammer type of loading observed in pipe networks. The methodology developed for calculating the annulus pressure transient taking into account the associated volume changes due to plastic deformation of the CT is reported[2, 3]. Since the Zr-2 CT material is anisotropic any plastic hoop strain also results in axial shortening[5] as described below.

4.1 Anisotropic Deformation of the Calandria Tube

4.1.1 Basic Equations of Plasticity Theory

Based on Hill's theory of anisotropic plasticity[6], the yield criterion for the case of a three dimensional stress state, in the three principal stress directions is given as:

$$F(\sigma_y - \sigma_z)^2 + G(\sigma_z - \sigma_x)^2 + H(\sigma_x - \sigma_y)^2 = 1 \qquad (6)$$

The anisotropy constants F, G and H are related to the uniaxial yield stresses in principal directions by

$$G + H = \frac{1}{x^2}, \quad H + F = \frac{1}{y^2}, \quad F + G = \frac{1}{z^2} \qquad \begin{matrix}(7)\\(a-c)\end{matrix}$$

The plastic strain increments de in the principal directions are related to the principal stresses as:

$$de_x = C\,[H(\sigma_x - \sigma_y) + G(\sigma_x - \sigma_z)]$$

$$de_y = C\,[F(\sigma_y - \sigma_z) + H(\sigma_y - \sigma_x)] \qquad \begin{matrix}(8)\\(a-c)\end{matrix}$$

$$de_z = C\,[G(\sigma_z - \sigma_x) + F(\sigma_z - \sigma_y)]$$

4.1.2 Application to Calandria Tube

For a tube under internal pressure, with fixed ends (Figure 4) the radial stress σ_z is approximately zero. Hence, the yield criterion and the incremental strains for the calandria tube are:

$$F\,\sigma_y^2 + G\,\sigma_x^2 + H(\sigma_x - \sigma_y)^2 = 1 \qquad \begin{matrix}(9)\\(a-c)\end{matrix}$$

$$de_x = C[H(\sigma_x - \sigma_y) + G\,\sigma_x]$$

$$de_y = C[F\sigma_y + H(\sigma_y - \sigma_x)]$$

56

Defining the ratio of anisotropy constants as:

$$\frac{H}{F} = R \qquad \text{and} \qquad \frac{H}{G} = P \tag{10}$$

the yield criterion and the yield stresses in Equation (6,7) can be written in terms of R, P, and H as:

$$\frac{\sigma_y^2}{R} + \frac{\sigma_x^2}{P} + (\sigma_x - \sigma_y)^2 = \frac{1}{H} \tag{11}$$

$$\frac{1}{x^2} = H(1 + \frac{1}{P}) \qquad \text{and} \qquad \frac{1}{y^2} = \frac{H(1 + \frac{1}{R})}{R} \tag{12 (a,b)}$$

When the two ends of the tube are fixed, the axial plastic strain increment (de_y) will be zero. Hence, from Equation (9 and 10) we have:

i.e., $\qquad \sigma_y = \dfrac{R \ \sigma_x}{1 + R}$ \hfill (13)

4.2 Estimation of Axial Force

The maximum axial force, F_a, during plastic deformation is obtained as:

$$F_a = \sigma_y \ 2\Pi \ rh \tag{14}$$

Using Equation (13) and substituting for hoop stress σ_x (= $\frac{P_i r}{h}$) the axial force is given by

$$F_a = 2\pi r^2 \ (\frac{R}{R+1}) \ P_i \tag{15}$$

To calculate the maximum axial force developed in the calandria tube from Equation (15) only the value of anisotropy constant R is required. The variation of maximum axial load F_a for various values of internal pressure P_i and anisotropy parameter R are shown in Figure 5.

4.3 Estimation of Permanent Plastic Axial Strain

An estimate of the axial shortening is obtained by considering the free axial deformation that is prevented by the end restraint. The free axial deformation is obtained by considering the anisotropic deformation of a closed ended free tube due to internal pressure.

The relation between the two principal stresses for this case is obtained by equilibrium consideration as:

$$\sigma_y = \frac{\sigma_x}{2} \tag{16}$$

57

The ratio of axial strain to the hoop strain is obtained from
Equation (9) and (16) as:

$$\frac{de_y}{de_x} = \frac{\frac{1}{R} - 1}{1 + \frac{2}{P}} \qquad (17)$$

Using this equation the calculated ratio of plastic strain in the
axial direction to that in the hoop direction is given in Table 3 for
various values of R and P.

5.0 COMPARISON WITH EXPERIMENTS

5.1 Comparison of Inextensional Deformation

The measured values of permanent bending deformations and the
inextensional deformation strain in the experiments are used to
validate the models developed for predicting the initial response.
From the measured plastic deflection in these tests the kinetic energy
imparted to the calandria tube is estimated from Equation (2). The
elastic strain energy associated with the bending of the pressure tube
along with the CT is also accounted for in the calculation of total
energy. Knowing the mass of the calandria tube the corresponding
impulse (I) delivered is calculated by using Equation (1). From the
pressure measurements and accelerometer readings the unbalanced load is
estimated to last about 2 ms (Δt) in all the tests. Based on this
time duration, the peak force (F_1) required to deliver the calculated
impulse is estimated as $I/\Delta t$. Knowing the value of force (F_1) the
expected inextensional strain $\epsilon_{\theta A}$ is calculated using
Equation (5). The estimated value of inextensional strain is seen to
be in reasonable agreement with the measured strains (see Table 1).

5.2 Comparison of Predicted Axial Forces and Strain Ratios

The measured axial force in the fixed-end tests with Zircaloy-2
calandria tubes are shown in Table 2 along with the measured plastic
strains and the estimated temperature increase in the CT during the
pressure transient. To compare these measured axial forces with those
predicted by the theory, it is necessary to correct these for the
thermal expansion of the calandria tube. The corrected values of the
axial force corresponding to the peak annulus pressure in each test are
shown in Figure 5. These results indicate good agreement between the
predicted and measured axial loads when the value of R equals 3 to 4.
Similarly the ratio of plastic axial strain to hoop strain measured in
Tests 7 to 9 agrees closely with the value predicted by Equation (17)
when the value of P equals to 2 to 3 and R equals 3 to 4 (Table 2).

6.0 CONCLUSIONS

Based on the observed deformation of the calandria tube in the full-scale tests performed, the response of the tube has been shown to be separable into two distinct stages. In the initial discharge stage (0-5 m/cs sec), the CT is subjected to a non-uniform pressure distribution which results in an impulse loading causing a permanent bending and inextensional deformation of the CT. In the later annulus pressurization stage observed axial loads in the CT and the axial shortening are shown to be the result of the anisotropic plastic deformation of the CT. The anisotropy factors R and P are estimated to be between 3 to 4 and 2 to 3 respectively for Zircaloy-2 calandria tube.

7.0 ACKNOWLEDGEMENTS

The Full-Scale Pressure Tube Rupture Program is funded via the COG-CANDEV agreement. The authors would like to acknowledge the efforts of the staff at Westinghouse Canada Inc. for performing the experiments.

REFERENCES

1.   Mosey, D., "Sudden pressure tube failure in Pickering Nuclear Generating Station unit 2", Nuclear Safety, Volume 26, No. 3, May-June 1985.

2.   Muzumdar, A.P., Hadallar, G.I., and Chase, R., "Experimental program to determine the consequences of pressure tube rupture in reactors". Paper presented at the International ANS/ENS Topical Meeting on "Thermal Reactor Safety", San Diego, California, February 1986.

3.   Muzumdar, A.P., Presley, J.K., and Kwee, M., "Simulation of CANDU pressure tube rupture experiments". Paper presented at the International ANS/ENS Topical Meeting on "Thermal Reactor Safety", San Diego, California, February 1986.

4.   Timoshenko, S. and Woinosky - Krieger, S. "Theory of plates and shells" McGraw-Hill Book Co. 1959.

5.   Mehan, R.L. "Effect of combined stress on yield and fracture behaviour of Zircaloy 2" J. of Basic Engineering, December 1961, p 499-512.

6.   Hill, R., "Mathematical theory of plasticity", Oxford University Press, 1971.

Table 1. Summary of test conditions and results from the full scale pressure tube rupture tests

Test No		1	2	3	4	5	6	7	8	9
Test Conditions										
Supply tank Pressure	MPa	11.6	9.2	8.2	10.5	7.5	8.5	8.5	8.6	7.7
Peak Annulus Pressure	MPa ~	22	9.05	7.6	9.0	9.8	8.7	9.7	9.3	8.5
Fluid Temperature	°C	290	301	295	307	255	256	251	271	251
Calandria Tube		5 mm thick stainless steel		2.5 mm thick stainless steel			prototype CANDU Zr-2 thickness 1.4 mm (tests 7,8,9 with fixed ends)			
Results										
Permanent Sag in the CT	mm	16	4.75*	13.7	N/A	9.6	15	8.2	11.8	N/A
Max Inextensional Strain on CT		740	1700	1500	N/A	500**	5600	4500	N/A	N/A
Calculated strain		735	–	1370	–	1281	4300†	4300†	–	–

N/A – Not Available
* – Value influenced by strain-hardening of tube from previous test
† – Estimated strain at limit load in inextensional bending
** – Value Suspect

Table 2. Test results and comparison for fixed ended Zr-2 calandria tube

	Test No.		
	7	8	9
CT temperature increase °C	100	100	100
Average plastic hoop strain % - ε_θ	0.75	0.75	0.60
Plastic axial strain ε_a%	-0.3	-0.5*	-0.2
Ratio of strain $\varepsilon_a/\varepsilon_\theta$	-0.4	-0.66	-0.30
Measured Net axial load kN	195	163	158

* Value suspected to be too large due to local strain asymmetry.

Table 3. Ratio of axial strain to circumferential strain for various anisotropy values

↓R/P→	2	3	4	5
2	-0.25	-0.3	-0.33	-0.355
3	-0.33	-0.4	-0.44	-0.476
4	-0.375	-0.45	-0.5	-0.536
5	-0.4	-0.48	-0.533	-0.571

Table 4. Notation

C	-	Constant defined in Equation 8
de	-	Incremental plastic strains (Equation 8)
dp	-	Permanent plastic bending displacement of the CT
e_a	-	Plastic axial strains
E	-	Youngs modules
F, G, H	-	Anisotropy constants in Equation (6)
F_1	-	The lateral load (see Figure 3)
F_y	-	The yield load of the CT
F_a	-	The axial force in the CT
h	-	Thickness of CT
I	-	Impulse delivered to the CT
KE	-	Kinetic energy
L	-	Length of CT between garter springs
m	-	Mass of the calandria tube
M_θ	-	Meridional bending moment
R, P	-	Ratio of anisotropy constants (Equation 10)
P_i	-	Internal pressure
r	-	Radius of CT
x, y, z	-	Co-ordinate directions (Figure 4)
X, Y, Z	-	yield stress in co-ordinate direction
V_e	-	Elastic strain energy of bending up to first yield.
ϵ_θ	-	Strain on the CT (Equation 5)
σ_x, σ_y, σ_z	-	Stresses in co-ordinate direction (Figure 4)
α	-	Coefficient of thermal expansion
ΔT	-	Temperature increment

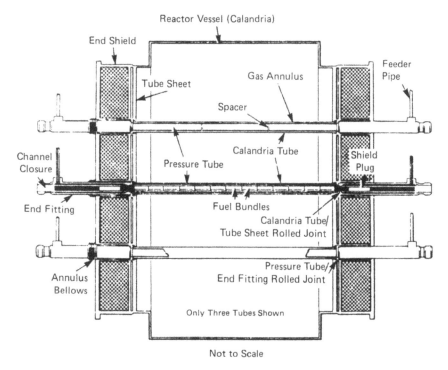

FIGURE 1
Schematic of CANDU Fuel Channels

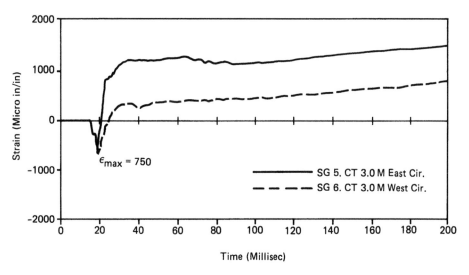

FIGURE 2
Strain Variation at the Side of the Calandria Tube in Pressure Tube Burst Tests

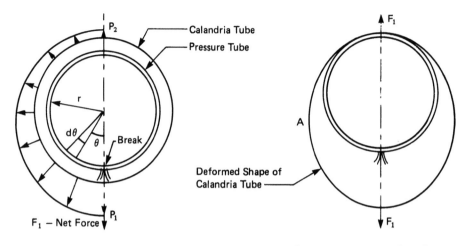

(a) Before Pressure/Calandria Tube Contact   (b) After Pressure and Calandria Tube Contact

FIGURE 3
Pressure Variation on Inside Surface of Calandria Tube

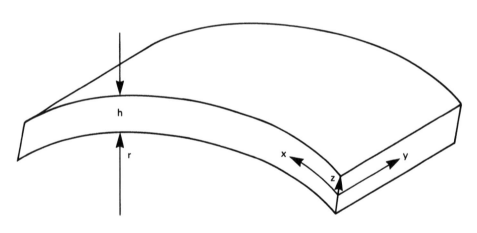

x, y and z	—	the Principal Stress Directions
x, y and z	—	Yield Stress in the Principal Stress Direction
σx, σy and σz	—	Principal Stresses

FIGURE 4
Geometry and the Principal Direction of a Tube Under Internal Pressure

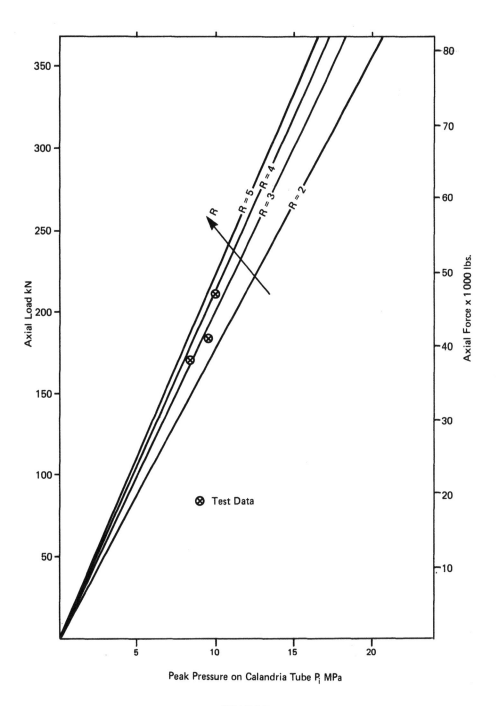

FIGURE 5
Axial Forces Developed in Calandria Tube

65

Blowdown and associated effects: Transient analysis II

Dynamic response of PWR vessel during a blowdown

D.Guilbaud

CEA-CEN Saclay, IRDI/DEMT/SMTS, Gif-sur-Yvette, France

1 INTRODUCTION

This paper presents a study of the behaviour of a 1300 PWR structures during the acoustical period of a blowdown accident, caused by a sudden break of the cold pipe at the outlet of the pump. The calculation gives displacements and pressures for many points of the vessel. Besides it gives the reaction forces in internal supports and also the loading level reached in the support device of the vessel.

This calculation was performed by the computer code TRISTANA, which uses a substructuration method on a modal basis (ref.1). This simplified method reduces computing size and time for tridimensional problems involving fluid structure interaction and sometimes localized non-linearities. It has already been validated during the calculation of the HDR blowdown (ref.2).

2 BASIC HYPOTHESIS OF THE CALCULATION

The decompression wave generated by the sudden break propagates along the discharge pipe and spreads into the downcomer. In this area, to simplify, the wave propagation is assumed to be hardly affected by the other loops. Therefore, only the vessel and the broken pipe are studied.

In this three dimensional problem, it is interesting to take advantage of the simplicity of each component: the broken pipe is a 1D structure and the vessel is an axisymmetrical one. Also it is worth while employing a sub-structuration procedure.

Moreover, in the acoustical phase of the LOCA, the internal movements are small and the structures are assumed to remain elastic. The fluid is still monophasic almost everywhere in the circuit and the flow has just started at the broken pipe. At this step, non-linearities, if they occur, are expected only at a few locations, for instance at the outlet pipe. So modal analysis can be used to represent the movement of each component.

In this study, the substructures are chosen link-free. At the connections, the substructures are isolated by the following boundary conditions:
- for mechanical links: $\vec{F} = 0$,
- for fluid links : $\overrightarrow{\text{grad}}\, p.\vec{n} = 0$

69

(no mass flow rate through the connection area).
So, the symmetry of the vessel can be preserved.

3 MODELIZATION

a) the discharge pipe:

As the pipe is concerned, only the movement of the fluid is represented by a set of acoustical modes of the closed pipe. The following
criteria is used to truncate the modal basis: the wave front must be
well represented by the modal basis. So the lower eigen period must
be about a quarter of the break opening time (the greatest eigenfrequency is closed to 4000 Hz).

b) the vessel and its internals

The vessel and the core barrel are modelized with axi-symmetrical
shells. The core is simply modelized by its upper and lower plates.
The set of fuel assemblies is represented by added mass. An analogous
procedure is used for the upper internals. Moreover the columns join-
ing the upper core plate and the upper support plate are reproduced by
an equivalent spring. The aim of this modelization is to reproduce
the transverse movements of the core barrel which are excited by the
decompression wave in priority.
   For the fluid we use the mesh drawn figure 3. Let us remark that
the set of fuel assemblies are represented by a rigid shell. The
fluid contained inside the core is modelized by an annular volume
located between this rigid shell and core barrel. This modelization
is sufficient to represent the plane wave propagation in the core. A
set of acoustical-mechanical coupled modes have been computed and fitted with some results of vibration tests performed in Paluel nuclear
power plant. (This procedure allows us to estimate the rigidity of
the hold-down spring which is difficult to calculate accurately).
   Due to the core barrel elasticity, the wave front is smooth in the
downcomer. Then, only the modes with eigenfrequency lower than 200 Hz
are used in the computation.

c) Fluid connection

Two fluid connections are employed:
- the first one, to connect the downcomer and the broken pipe,
- the second one, to introduce an acoustical impedance which controls
  the wave reflexion on the two phase zone localized at the break.

d) Mechanical connections

The upper core plate supports and the core barrel radial support are
modelized by mechanical links. During the blowdown, the gap between
the structures are neglected, so the links are permanent ones.

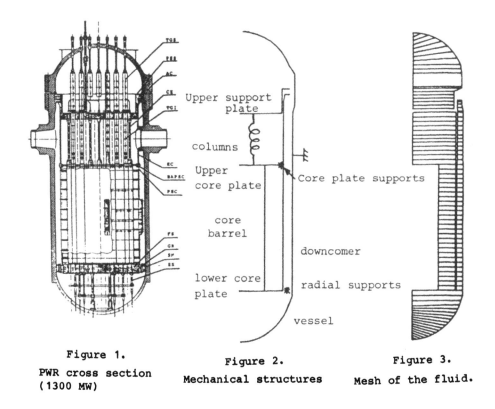

Figure 1.
PWR cross section
(1300 MW)

Figure 2.
Mechanical structures

Figure 3.
Mesh of the fluid.

e) Break representation

An acoustical impedance Z is used to simulate the wave reflexion on
the two phase zone localized at the break. Moreover, this impedance
allows to control the mass flow rate at the break. It can be seen
that the mass flow rate reaches a maximum value. So, a simple way to
adjust this impedance is to choose the one which gets back this maximum value. A thermohydraulic calculation has been performed with the
PLEXUS code. In this computation, only the fluid is concerned. It is
chosen homogeneous and in thermoequilibrium. The critical flow is
given by the Moody model.

f) Loading

The break creates:
- on the fluid outlet pipe:
  an acoustical pressure source such that the pressure inside the pipe
  reaches the saturation pressure level,
- on the broken pipe section:
  a force due to the tension release in the pipe. The resulting
  efforts are applied on the junction with the vessel.
  The loading time history is assumed to be a simple step reached in
one millisecond.

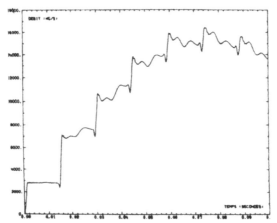

Figure 4. Mass flow rate at the break.

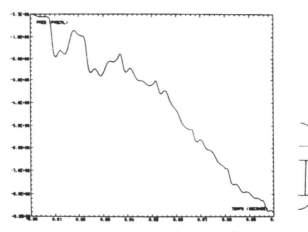

Figure 5. Pressure at the junction between
the vessel and the broken pipe.

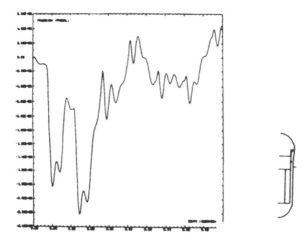

Figure 6. Differential pressure across core
barrel in front of the junction.

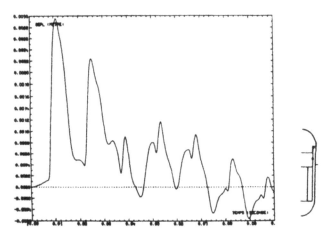

Figure 7. Radial force on a upper core plate
support (90°).

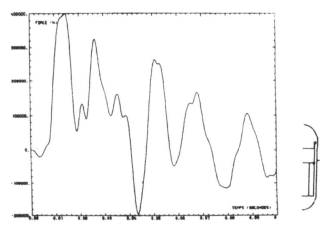

Figure 8. Radial displacement of the core
barrel in front of the junction.

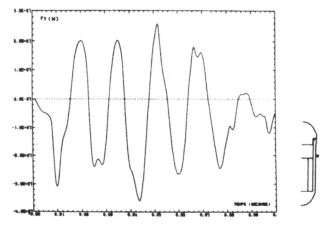

Figure 9. Resulting force on the vessel
support in the y axis.

73

4 DISCUSSION OF THE BLOWDOWN CALCULATION RESULTS

The maximum mass flow rate reached is closed to 16,000 kg/s, the value given by the Plexus code (fig. 4).

The pressure decreases globally in the downcomer (fig. 5) with a characteristic time associated with the resonance frequency of the Helmoltz system: vessel internal volume/blowdown pipe.

When the decompression wave arrives in the downcomer the pressure starts to decrease (20 10^5 Pa).

When the pressure acting in the core barrel decreases, the core barrel starts moving towards the blowdown nozzle and the shell modes are excited.

The largest displacements are those of the core barrel in the neighbourhood of the blowdown nozzle (fig. 7). The maximum displacement is about 3 mm.

The transverse movement of the core barrel creates reaction forces on the support.

The maximum efforts observed on the upper core plate supports are tangential ones. They are closed to 2 10^6 N.

The maximum efforts observed on the radial supports are radial efforts produced by the contraction of the vessel. They are closed to 4.5 10^6 N. It seems that no plastification occurs in the supports.

The calculation also provides us with an estimation of reaction forces applied on the support device of the vessel.

The maximum force is 3.6 10^7 N and the maximum moment is closed to 27 10^6 NxM.

5 CONCLUSION

This substructuration method reduces significantly the number of degrees of freedom. Its requires few calculation compared with a finite element program dealing with fluid structure interaction in a three-dimensional geometry. However, its implementation necessitates as a preliminary a correct approach of the physical phenomena involved. Particularly, it is very important to choose a modal basis giving a good representation of those phenomena. Besides, in our case, a fitting of the model by available experimental results provide us with a reliable modal basis.

With regard to the results obtained here, the assumptions seem work reasonably well. Specially the single phase flow hypothesis remains valid during the three-quarter of the whole calculation time.

The internal modelization, somewhat difficult to perform, seems sufficient to reproduce the core movement and therefore, forces applied on upper core plates supports and on radial supports.

It is interesting to remark that efforts, excluding the compression ones due to vessel contraction are maximum within the very first instants of the accident. This point confirms the importance of the acoustical studies of the LOCA.

6 REFERENCES

(1) Guilbaud, D., Jeanpierre, F., Gibert, R.J., "A substructure method to compute the 3D fluid-structure interaction during blowdown". SMIRT 7, Chicago, 1983.

(2) Guilbaud, D., "Calculation of a HDR blowdown test using a sub-structure method.
SMIRT 8, Brussels, 1985.

(3) Wolf, Schumann, Scholl, "Experiment and analytical results of coupled fluid structure interaction during blowdown of the HDR vessel", SMIRT 7, Chicago, 1983.

(4) Jeanpierre, F., Gibert, R.J., Hoffmann, A., Livolant, "Fluid structure interaction. A general method used in the CEASEMT computer programs", SMIRT 5, Berlin, 1979.

(5) Axisa, F., Gibert, R.J., "Non linear analysis of fluid structure coupled transients in piping systems". ASME PVP Conference, Orlando, 1982, PVP63, p. 15.

Effect of finite-length CRD housing model on the response of BWR piping systems

M.Y.Lee, A.L.Unemori & M.Chatterjee
ASD International Inc., San Francisco, Calif., USA

T.Yoshinaga & N.Gotoh
Hitachi Works, Hitachi Ltd, Ibaraki-ken, Japan

ABSTRACT

Transient response of the Control Rod Drive (CRD) piping system subject
to start-up SCRAM conditions is calculated by the Method of Characteristics using a new mathematical modelling of the CRD housing with a
finite, time-dependent piston travel length. Both the compressibility
of the flow characteristics within the CRD housing and the dynamic
effect of the CRD piston travelling as a moving boundary condition are
considered in the present analysis. Calculation results are confirmed
by previous experimental data in predicting both the SCRAM event evolution and the pressure peak occurrences.

1. INTRODUCTION

The CRD hydraulic unit plays a major role in providing quick and safe
reactor core control for nuclear power plants. Existing evidence
strongly indicates the need for a good shutdown policy for the reactor.
During a reactor SCRAM operation, the control rods must be quickly
inserted to reduce the power level in the reactor. This is achieved by
applying a large pressure differential across the upstream and downstream faces of the drive piston. Severe hydraulic transient water
hammer phenomena can thereby occur in the CRD piping system.

In the analysis of the transient response of CRD piping systems subject
to SCRAM conditions, a lumped-parameter model of the CRD housing and
piston is frequently used (Watanabe, 1983). Under this assumption, the
CRD housing and piston is considered to be inelastic and to contain an
incompressible fluid. It infers that the elasticity of this part of
the piping system is not nearly as important as its inertia in determining the transient response. However, since the travel length of the
piston inside the CRD housing is much greater than most of the pipe run
segments in the CRD piping system, a compressible flow model for the
CRD housing with a finite, time-dependent piston travel length using
the Method of Characteristics (Wylie, 1978) is considered to be necessary to maintain a continuous fluid model. For this reason, a new
finite-length CRD housing model which takes into account both the flow
compressibility within the CRD housing and the dynamic effect of the
CRD piston travelling as a moving boundary condition is developed.

The finite length CRD housing model is implemented into the PIPEFLOW

77

(ASD International, 1986) computer program to perform this waterhammer transient analysis for a typical CRD piping system subject to start-up SCRAM conditions. The analysis covers event durations, discharge from the accumulator, insert/withdraw SCRAM valve openings and the response due to stoppage of the CRD piston as dictated by the action of the CRD buffer assembly. Analysis results are confirmed by previous experimental data for a 1100 MWe BWR in predicting both the SCRAM event evolution and the pressure peak occurrences.

2. MODELLING OF CRD MECHANISM

Figure 1 shows a typical CRD piping system. Upon arrival of the SCRAM signal, the insert line discharge supplied by the accumulator will exert a high pressure at the bottom side of the CRD piston. When the pressure differential across the piston increases to the point where it can overcome the inertia force and the associated friction forces of the piston, the control rod begins to accelerate upward. Since the SCRAM time depends strongly on the insertion speed of the rod, which is directly related to the pressure differential across the upstream and downstream faces of the CRD piston, it is believed that accurate prediction of this pressure differential will have a dominant effect on the overall response of the CRD piping system under SCRAM operation.

A finite-length CRD housing is modelled as two additional pipe segments adjacent to the drive piston as shown in Figure 2. An adjustable grid generation scheme which allows instantaneous variation in the integration points in these two pipes as the CRD piston travels has been adopted. Boundary conditions at both ends of the CRD housing are used to couple with the Characteristic equations and the dynamic equation of the CRD piston at each new CRD piston position during its entire length of travel. Pressures on the upstream and downstream faces of the CRD piston are thereby calculated <u>exactly</u> at their instantaneous positions as the piston travels within the CRD housing.

The governing equations of the CRD piston motion, in terms of the pressure head, H, and the flow discharge, Q, can be written as

$$\gamma(A^u H^u - A^d H^d) - mg - F_b - A^r p^r = m \frac{dV}{dt} \qquad (1)$$

where

V = flow velocity = Q/A

t = time variable

γ = specific weight of water

A^u = area of CRD housing at upstream face of piston

A^d = area of CRD housing at downstream face of piston

H^u = pressure head inside CRD housing at upstream face of piston

H^d = pressure head inside CRD housing at downstream face of piston

m = mass of CRD piston

g = gravitational constant

F_b = sum of all frictional and resistive forces

A^r = contact area between the Reactor Pressure Vessel (RPV) and the upstream face of the CRD piston

P^r = pressure of RPV

Furthermore, the continuity equation requires that

$$\frac{Q^u}{A^u} = \frac{Q^d}{A^d} \tag{2}$$

where

Q^u = flow discharge at the upstream face of the piston

Q^d = flow discharge at the downstream face of the piston

Finally, the well-established Characteristic equations (Wylie, 1978) can be written as

$$C^+ \begin{cases} \dfrac{g}{a}\dfrac{dH^u}{dt} + \dfrac{dV}{dt} - \dfrac{g}{a}V\sin\alpha + \dfrac{fV|V|}{2D} = 0 & (3) \\[2ex] \dfrac{dx}{dt} = V + a & (4) \end{cases}$$

$$C^- \begin{cases} -\dfrac{g}{a}\dfrac{dH^d}{dt} + \dfrac{dV}{dt} + \dfrac{g}{a}V\sin\alpha + \dfrac{fV|V|}{2D} = 0 & (5) \\[2ex] \dfrac{dx}{dt} = V - a & (6) \end{cases}$$

where

a   =   speed of pressure pulse

α =   inclination angle of the pipe

f   =   Darcy-Weisbach friction factor

D   =   diameter of the pipe

x   =   distance along the pipe

After elaborate algebraic manipulations, Equations (1) - (6) are solved simultaneously for the four unknowns, namely, H^u, H^d, Q^u and Q^d. Details regarding the numerical procedures of Finite Difference approximations and the Modified Newton-Raphson direct integration method, including second-order friction terms can be found in Reference 3.

3. RESULTS, DISCUSSION AND EXPERIMENTAL VERIFICATION

To test and validate the numerical accuracy of the present finite-length CRD housing model, the PIPEFLOW program has been used to perform a transient waterhammer analysis of a CRD piping system (Figure 1) under start-up SCRAM conditions. The underlying CRD piping system is approximately 80 m in length, and consists of pipe sections of sizes 20A, 25A and 32A.

The results of the analysis are presented in the form of plots of pressure head time histories at selected locations in the insert and withdraw pipe lines. Figure 3 shows pressure head at a point located after the insert SCRAM valve in the insert line. It indicates that the pressure rise due to the insert SCRAM valve opening is significant for points near the accumulator. Figure 4, which corresponds to a point located near the intersection of the CRD housing and the insert pipe line, shows clearly that the movement of the piston has a relieving effect on the upstream pressure by increasing the available volume underneath the piston. It also shows the oscillation decay and reflection of the pressure waves due to the movement of the CRD piston. More important, it shows that the buffer action and the stoppage of the piston severely changes the response of the system.

Figure 5, which corresponds to a point located downstream of the CRD piston in the withdraw line, shows that the pressure rises rapidly when the CRD piston first starts to accelerate (which occurs when the pressure wave due to the insert SCRAM valve opening first arrives at the piston). However, the pressure falls quickly and even drops down to vapor pressure as soon as the piston stops, and stays at a low pressure (liquid column separation) for the rest of the transient.

Analysis results are confirmed by previous experimental data in predicting both the SCRAM event evolution and the pressure peak occurrences. Figure 6, which shows the experimentally-measured pressure time history at the intersection of the insert pipe line and CRD housing, should be compared with the analytical calculation of Figure 4. As can be seen, the overall correlation between the analytical solution and the experimental data is very close, in spite of possible discrepancies in the material properties and measurement point locations in the experimental set-up.

4. SUMMARY AND CONCLUSIONS

Major observations and conclusions reached from the present study of the effects of the finite-length CRD housing on the response of a BWR power plant CRD piping system are best summarized as follows:

o Pressures in the pipe segments near the CRD housing are, in general, much higher than those away from the CRD housing

o Pressures in the insert line are, in general, much higher than those in the withdraw line

o Pressure rises due to the CRD piston stoppage are greater than that due to the insert SCRAM valve opening for pipe segments near the CRD housing in the insert line

o Significant parts of pressure changes in the withdraw line are only due to the opening of the insert SCRAM valve

o Water column separation is predicted near the CRD housing in the insert line for a short duration, but in the withdraw line an extensive vapor formation is predicted over most of the transient

o The inertia force due to the weight of the CRD piston is not a dominant factor in the overall response, whereas the pressure differential across the upstream and downstream faces of the CRD piston is

REFERENCES

Watanabe, Y., and Y. Motora, "Analysis of CR Scrammability Characteristics on the Condition of the Forced Vibration of Fuel Assemblies", 7th International Conference on STRUCTURAL MECHANICS IN REACTOR TECHNOLOGY, August, 1983

Wylie, E. B., and V. L. Streeter, FLUID TRANSIENTS, McGraw-Hill Book Company, New York, 1978

ASD International, Inc., PIPEFLOW: COMPUTER PROGRAM FOR WATERHAMMER TRANSIENT FLOW ANALYSIS OF PIPING SYSTEMS, User's Manual, Version 1.2, May, 1986

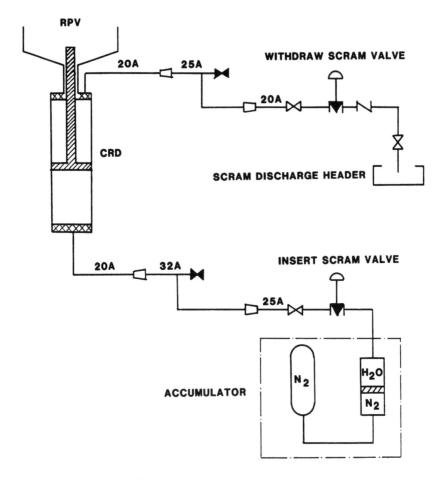

FIGURE 1   TYPICAL CRD PIPING SYSTEM

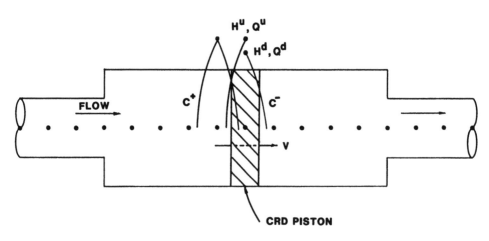

CRD PISTON

FIGURE 2   SCHEMATIC DIAGRAM OF FINITE LENGTH CRD HOUSING USING METHOD
OF CHARACTERISTICS

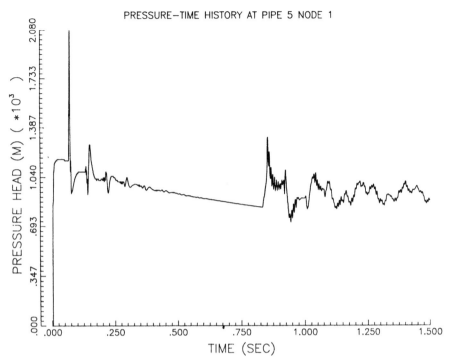

FIGURE 3 CALCULATED PRESSURE TIME HISTORY AT POINT AFTER INSERT SCRAM VALVE IN THE INSERT PIPE

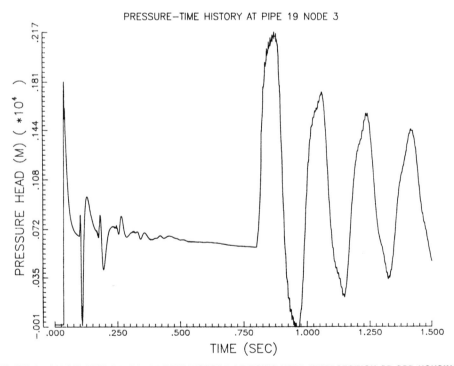

FIGURE 4 CALCULATED PRESSURE TIME HISTORY AT POINT NEAR INTERSECTION OF CRD HOUSING AND INSERT PIPE

PRESSURE–TIME HISTORY AT PIPE 21 NODE 1

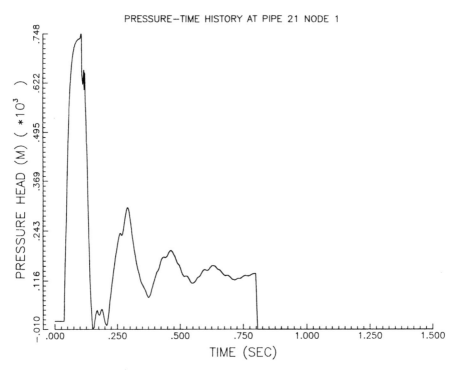

TIME (SEC)

FIGURE 5 CALCULATED PRESSURE TIME HISTORY AT POINT DOWNSTREAM OF CRD PISTON IN THE WITHDRAW PIPE

(kg/cm²)

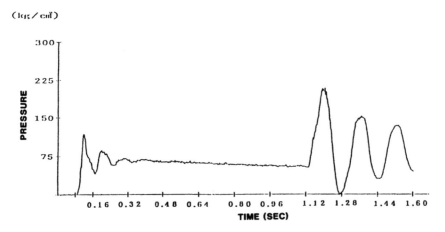

TIME (SEC)

FIGURE 6 EXPERIMENTAL RESULT OF PRESSURE TIME HISTORY AT INTERSECTION OF CRD HOUSING AND INSERT PIPE

84

Elastic and elastic-plastic behaviour of a piping system during blowdown – Comparison of measurement and calculation

W.Petruschke & G.Strunk
Rheinisch-Westfälischer Technischer Überwachungs-Verein e.V., Essen, FR Germany

1 INTRODUCTION

As part of the German HDR Safety Program, blowdown tests were performed /1/, /2/, /3/.

Fig. 1 shows the experimental loop on which the blow-down tests T21.1 and T21.3 were conducted. In the test T21.1 the safety-valve had an optimized damping characteristic, while it was almost undamped in the test T21.3. The undamped characteristic was used to induce very high pressure waves so that the piping response would also be on a higher level up to the plastic range.

Comparisons of former blowdown tests and corresponding precalculations have shown that it is difficult to distinguish in the set of modelling parameters those which are significant for possible deviations between test and analysis. For this reason some preliminary tests were conducted to check the accuracy of the structural model:

test	data compared
- static loading by single horizontal and vertical forces	- displacements, stresses
- snap-back test	- natural frequencies, mode shapes

With the informations from these tests the structural model could be optimized so that the precalculated elastic piping response would additionaly be influenced only by the pressure time histories and the method of application of the fluid-dynamic forces to the structural model.

2 STRUCTURAL MODEL

For simulation of the preliminary tests and for the precalculations of T21.1 the finite element program STARDYNE was used. The piping was modelled by beam elements with pipe cross-section using flexibility factors according to the Karman theory for the elbows, (Fig. 2). According to the mass-ratio the reactor pressure vessel could be decoupled from the piping, and so it could be neglected in the analysis. Only the flexibility of the RPV-nozzle was considered by

85

introducing local springs at the end of the pipe. The material behaviour was assumed to be linear elastic. For the dynamic analysis a damping value of 2 % was considered.

For the precalculations of T21.1 the load assumptions were given by the measured pressure time histories near the elbows. As a first step a linear interpolation /4/ was made from the measuring points to the centre points of the elbows. As a second step the differential forces of successive elbows were evaluated and applied to the structure between the corresponding elbows.

For the precalculations of the second blowdown experiment T21.3 the finite element piping program NONPIPE was used. This has the capability to simulate elastic-plastic material behaviour.

By using the same topology as for the test T21.1 one calculation was performed with elastic and one with elastic-plastic material behaviour.

3 MODEL CHECK WITH PRELIMINARY TESTS

In the static test the piping was displaced by a concentrated force in the lower part of the vertical line of the piping. Fig. 3 and Fig. 4 show the comparison of the measured and calculated deflections and stresses. The calculated stresses at the elbow are based on the slightly modified detailed analysis procedure of ASME /5/.

The first natural frequencies and mode-shapes are measured from the free vibrations of the piping, which was excited in a snap-back test. In Fig. 5 the first 5 natural frequencies are compared with the measured data.

A comparison of the measured and calculated data shows that the agreement of the static deflections and stresses, natural frequencies and mode shapes is quite good. Parameter variations with regard to the boundary conditions and the flexibility factors of the elbows did not change the overall agreement.

Thus for the further calculations the prescribed piping model was used without any modification.

4 BLOWDOWN TEST T21.1

This blowdown test was planned with an optimized damping characteristic of the valve, so that the piping response would be in the elastic range. But from the evaluation of the measured data it was possible to see that at two local positions plasticity had occured. Nevertheless the overall behaviour of the piping was nearly elastic so that the measured and calculated data are comparable. Fig. 6 shows the comparison of a typical displacement time history. The agreement of the two time histories up to 0.2 sec is very good; after this a slight shift of the fundamental frequency and an overestimation of the amplitudes in the precalculation can be seen.

The reason for this is established using the measured pressure time histories which include the natural frequencies of the piping system caused by fluid structure interaction. This frequency content leads to a pseudo resonance of the piping model.

Fig. 7 shows the maximum hoop stress time history in the centre of elbow 1. The stress calculation procedure is the same as used for the static precalculations, which leads to a reasonable good approximation of the measured data.

5 BLOWDOWN TEST T21.3

This second blowdown experiment was planned in such a way that the piping response would be at a higher level up to the plastic range.

For the simulation of this experiment, in contrast to T21.1 precalculated pressure time histories related to the elbow centre points were used so. Thus no special interpolation technique was necessary.

With the intention of qualifying simplified elastic-plastic calculation methods and of making a best estimate analysis, two precalculations were performed.

5.1 Simplified elastic-plastic precalculation

The aims of the present analysis were to describe the overall behaviour of the piping as characterized by the displacement, and the local behaviour as characterized by stresses and strains. In cases where plasticity occurs only in local regions, it was possible to adopt a linear elastic approach for simulation of the overall behaviour. Therefore as a first step the linear-elastic response of the piping was calculated. In a second step reduced Youngs moduli were introduced for those elements which were loaded over the yield point. Assuming that the overall data thus calculated represented a reasonable approximation, some additional studies were conducted on this basis for computing stresses and strains of the highly stressed parts of the piping.

Thus an additional static FEM-analysis for the elbow near the RPV was conducted using shell elements with material and geometrical nonlinearities. The results from this analysis were relationships between the rotation of the cross-sections at the ends of the elbow and the strains at different section points. On the basis of the time-dependent rotations of the elbow the strain time histories were calculated at these section points.

5.2 Elastic-plastic precalculation

The second precalculation was planned as a "best estimate analysis" taking into account the elastic-plastic material behaviour. For reasons of economy and in order to qualify more simplified approximations, pipe elements with nonlinear moment-curvature and moment-twist relationships were used for the model, and these were suitable for simulating the nonlinear material behaviour and the energy dissipation during cyclic loading. These elements are very effective for describing the global behaviour of the structure such as displacements and strains in beam sections. Compared to the more sophisticated finite element technique using incremental material laws a small error arises from the assumption that in the case of yielding, bending and torsional moments are decoupled. Because of the assumptions related to the beam theory only "beam strains" can be calculated.

5.3 Results

Fig. 8 shows a comparision of the measured displacements and the results of the two calculations. The results of the simplified elastic-plastic analysis still show a sufficient agreement with respect to the maximum displacements. But after 0.1 sec the frequency content of the measured data is quite different, which is caused by the effects of plasticity.

The agreement of the elastic-plastic calculation is significantly better. For the other measuring points the same tendency can be seen with which in general the percentage deviation is greater at locations with smaller deflections. In elbow 1 the expected high strains did not occur, and so the strains remained nearly elastic. For this reason the strain amplitudes of the elastic-plastic calculation were derived from the detailed stress analysis prodecure according to /5/ using the moment-time histories. The results are compared in Fig. 9 where the elastic-plastic calculation is also the better approach, while the simplified analysis still gives reasonable results. In Fig. 10 the strains of the elastic-plastic calculation in the lower part of the piping are compared with the measured data. Reasons for the greater deviations have not yet been completely clarified. One reason may be local effects coming from the measured data because the strain gauges were only applied at one point on the circumference of the pipe section.

An additional analysis with the measured data is planned in order to study the influence of the precalculated pressure time histories.

6 CONCLUSIONS

The investigations according to the system identification show that the piping model using beam theory and flexibility factors according to the Karman theory are adequate for evaluating natural frequencies, mode shapes, static displacements and stresses.

The same accuracy can be seen by comparing the piping response due to blowdown within the elastic range

The simplified elastic-plastic analysis in general overestimates the maximum amplitudes while the frequency content is not simulated very well. For practical purposes, it can be an adequate tool in many cases.

The elastic-plastic analysis is the most expensive procedure but gives also the best results. The use of beam elements with multilinear moment-curvature relationsships results in a good approximation for the global behaviour (displacements). The strains according to this theory only include the beam deformation modes.

References

/1/ 10. Statusbericht des Projektes HDR-Sicherheitprogramm, PHDR-
     Arbeitsbericht 05.27/86
/2/ Versuchsprotokoll Blowdownversuch mit Ventilschließen, Versuchs-
     gruppe RORB, T21.3 PHDR-Arbeitsbericht 2.205/85
/3/ Quick-Look-Report, Versuche T21.0, T21.1, T21.3
     Technischer Fachbericht PHDR 58-86
/4/ Piping Response Due to Blowdown; Significant Parameters for a
     Comparison of Experimental and Analytical Results
     F. Bietenbeck, W. Petruschke, H. Wünnenberg
     SMIRT 8, F1 3/7
/5/ ASME III, Div.1, NB 3685,
     ASME, New York, 1983

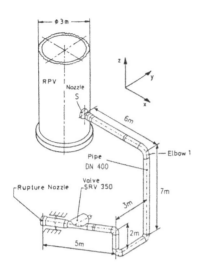

Fig. 1 Experimental Loop

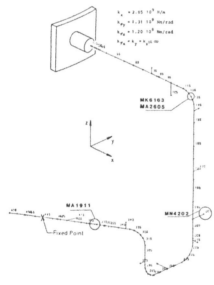

Fig. 2 Piping Model

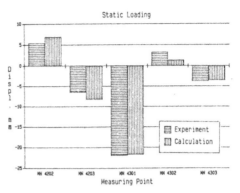

Fig. 3 Static Deflections

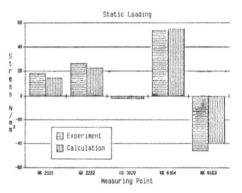

Fig. 4 Static Stresses

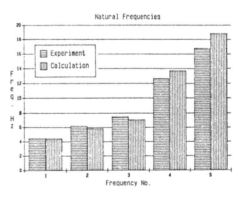

Fig. 5 Natural Frequencies

89

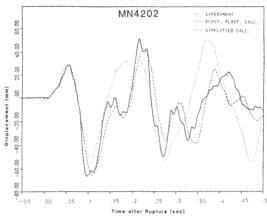

Fig. 8 T21.3: Displacement Time History

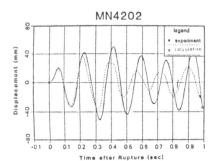

Fig. 6 T21.1: Displacement
Time History

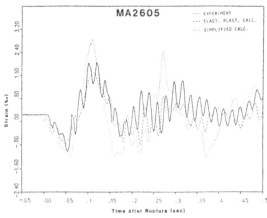

Fig. 9 T21.3: Strain Time History

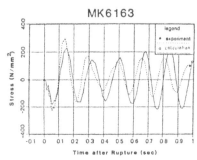

Fig. 7 T21.1: Stress Time
History

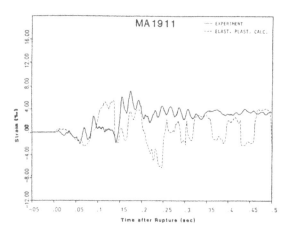

Fig. 10 T21.3: Strain Time History

Elastic-plastic response of a piping system due to simulated double-ended guillotine break events

K.Kussmaul & H.Diem
Staatliche Materialprüfungsanstalt (MPA), University of Stuttgart, FR Germany
H.Hunger & G.Katzenmeier
Project HDR-Safety Program, Kernforschungszentrum Karlsruhe, FR Germany

1 INTRODUCTION

In the design criteria for nuclear power plant components of light water reactors of the Federal Republic of Germany (FRG) a double-ended guillotine break of a feedwater line outside the reactor containment is postulated (Fig. 1) /1/. In case of a blowdown induced by a double-ended guillotine break in a boiling water reactor the feedwater check valves (SRV) serve the purpose of closing so quickly that the consequences of the accident are limited. The blowdown first produces a pressure relief wave within the piping followed by high pressure peaks (water hammer effect) as a result of the quick closure of the check valve. The differential pressures in the individual pipe sections, especially in front of and behind the pipe bends, give rise to fluid dynamic forces in the piping which load the structure considerably. To show the load bearing and plasticizing capacity of high ductile pipework materials nowadays used in reactor technology /2/, three blowdown tests were performed at the decommissioned Heißdampfreaktor (HDR) near Franfurt. The experiments reported in this paper are part of Phase II of the HDR Safety Program which is sponsored by the Ministry of Research and Technology of the Federal Republic of Germany /3, 4/. These investigations complete and continue former tests which had been performed in the scope of Phase I of the HDR Safety Program /5/.

2 EXPERIMENTAL FACILITY

To be able to perform blowdown experiments with high structural loading a flexible piping system, ND 400, was built up at the decommisioned HDR-Plant. The basic design concept was similar to that of a typical feedwater line in a boiling water reactor having two fix points and no other supports (Figures 2 and 3). The piping system had a length of about 23 m including four 90°-pipe bends and was mainly made of the low alloyed ferritic material 15 NiCuMoNb 5 with yield strength $R_{p0.2,\vartheta}$ of 467 MPa. Special parts such

as the rupture nozzle, a measurement ring and three
reducers required for reasons of feedwater line
construction and test performance consist of the low
alloyed ferritic material 15 Mo 3 with yield strength
$R_{p0.2,9}$ of 235 MPa. The feedwater check valve is
manufactured from cast steel GS C 25 (Fig. 4).

In order to concentrate the high loading on the RPV nozzle
area, the pipe section connected to it was reduced in wall
thickness to about 16 mm. The rest of pipework had about
20 mm wall thickness, the pipe elbows had an average wall
thickness of 27 mm.

The valve used for testing was a feedwater check valve
(ND 350) with hydraulic end damping. Its closing
characteristic was successively varied during three
experiments in order to increase the loading. The damping
characteristic of the valve as adjusted in test T21.1 was
simular to that of feedwater check valves of boiling water
reactors in operation. As undamped feedwater check valves
usually are not employed in nuclear power plants of the
FRG, the damping variation served only for increasing the
loading. Nevertheless the test can be considered from a
conservative point of view as worst case conditions to show
the safety margin.

The initial test conditions in the pressure vessel of
7 MPa and 285 °C in test T21.1 resp. 9 MPa and 305 °C in
the tests T21.3 and T21.4 were similar to those in a
boiling water reactor. The examined pipework itself
contained subcooled water at 220 °C.

The blowdown was initiated by blasting off two burst discs
at the rupture nozzle. The rupture opening time of about
two to three milliseconds corresponds to that of a natural
pipe rupture. In order to be able to record the fluid and
structural dynamic sequence 185 measuring transducers in
total had been attached to the piping system for recording
pressures, temperatures, deflections and strains. In order
to demonstrate the plasticizing capacity of pipe bends, one
elbow had been equipped with 76 high temperature strain
gages. Strain gage measurement had also been performed at
two positions at the inner surface, which usually is
impossible to control experimentally in working plants.

3 EXPERIMENTAL RESULTS

In the three blowdown experiments performed in the scope of
Phase II of the HDR Safety Program different closing times
of the feedwater check valve and pressure responses were
obtained as a function of the damping characteristics of
the valve (Figures 5 and 6). The piping responded to the
pressure relief waves by an upward movement and to the
pressure peaks occurring during valve closure by a downward
movement where the second effect was always dominating. In
all three experiments noticeable plastifications were
measured in measuring cross sections located at the above
mentioned reducers near the fix points subjected to the
lower yield strength of the material 15 Mo 3 in comparison
to the high-strength steel 15 NiCuMoNb 5 (Figures 7 and 8).

92

As the test had been performed one after the other, possible strain hardening effects have to be considered in evaluating the measured data.

The maximum loading on the piping due to structural movements was dependent on the relation between pressure relief wave and pressure peak on the one hand, and on the ratio of valve closure time to mass flow development on the other hand. In the experiment T21.1 (optimized valve damping) closing of the valve was terminated at a point of time when the mass flow had already reduced almost to zero. The value of the pressure peak (10,1 MPa) in this case attained only 1.5 times the operating pressure. The maximum loading in test T21.1 occurred at the reducers made from 15 Mo 3, still during closing of the valve but already during the downwards movement of the pipework system, and was due to dominating bending effects. By contrast, the valve closing in experiment T21.3 (no damping) was completed in a phase where the mass flow still continued to increase. The pressure peak of 28,6 MPa attained about 3.2 times the operating pressure. In this experiment the "water hammer effect" caused plastifications both in the first and in the second phase of pipe movement. In the first phase of movement the yield strength at test temperature was exceeded at all points where positive strains caused by internal pressure and the bending load were superimposed.

The bending moments and the loading on the pipe elbow No. 1 remained relatively low in the experiments T21.1 and T21.3 so that the yield strength of the material was exceeded only on the inner surface of the bend flank in test T21.3.

In the experiment T21.4 the attempt was made to delay valve closure by initial damping until the two phases of movement of the piping system were tuned to each other and a high bending moment occurred in the area of elbow No. 1. This measure caused the "water hammer effect" to increase and led to a pressure peak of 31,5 MPa ($\cong$ 3.5 times the operating pressure). Plastifications were nearly identical with those noted in the experiment T21.3 only the elbow underwent higher loading. In Figure 9 measured circumferential strains at the bend flank of elbow No. 1 are represented as functions of load history. This plot gives an impression of the pronounced elbow ovalization and show that the inner surface is loaded higher than the outer surface. It is also shown that the elbow is loaded over a large range in the same magnitude. This effect is important with respect to a possible failure of a precracked bend /6/.

The stress state in the bend area at the time of maximum loading is clearly visible from the plots of fictive-elastic equivalent stresses according to the Mises-Hencky-Criterion (Figure 10). The stresses are calculated by the following equations from measured strains at the inner and outer surface:

$$(1) \qquad \sigma_{circ.} = \frac{E}{1-\mu^2} \left(\varepsilon_{circ.} + \mu\, \varepsilon_{long.} \right)$$

$$(2) \qquad \sigma_{long.} = \frac{E}{1-\mu^2} \left(\varepsilon_{long.} + \mu\, \varepsilon_{circ.} \right)$$

$$(3) \qquad \tau_{cl} = \frac{E}{1+\mu} \left(\varepsilon_{45°} - \frac{\varepsilon_{circ.} + \varepsilon_{long.}}{2} \right)$$

$$(4) \qquad \sigma_{eq.} = \sqrt{\sigma_{circ.}^2 + \sigma_{long.}^2 - \sigma_{circ.}\cdot\sigma_{long.} + 3\,\tau_{cl}^2}$$

It could be shown that the influence of shear stresses on the magnitude of $\sigma_{eq.}$ was negligible. The grafic representation in form of single plane orthomorphic projection of the elbow surface (Figure 10) shows that in test T21.3 only at the inner surface the yield strength of the elbow material 15 NiCuMoNb 5 had been exeeded, where as in test T21.4 a narrow plastic zone (shaded area) developed from the onset to the end of the bend. As had been expected, plastification was stronger on the inner surface /6/. A detailed description of load-history dependend formation of plasticized zones on the outer as well as on the inner surface of pipe bends will be presented in /7/.

In neither experiment involving a non-damped valve the internal pressure did not play any role in elbow stressing at the time of maximum loading. The energy of motion introduced into the piping system by blowdown loading was dissipated above all by work done in plastic deformation at two reducers made from the material 15 Mo 3. Figure 11 shows the remnant strain summed up in the course of the three experiments. It should be mentioned here that despite three times of plastification no loss of integrity has been observed on the piping with a summed up remnant strain of 2.9 %.

4 ANALYTICAL DESCRIPTION OF PIPING BEHAVIOUR

The results of the piping system during the test with undamped valve closing was analyzed with the ABAQUS finite element-code using special pipe "PIPE 31" and pipe bend elements "ELBOW 31" /8/. The loading of the finite element model was generated by measured pressure values at different locations within the test pipe. The results of measurement are compared with the results of the non-linear computation for the experiment T21.3. The computation involving modelling of real stress/strain representations at test temperature of the materials 15 Mo 3 and 15 NiCuMoNb 5 described the plastifications in the nozzle zone in a good way as shown in a grafic representation of the strain history at a measurement point (Q1) near the RPV in Fig. 12. An analysis of the cross-sectional deformation behaviour (ovalization) in the center of elbow No. 1 at the time of maximum loading (t = 0,1 s) is shown in Figure 13 and compared with the measured values. In the

94

computation two points of time, 6 milliseconds apart, are
indicated which give an impression of the dynamic process
of deformation. The agreement between the measurement and
the calculation can be considered as very good.

5 SUMMARY AND CONCLUSIONS

From the blowdown experiments performed on the HDR
feedwater line with feedwater check valve the conclusion
can be drawn that high transient loads of up to plastic
strains of 3 %, acting on an initially integer piping
system, can be sustained without loss of integrity for a
low number of load cycles due to the plasticizing capacity
of the pipework materials nowadays used in reactor
technology. In the experiments carried out with ferritic
piping of ND 400 pressure peaks up to about 31,5 MPa were
achieved which resulted in excessive strains of up to 3 %.
By nonlinear finite element computations (ABAQUS) it was
possible to describe the elastic-plastic behaviour of the
piping in a good approximation.
On account of the safety margins proved in the
experiments, potential inaccuracies in theoretical
structure analyses are accommodated so as to be on the safe
side. On the other hand, it appears that designing
pipework with reference to elastic stress categories does
not adequately take into account the actual reserves of the
pipework material.

REFERENCES

1 Sicherheitstechnische Regel des KTA, Komponenten
  des Primärkreises von Leichtwasserreaktoren
  KTA 3201.2

2 Kussmaul. K.: Die Gewährleistung der Umschließung
  atw, pp. 354/361, July/August 1978.

3 HDR-Safety Program Phase II,
  PHDR-Report No. 05.19/84, Kernforschungszentrum
  Karlsruhe, January 1984

4 Diem, H., Hunger, H.: Überelastische Rohrverformung
  unter scharfer Druckstoßlast bei Blowdown mit unge-
  dämpftem Ventilschließen,
  10. Statusbericht HDR-Sicherheitsprogramm,
  PHDR-Report No. 05.27/86. Karlsruhe, December 1986,

5 Müller-Dietsche, W., Katzenmeier, G.: Reactor Safety
  Investigations at the "Heißdampfreaktor" Karlstein (HDR),
  Final Report Phase I, Technical Report No. 60/85,
  Kernforschungszentrum Karlsruhe, July 1985

6 Kussmaul, K., Diem, H., Blind, D.: Deformation and
  Failure Behaviour of Elbows, Experimental Results
  and Analytical Predictions,
  1986 ASME Joint PVP and CED Conference,
  July 1986, Chicago, Illinois

7 Kussmaul, K., Diem, H., Blind, D.: Investigations on
  the Plastic Behaviour of Pipe Bends.
  To be presented at the 1987 ASME PVP Conference,
  June 28 - July 2, 1987, San Diego, California.

8 Kussmaul, K., Kerkhof, K., Blind, D., Sauter, A.:
  System Response of a Feedwater Line During Blowdown
  Loading.
  To be presented at the 1987 ASME PVP Conference,
  June 28 - July 2, 1987, San Diego, California.

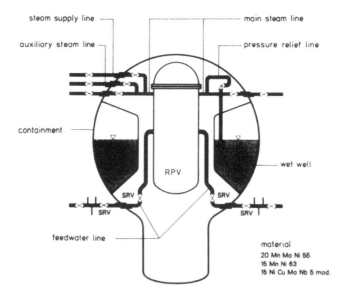

Fig. 1. Pressure retaining boundary of a BWR

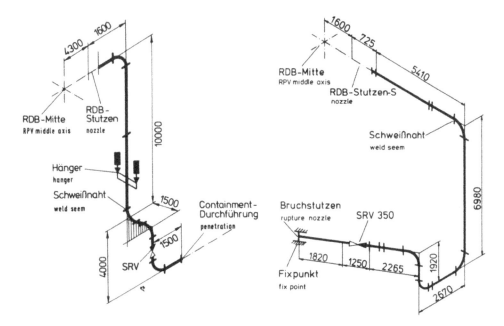

Fig. 2. Typical feedwater line
of a BWR

Fig. 3. Feedwater line of
HDR-experimental plant

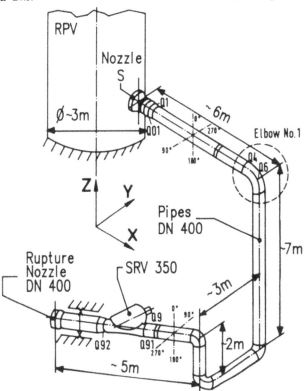

Fig. 4. Investigated test piping system
with main parts

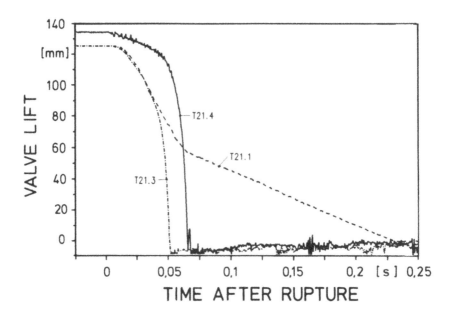

Fig. 5. Closing characteristics of the feedwater check
valve in the three experiments

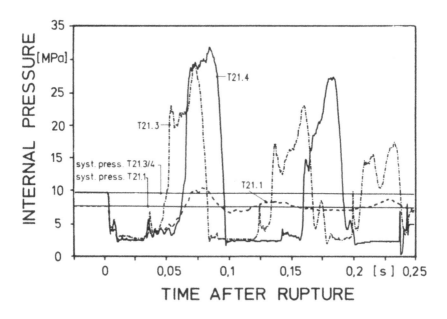

Fig. 6. Pressure responses at a point near the rupture
nozzle due to different closing characteristics
of the feedwater check valve

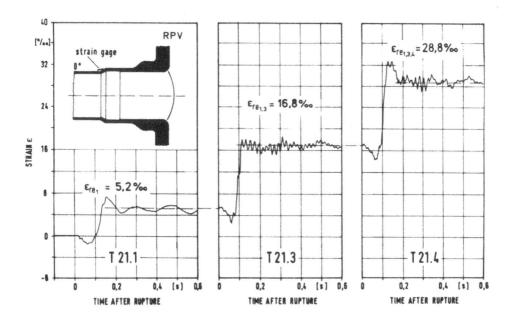

Fig. 7. Longitudinal strain histories at a point near the RPV-nozzle
(Q01) and summed up remnant strains

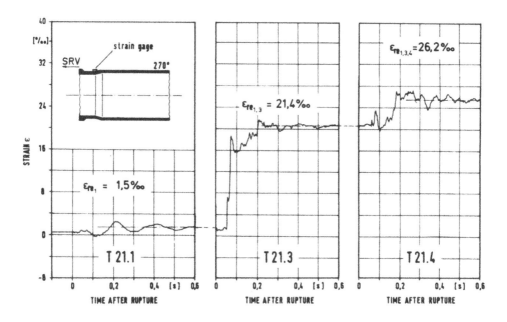

Fig. 8. Longitudinal strain histories at a point near the SRV (Q9)
and summed up remnant strains

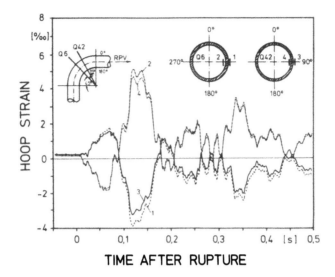

Fig. 9. Circumferential strain histories at two points at the flank of elbow No. 1 in experiment T21.4

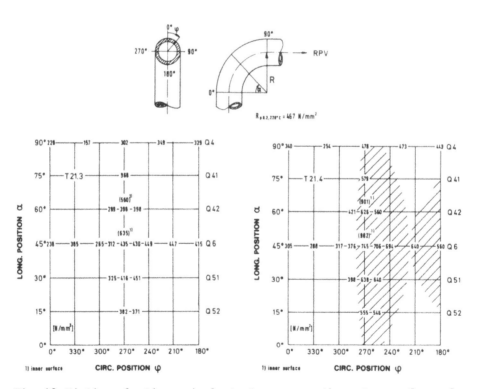

Fig. 10. Fictive-elastic equivalent stresses on the outer surface of elbow No. 1 at the time of maximum loading in the experiments T21.3 and T21.4

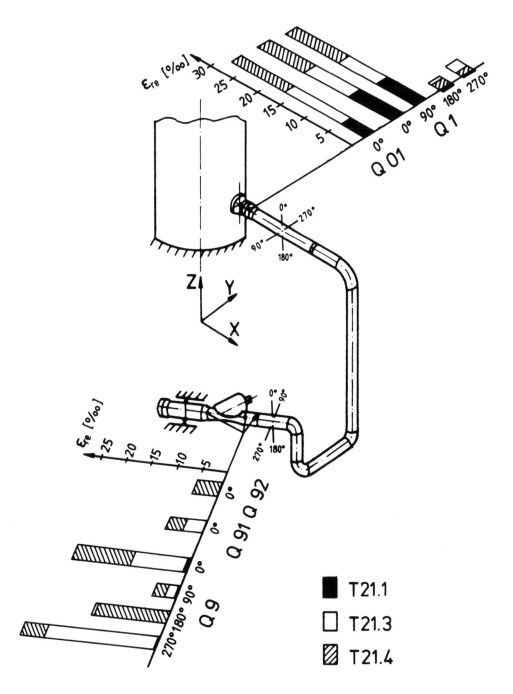

Fig. 11. Summed up remnant longitudinal strains at the maximum
loaded points in the course of the individual experiments

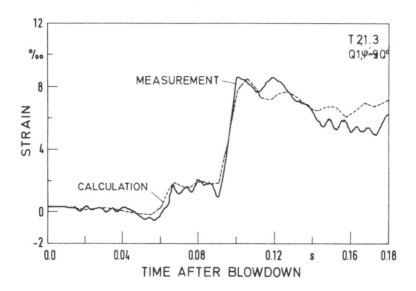

Fig. 12. Calculated and measured longitudinal strain
history at a point near the RPV-nozzle (Q1)

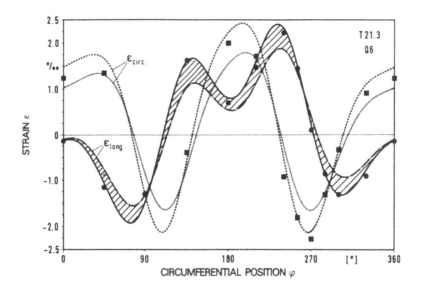

Fig. 13. Comparison of measured and calculated strain
distributions in the midsection of elbow No.1 at
the time of maximum loading in test T21.3

102

Pipe loads in the case of a safety valve blowdown with phase change

W.Zeitner & R.Puzalowski
Kraftwerk Union AG, Erlangen, FR Germany
U.Simon
Kraftwerk Union AG, Karlstein, FR Germany

1 INTRODUCTION

For design of piping systems the dynamical response of the structure due to flow induced forces under normal and accident conditions has to be known.

As experimental tests can not provide all stresses, strains and displacements during transient loadings, in many cases the underlying analysis has to be carried out by a computer simulation of the piping system based on an appropriate numerical structural model.

This approach is of particular interest in design of nuclear reactor components, where also hypothetical accident conditions have to be considered.

As an example in the present paper we discuss the safety- and relief-valve systems of a KWU pressurized water reactor that functionally protects the primary reactorsystem against overpressure. This facility consists essentially of the pressurizer and the main blowdown pipes to the safety- and relief valves.

The transient loadings for this piping system emerge from flow induced forces during valve operation under normal conditions (steam discharge) and under accident conditions (subcooled water and 2-phase mixture discharge). The latter situation may occur in a few cases of anticipated transients without reactor shut down (ATWS).

On a fullscale KWU-test facility /1/ blowdown experiments were carried out to get knowledge about some physical parameters, to improve the fluiddynamic code for calculating load functions and to correct the structural model.

Which the load functions as input the resulting motions of the structural model were calculated.

2 THE PROBLEM AND THE APPROACH TO SOLUTION

The strains induced in the piping by the blowdown process have been evaluated earlier for the loadcase steam blowdown in some power plants. Insufficient knowledge in evaluating hydrodynamic loading and the thereby induced vibration of the piping had to be dealt with using conservative assumptions eg. the rate of valve stroke or the structural damping.

Another point of uncertainty is the stiffness of supports. Whereas
for static loadings the assumption of very stiff supports normally
leads to conservative results, in the dynamic case no such simple
rule holds.

From a series of blowdown experiments of the test facility, see Fig.1,
one typical test was chosen, showing steam blowdown followed by subcooled water-steam mixture.

Measurements of pressure, forces and displacements serve to adjust
the hydrodynamical code and to correct the structural dynamical model.

3 EXPERIMENTS CARRIED OUT

In a first series of experiments, tests were performed with a very
stiff and rigid support of the pipe. The pipe was supported in its
horizontal and vertical axial direction, and the support loads as
well as the in-pipe pressures were measured. In a second series
of experiments the pipe supports were changed to allow for considerable
pipe vibration, comparable to the situation in a power plant. The
displacements, the snubber forces and the in-pipe pressures were
measured, see Fig.2.

The first series of tests allowed a verification of the hydrodynamic
codes described in chapter 4, the second series of tests allowed a
correction of the structual dynamic parameters described in chapter 5.

4 ESTIMATE OF HYDRODYNAMIC LOAD FUNCTIONS

The hydrodynamic load functions which are input for the structural
analysis have been determined with the postprocessor TRAFO from
 fluiddynamic data generated with the Transient Reactor Analysis Code
TRAC-PF1 by conducting post test calculations of the experiments.

While in single-phase blowdown (steam or subcooled water) the force
amplitudes depend essentially on the valve stem velocity during valve-
opening and closing-operation, in two-phase mixture blowdown they
are very strongly affected by the degree of subcooling of the water-
front that follows the initial steam discharge and hits the valve
orifice.

In an earlier paper /2/ we have already reported on the single phase
post-test calculation stressing the fact, that due to much smaller
valve stem velocities under water conditions the force amplitudes are
only slightly higher compared to those from a steam blowdown.

Conducting post-test calculations of the two-phase mixture experiments,
we found that the code TRAC-PF1 as it stands was not capable to simulate
the steep void gradients in the safety line caused by the propagating
steam-water-interface in the various cases.

In order to fit the experimental data with respect to pressure gradients and the resulting measured pipe-support forces in the two-phase
mixture tests, the code needed to be updated in its interphase correlations. With these corrections (for details see /3/) the experimental
data could be reproduced with fairly close agreement.

As an example fig.3 shows the valve inlet pressure of a post-test
calculation compared with the experimental data of a two-phase mixture
blow-down test with 15 K subcooled water. In fig.4 the associated
normalized force time-history F/F_{max} for the longest straight pipe
segment of the safety-valve line is given. Since in the region of

moving mixture front the calculated in-pipe forces are about 20 %
too low compared with the experimental results, a scaling of the
loadfunctions by a factor of 1.2 restricted to this region was performed.

5 RESPONSE OF PIPING DUE TO HYDRODYNAMIC LOADING

The analysed part of the test facility comprehends the two vessels
simulating the pressurizer, the blowdown piping, the safety valve
with the attached steam dome, the piping to the relief tank and parts
of the feeding pipe. These components were transformed to a finite
element beam model with the computer code KWUROHR. The four snubbers
were represented by linear springs. Stiffness of steamdome supports
and the vessel support were calculated using simple beam equations.
However the vessel support stiffnesses associated with rotation are
in doubt due to incertitudes in the fastening of the double-T-beams
to the concrete. Therefore experimental results were used to fit these
stiffnesses.
    Along the two greater straight parts of the blow-down piping the
stiffnesses in axial direction at the nodes 55 and 68 were measured.
Furthermore from the displacement signals W9 and W10, see fig.5 and 6,
an eigenfrequency of about 3,4 Hz was estimated. The vessel support
stiffness mainly affects both the piping stiffness at node 55 and the
mentioned eigenfrequency, which is associated with a motion in axial
direction at node 55. From the table below you can see that in this
point a certain degree of optimization is reached, since the ratios
of calculated to experimental values equal one.
    The response of piping was determined by direct integration of
the equations of motion including the hydrodynamic functions mentioned
above. One parameter till open was the degree of critical damping.
From the whole recorded time histories at W9 and W10 a value of about
5 % can be derived. For the whole structure this value seems to be
somewhat to high. In a first run of integration a value of 1 % critical
damping results in a mean value of the ratio calculated to experimental
range of all signals of 1.05. However, 3 % critical damping seems to be
a good compromise, leading to the results listed in the table below.
As a further example fig.5 and fig.6 show the calculated time histories
for the signals W9 and W10 in comparison to experimental results.

6 CONCLUSION

Steam blowdown with mixture of subcooled water represents a very
complex loading for the pressurizer-safety valve system. Having
adjusted the hydrodynamic loads and structural parameters to experimental data, a computer model is available to calculate similar load
cases. On the other hand, our experience is confirmed that in the
case of complex loadings a reliable evaluation of strains and stresses
in a piping is only possible based on experiments.

REFERENCES

/1/ Simon, U., A.Knapp, W.von Rhein, F.Agemar, W.Hofbeck & R.Puzalowski.
    1986. Untersuchungen zur Funktionssicherheit der Druckhalter-Sicher-
    heitsventile beim Abblasen von heißem Druckwasser und vollem Massen-
    durchsatz in Originalgeometrie. Förderungsvorhaben BMFT 1500 636/7.

/2/ Puzalowski, R. & U.Neumann. 1986. Numerical simulation of self-
    actuating valves and its application. lst.Int.Multiphase Fluid
    Transients Symposium. ASME, FED-Vol.41, p.85

/3/ Puzalowski, R & U.Neumann. 1987. Interfacial mass, heat and momen-
    tum transfer correlations in fast propagating two-phase mixture fronts.
    To be published in Proc.of 1987 ICHMT Int.Symposium on Transient
    Phenomene in Multiphase Flow.

TABLE

	exp.value	calc.value	ratio calc/exp.
Stiffness in N/mm			
node number 55	4400	4260	0.97
" " 68	7200	8110	1.13
Eigenfrequency in Hz			
Signals W9, W10	3,4	3,58	1,05
Displacement ranges in mm			
W9	8,4	6,94	0,83
W10	12,4	8,46	0,68
W12	26,0	22,9	0,88
W13	8,0	5,55	0,69
W14	9,5	7,76	0,82
W15	11,4	10,5	0,92
W16	6,0	6,93	1,16
Force ranges in KN			
D8	92	57,3	0,62
D9	36	44,3	1,23
D10	58	65,2	1.12
D11	17	16,9	0.99

$$\text{Total mean value} \qquad \frac{\sum_{i=1}^{n} \frac{calc}{exp}}{n} = 0,936$$

$$\text{Total root mean square value} \qquad \sqrt{\frac{\sum_{i=1}^{n}(1-\frac{calc}{exp})^2}{n}} = 0,195$$

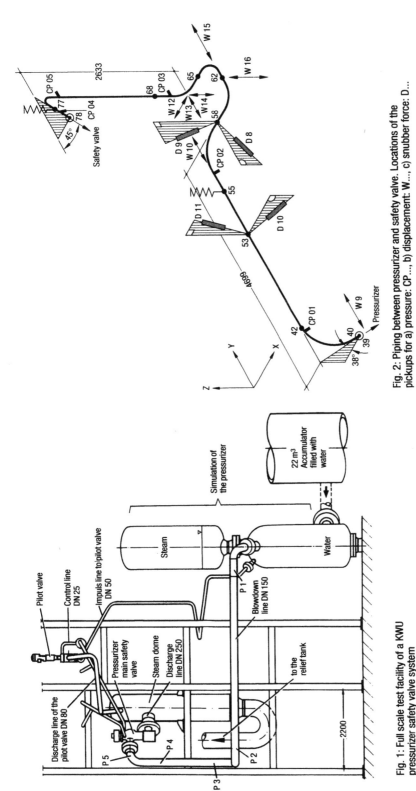

Fig. 2: Piping between pressurizer and safety valve. Locations of the pickups for a) pressure: CP ..., b) displacement: W ...., c) snubber force: D ...

Fig. 1: Full scale test facility of a KWU pressurizer safety valve system

107

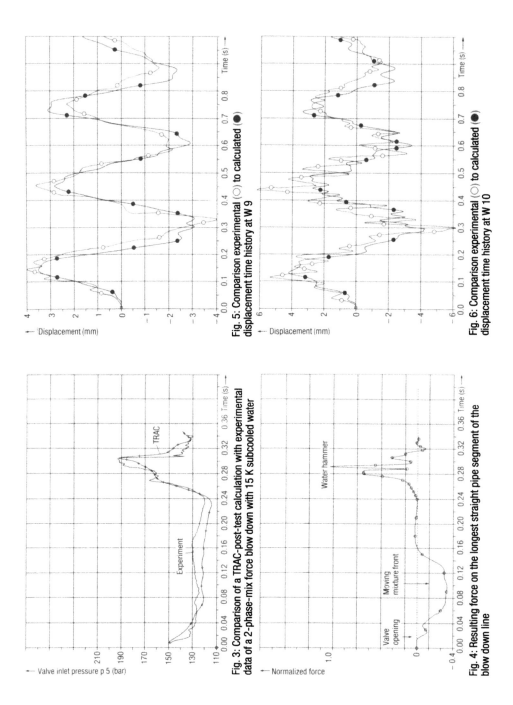

Fig. 5: Comparison experimental (○) to calculated (●)
displacement time history at W 9

Fig. 6: Comparison experimental (○) to calculated (●)
displacement time history at W 10

Fig. 3: Comparison of a TRAC-post-test calculation with experimental
data of a 2-phase-mix force blow down with 15 K subcooled water

Fig. 4: Resulting force on the longest straight pipe segment of the
blow down line

108

Blowdown thrust force and pipe whip analyses of large diameter pipes under LWR LOCA conditions

I.Hashiguchi, K.Kato, S.Ueda, T.Isozaki & S.Miyazono
Japan Atomic Energy Research Institute, Ibaraki-ken

1. Introduction

An instantaneous pipe rupture is postulated as a hypothetical accident in the design of the nuclear power plants. The ruptured pipe at double ended guillotine break would move rapidly and impact surrounding structures. This dynamic motion is called a pipe whip. Pipe whip restraints and other protective structures are installed to limit the pipe whip motion in the nuclear power plants. In order to design such protective structures, it is necessary to evaluate the blowdown thrust force for a ruptured large diameter pipe under the loss of coolant accident (LOCA) conditions. The present paper shows
(1) the verification of the thermal-hydraulic analysis code RELAP4/MOD6 and its post-processor BLOWDOWN for the blowdown thrust force analysis under both BWR and PWR LOCA conditions by comparison between analytical and experimental results,
(2) pipe whip analysis of an 8 inch diameter pipe using the general purpose finite element code ADINA with analytical blowdown thrust force under PWR LOCA conditions, and
(3) blowdown thrust force analyses for various diameters of pipes under both BWR and PWR LOCA conditions.
The jet discharge and the pipe whip tests were carried out using 4, 6 and 8 inch pipes at the Japan Atomic Energy Research Institute (JAERI). BLOWDOWN was developed to calculate blowdown thrust force at JAERI in 1981.

2. Verification of RELAP4/MOD6 and BLOWDOWN

2.1 Experimental and analytical conditions using 8 inch pipe

Figure 1 shows a typical schematic setup of the jet discharge test with an 8 inch pipe at JAERI. The inner diameter of the test pipe is 190.9 mm at BWR test, 170.3 mm at PWR test. The volume of the pressure vessel and the pressurizer is 4.0 m³ and 0.5 m³ respectively. The pressurizer is not used in the experiment under BWR LOCA conditions. Values of 6.76 MPa and 285 °C were settled in the pressure vessel for initial pressure and temperature respectively under BWR LOCA conditions, similarly 15.42 MPa and 325 °C under PWR LOCA conditions. The instantaneous guillotine break was simulated by break of the rupture disk with an electric arc. The Henry-Fauske/homogeneous equilibrium model(HEM) was used as a critical flow model. This is a combination of Henry-Fauske model and HEM with a specified transition from one to the other. The discharge coefficient C_D is selected as 1.0 in all the calculation/1/.

2.2 Comparison of analytical and experimental results

It is necessary to define control volumes in the calculation using RELAP4/MOD6. The test pipe was divided into about 50 volumes and the length of a volume is 300 mm in both BWR and PWR analyses. Calculations using rough models(the length of a volume is 1000 mm) were also performed and analytical results agreed well with experimental ones in BWR analysis. In PWR analysis using this rough model, agreement was not found between analytical and experimental results. The minimum number of the volume to obtain good results is variable with change of calculating conditions in the analysis by RELAP4/MOD6. The distribution of initial temperature in the test pipe is an important parameter. To obtain good agreement between analytical and experimental results, it is necessary to use precise temperature distribution in the analysis by RELAP4/MOD6.

Figure 2 shows the comparison of blowdown thrust forces between analyses and experiments. Under BWR LOCA conditions, the experimental peak value is about 188 kN and it occurs at 283 ms after the jet discharge initiation. On the other hand, the analytical peak value is 175 kN and it occurs at 142 ms. The difference of the peak values is 6.9 %. Under PWR LOCA conditions, the experimental peak value is about 320 kN and it occurs at 150 ms after the jet discharge initiation, and the analytical peak value is 279 kN and it occurs at 190 ms. The difference of the peak values is 12.8 %. It is verified that analytical results agree well with experimental ones under both BWR and PWR LOCA conditions.

3. Pipe whip analysis of 8 inch diameter pipe under PWR LOCA conditions

The evaluation of the dynamic interaction between the test pipe and the restraint was performed about an 8 inch diameter pipe under PWR LOCA conditions. The beam element was utilized for representation of the test pipe, the truss element for two restraints. The overhang length is 500 mm. The analytical blowdown thrust force history from RELAP4/MOD6 and BLOWDOWN was used as input data for ADINA.

Figure 3 shows a comparison of sum of two restraint force histories between analysis and pipe whip test. A good agreement is found in experimental and analytical restraint forces, especially the maximum value, until 100 ms after the jet discharge initiation.

4. Blowdown thrust force analyses of large diameter pipes

Following pipes were adopted to examine the relation between the flow sectional area and the blowdown thrust force : 4, 8, 16, 24 and 28 inch diameter pipes under BWR LOCA conditions, 20, 24, 28, 32 and 34 inch diameter pipes under PWR LOCA conditions.

Table 1 shows the initial conditions for analyses. The volume of the pressure vessel and the pressurizer corresponds to one of 1100 MW power plants in Japan. The initial conditions of inner fluid is the same as experiments at JAERI. Figure 4 and 5 present blowdown thrust force histories from these analyses. Figure 6 shows the relation of the thrust coefficient and the flow sectional area. The thrust coefficient C_T is defined as

$$C_T = F_T/A(P_0-P\infty).$$

In this equation, F_T is the peak value of blowdown thrust force, A the sectional flow area, P_0 the initial pressure in the pressure vessel and $P\infty$ the atmospheric pressure. It is revealed in this figure that C_T increases with the flow sectional area under BWR LOCA conditions and it decreases with increasing flow sectional area under PWR LOCA conditions. The trend of C_T in the BWR analyses comes mainly from the reduction of friction in the analyzed pipe. In the PWR analysis, this is because decreasing of peak value of the pressure at pipe exit and the rapid change of the inner fluid from subcooled state to saturation in the pressure vessel occur with increasing flow sectional area. These relations are shown in Table 2 and Figure 7. Table 2 shows the peak value of pressure at pipe exit for several diameter pipes and Figure 7 shows the pressure history in the pressure vessel under PWR LOCA conditions.

5. Conclusions

The following conclusions are obtained in the present paper.
(1) RELAP4/MOD6 and BLOWDOWN are useful for the blowdown analysis under LWR LOCA conditions.
(2) RELAP4/MOD6, BLOWDOWN and ADINA are useful for the pipe whip analysis under PWR LOCA conditions.
(3) The thrust coefficient increases with the flow sectional area under BWR LOCA conditions, and it decreases with increasing flow sectional area under PWR LOCA conditions.
(4) The thrust coefficient has a value of 0.86 to 1.00 under BWR LOCA conditions, 0.80 to 0.94 under PWR LOCA conditions.

Acknowledgements

This work was performed under the contract between the Science and Technology Agency of Japan and Japan Atomic Energy Research Institute to demonstrate the safety for pipe rupture of the primary coolant circuits in nuclear power plants. The authors would like to appreciate the members of Committee on the assessment of Safety Research for Nuclear Research Structural Components at JAERI (Chairman : Prof. Emeritus. Y.Ando, University of Tokyo) for their fruitful comments. Also they would like to appreciate Mr. K. Sato, Director of Nuclear Safety Research at JAERI, and Dr. T. Shimooke, Dupty Head of Department of Nuclear Safety Research at JAERI, for their great supports.

References

/1/ Yano, T., Miyazaki, N. & Isozaki, T. 1982. Transient analysis of blowdown thrust force under PWR LOCA. Nucl. Engrg. Des. 75: 157-168.
/2/ Sekiya, H., et al. 1985. Experimental and analytical studies of 8 inch pipe whip tests under BWR LOCA conditions. 8th Trans. of the SMiRT.

/3/ Miyazaki, N. & Akimoto, T. 1983. Blowdown force analysis of piping system under LOCA conditions using BLOWDOWN code. Nucl. Engrg. Des. -76: 121-135.
/4/ Miyazaki, N., et al. 1983. Experimental and analytical studies of 4-inch pipe whip tests under PWR LOCA conditions. 7th Trans. of the SMiRT.

Table 1. Conditions to calculate large diameter pipes.

	BWR	PWR
initial pressure in pressure vessel (MPa)	6.76	15.42
initial temperature in pressure vessel (°C)	285	325
volume of pressure vessel (m³)	639.11	167.26
volume of pressurizer (m³)	——	51

Table 2. Peak values of pressure at pipe exit under PWR LOCA conditions.

PWR					
diameter of pipe	20 inch	24 inch	28 inch	32 inch	34 inch
peak value of pressure at pipe exit (MPa)	11.90	11.31	10.96	10.69	10.52

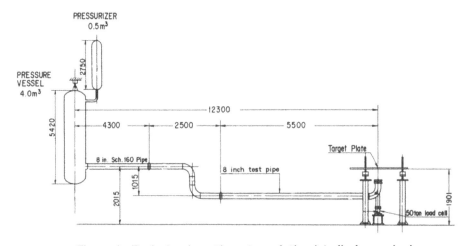

Figure 1. Typical schematic setup of the jet discharge test.

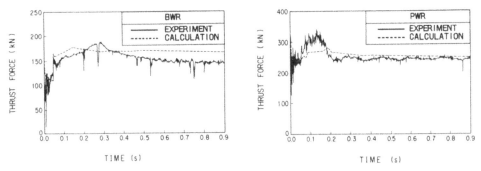

Figure 2. Comparison between analytical and experimental results of the blowdown thrust force.

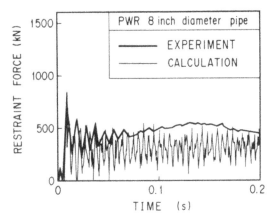

Figure 3. Sum of two restraint forces about an 8 inch test pipe.

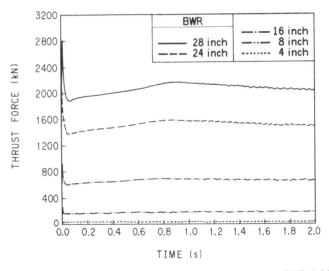

Figure 4. Analytical results of blowdown thrust force under BWR LOCA conditions.

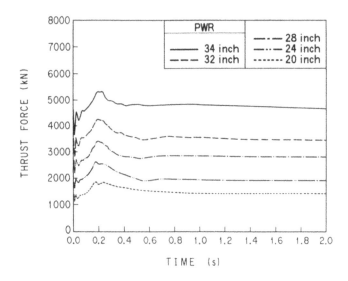

Figure 5. Analytical results of blowdown thrust force under PWR LOCA conditions.

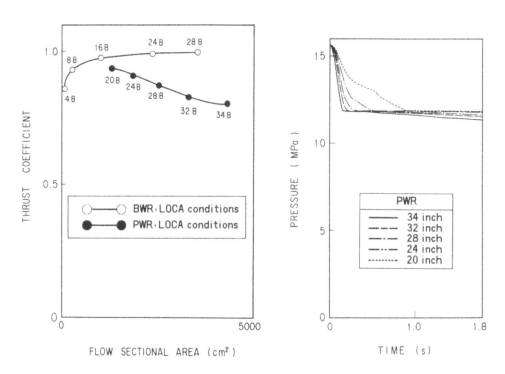

Figure 6. Analytical results of thrust
coefficient.

Figure 7. Analytical results of pressure
history in the pressure vessel
under PWR LOCA conditions.

A multi-step approach for evaluation of pipe impact effects

J.M.Vazquez Sierra
Nuclenor, Santander, Spain

J.Marti
Principia España, S.A., Madrid, Spain

R.Molina
Iberinsa, Madrid, Spain

1 INTRODUCTION

The licensing of new and requalification of existing plant requires the consideration of effects arising from postulated breaks in high-energy lines. If the resulting jets or whipping pipes affect equipment or components (with safety-related functions in relation with the postulated break), their structural integrity and functionality has to be guaranteed. This can be achieved either by demonstrating sufficient ruggedness, or by obviating the problem with hardware (restraints, screens, deflectors, etc). The present paper is orientated towards the first solution.

A methodology has been developed and applied to the requalification of high-energy piping at the Santa María de Garoña NPP in Spain. It provides techniques for evaluation of pipe-whip and jet effects on various structures inside the containment: containment liner, pedestal, shield wall, pipes and penetrations. Items of little structural strength (such as cables, conduits, etc) were excluded from this approach for obvious reasons.

The methodology developed decomposes the evaluation efforts in various levels of decreasing conservatism and increasing complexity and cost. Higher evaluation levels are applied only to interactions where lower levels fail and where there is a reasonable expectation that more realistic analyses might demonstrate acceptability. The evaluation levels are described in this paper for pipe-on-pipe impact interactions. This case has been selected because it is common to all plants and because, from the information presented, other applications (targets other than pipes, jet forces rather than impacts) are relatively straightforward.

2 GENERAL METHODOLOGY

Advances in computer hardware and software have for some time allowed conducting three-dimensional calculations of pipe-on-pipe impact (see, for example, Martí et al, 1983; Mackay et al, 1984). However, that process is expensive and very often unnecessary. Hence, the emphasis must be placed in finding simplified, yet conservative, procedures which allow reserving full 3-D analyses for the cases in which they became unavoidable.

Many researchers have worked on such simplified procedures (see, for instance, Enis et al, 1980; Clendening et al, 1983; Prinja et al, 1984) but the complexity of the problem and number of variables of interest make it difficult to arrive at solutions of generalised usefulness.

The essence of the lower level procedures proposed here is the assumption that the effective missile energy is dissipated plastically in the interaction. The effective energy of the missile is defined as the part of its kinetic energy which corresponds to the velocity change effected by the impact. The assumption can be formulated as follows:

(1) $$E = \int_{o}^{d} F(x) \, dx,$$

where E is the effective missile energy, F(x) is the force of interaction developed as a function of the relative missile-target displacement x, and d is the relative displacement for which E is dissipated.

The structural survivability of the target is then assessed judging whether it can sucessfully sustain F(d) as a static load. The method is conservative but is useful for quick dismissal of many safe interactions.

The procedure hinges on the determination of F(x), that is, the function relating the forces developed to the interpenetrations produced. The derivation of F(x) is discussed in the next section for the case of pipe-on-pipe impact.

3 DETERMINATION OF F(x) FOR PIPE-ON-PIPE IMPACT

When considering the collapse of a pipe, two different mechanisms contribute relative displacements (see Fig.1):

1. Local collapse, which occurs as a progressive ovalisation of the section directly involved; its occurrance is unrelated to possible displacements of the pipe axis.

2. Structural collapse, which appears as a progressive displacement of the pipe axis from its initial position; such deformations may occur independently of whether the local section changes shape are not.

The two mechanisms clearly operate in series. As a first approximation they can be considered to be mutually independent. There are however some cross-influences, eg: large ovalisations facilitate the formation of a structural hinge at that location; and large axis displacements require some ovalisation to take place.

Notice that, although local collapse will not occur in a pipe of greater schedule, structural collapse might take place if the combination of dimensions is adequate for it. In that sense, the "equal schedule argument" for qualifying pipe-on-pipe impact events need not be conservative in all cases.

116

3.1 Local collapse

Following the Enis et al (1980) approach, built on the experimental results of Peech et al (1977), local collapse
can be represented with an expression of the type:

(2) $F = F_o + Kx$,

where F_o and K are functions of the pipe dimensions and material characteristics. The equation neglects initial elastic deformations and is depicted in Figure 2.
   The values of F_o and K are derived based on semiempirical
correlations which include the hypothesis that local collapse is resisted by two additive mechanisms:
   a) ring crushing, which generates a force proportional to
the crushed length
   b) strengthening contributed by the rest of the pipe outside the crushed length, an effect which is independent of
the crushed length.
   The detailed expressions can be found in the referenced
paper. Local deformations must be bound by a limit of what
is admissible. Based on experimental results (Clendening et
al, 1983) and analytical considerations (Prinja et al,
1984), it seemed appropriate and conservative to limit local collapse to decreases in the section diameter below
30%.

3.2 Structural collapse

From the viewpoint of structural collapse, a pipe can be
considered as a beam provided with some support conditions
and subjected to a concentrated load. As the load increases, elastic deformations form until a first plastic
hinge develops. This produces a knee in the force-displacement relationship which describes structural collapse.
   Further increases in the load result in activation of
hardening at the hinge and, eventually, in the formation of
a second hinge. The process continues until unacceptable
plastic strains occur at one of the plastic hinges formed.
Based on licensing precedents, the maximum acceptable
strain was conservatively taken as 45% of the ultimate uniaxial tensile strain of the material.
   The mathematical details are omitted here for lack of
space but they are not too difficult to derive, possibly
with the exception of the incorporation of hardening. Figure 3 presents a typical force-displacement representation
of structural collapse.

3.3 Combination of effects

The local and structural collapse mechanisms of both missile and target pipes act as non-linear springs operating
in series. At each value of the force F, the global tangent
stiffness is:

117

(3) $K = 1/\Sigma(1/K_i)$,

where K_i is the tangent stiffness of mechanism i at that
value of the force.
   This combination procedure permits generating a force-
displacement relationship F(x), as sought, for representing
the energy dissipation characteristics.

4 EVALUATION OF INTERACTIONS

The above relationship can then be applied to the qualification of pipe-on-pipe interactions. If the dissipation of
the effective kinetic energy exceeds one of the allowable
limits, the interaction cannot be qualified. Those limits
are imposed to local deformations (section 3.1), structural
deformations (section 3.2) and, possibly, to losses in
cross-section of the target pipe if it must maintain its
function past the impact even.
   Three-dimensional non-linear dynamic analyses were conducted to confirm the above procedures and indicated that
the ability to dissipate energy was being underestimated.
Actual values were at least 30% and, possibly, as much as
100% greater than calculated.
   The proposed procedure was used as a first level qualification tool. It led to the immediate qualification of about
80% of the almost 400 interactions resulting from high-
energy line breaks that required structural assessment
inside the containment of Garoña NPP.
   When interactions could not be qualified in this way, but
the marging of exceedance was sufficiently small, progressively more sophisticated approaches were utilised, including 3-D dynamic calculations.
   An example of the latter can be seen in Fig. 4. It is an
8in Sch80 pipe continuously supported along two generators
106° apart and impacted by a 10in Sch100 missile. The analysis was carried out with PR3D. The figure shows the deformations of the target pipe when a 40% decrease of the diameter had been reached.
   It is interesting to mention that the combination of the
above procedures, together with detailed safety analyses of
the Garoña plant, allowed evaluating all interactions resulting from high energy breaks inside containment without
recourse to hardware modifications.

5 CONCLUSIONS

A flexible procedure was developed for evaluation of interactions resulting from postulated high-energy line
breaks inside containment.
   The procedure includes steps of increasing sophistica-
tion. It starts with inexpensive energy dissipation assessments for very safe interactions and proceeds to 3-D nonlinear dynamic analyses when warranted.
   The simplest approaches allowed qualification of 80% of
interactions at Garoña NPP. The implementation of the pro-

cedures proposed here, combined with detail safety analyses of the plant, permitted qualifying all resulting interactions without impoing plant design changes.

REFERENCES

Enis, R.O., Bernal, D.B. and Burdetter, E.G. (1980) "A Design Guide for Evaluation of Barriers for Impact from Whipping Pipes", 2nd ASCE Conference on Civil Engineering and Nuclear Power, September.

Clendening, W.R., Saari, K., Wong, H., Masuda, N. and Miyashita K. (1983) "Pipe Whip Experiments Involving Impacts between Pipes", 7th SMiRT, Chicago, Paper F2/2.

Mackay, D.M., Martí, J. and Prinja, N.K. (1984) "Three-Dimensional Analysis of Pipe-on-Pipe Impact and Fracture", 3rd International Conference on Numerical Methods in Fracture Mechanics, Swansea, U.K., March 26-30.

Martí, J., Kalsi, G.S. and Prinja, N.K. (1983) "Three-Dimensional Non-Linear Analysis of Pipe-to-Pipe Impact", ASCE Engineering Mechanics Division Specialty Conference, Purdue University, West Lafayette, Indiana, USA, May 23-25.

Peech, J.M., Roemer, R.E., Pirotin, S.D., East, G.H. and Goldstein, N.A. (1977) "Local Crush Rigidity of Pipes and Elbows", 4th SMiRT, San Francisco, California, August, Paper F3/8.

Prinja, N.K., Day, B.V. and Parker, J.V. (1984), "Damage Criteria for Pipe on Pipe Impact", International Conference on Structural Impact and Crashworthiness, London, 16-20 July, Vol. 2, Elsevier Applied Science Publishers, London.

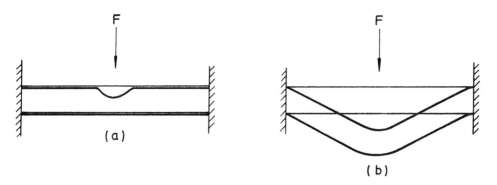

Fig. 1 Pipe collapse mechanisms a) local b) structural

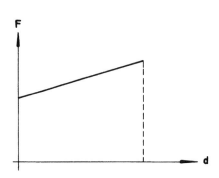

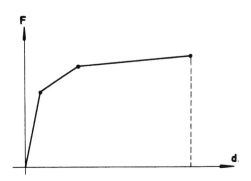

Fig. 2 Force displacement
relationship for local
collapse. Displacement
limited to 30% of diameter

Fig. 3 Force displacement
relationship for structural
collapse. Limited by
development of 45% of
ultimate strains in a hinge

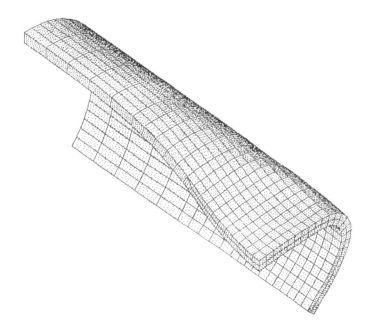

Fig. 4 Deformations of longitudinally supported target pipe
after loss of 40% of local diameter

Pipe-whip experiment and numerical analysis

N.Chiba
Mechanical Engineering Research Laboratory, Hitachi Ltd, Tsuchiura, Japan
N.Sueyoshi, T.Matsunobu & T.Wadayama
Hitachi Works, Hitachi Ltd, Japan
J.Kaneko
Power Reactor and Nuclear Fuel Development Corporation, Tokyo, Japan

1 INTRODUCTION

In a postulated guillotine-break of a pressure-tube-type heavy-water reactor's primary piping, there is a possibility that the ruptured pipe
accelerated by escaping pressurized fluid will whip and form a hinge on
itself. There is also a possibility that the ruptured pipe will strike
the adjacent pipe causing severe local damage to the impacted zone.
These situations may occur because the reactor consists of numerous,
closely packed, small diameter pipes.

In order to assess the integrity of the primary piping system, two
types of impact experiments, whip tests and pipe-to-pipe impact tests,
were carried out with pipes of the same material and same size used in
the reactor. Test pipes were accelerated by detonating powder installed
at the free end of the pipes. The powder's thrust force simulates the
blowdown force in a postulated pipe break accident.

Finite element analsis was also carried out to simulate the dynamic and
highly nonlinear pipe behavior.

2 TEST CONDITIONS

Two types of impact tests were carried out: (a) unrestrained whip test,
and (b) pipe-to-pipe impact test. In the unrestrained whip test for the
inlet piping in the reactor's primary system, a 2-inch schedule 80 pipe,
made of type 316 stainless steel, 4 m in length with one end fixed on the
floor, was accelerated by the detonation of an explosive charge at the
free end under room temperature conditions. The explosive used was chosen so that thrust force and duration time were 12.3 kN and 100 ms, respectively. Thrust force magnitude was determined from a blowdown analysis of the reactor's inlet piping in a postulated pipe rupture utiliz-
ing an envelope curve of the thrust history; Duration time was determined so that thermal reaction from the explosive would be finished before the pipe strikes the floor.

In the pipe-to-pipe impact test for the outlet piping in the reactor's
primary system, a 3-inch schedule 80 pipe, made of type 316 stainless
steel, 5.5 m in length with both ends simply supported, was impacted by
another pipe of the same size and material, 5 m in length, charged with
an explosive at the free end. The gap distance between impact and target
pipe was 70 cm, and the test was carried out under room temperature conditions. The explosive was chosen so that thrust force and duration time

121

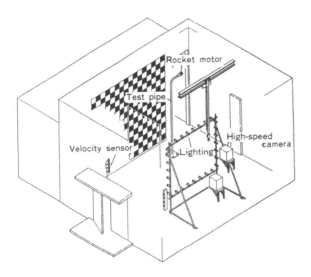

Figure 1. Test facility and instrumentation for 2-inch whip test.

Table 1. Test conditions

(a) Whip test

Pipe					Load
Outer diameter	Thickness	Material	Length	Support	
61.0mm	6.3mm	316s/s	4 m	Cantilever	12.3kN ×100ms

(b) Impact test

	Outer diameter	Thickness	Material	Length	Support	Load
Impact pipe	89.1mm	8.25mm	316s/s	5m	Cantilever	59kN ×27ms
Target pipe	89.5mm	8.4mm	316s/s	5.5m	Simply supported	——

would be 59 kN and 27 ms, respectively. Thrust force magnitude was determined from blowdown analysis of the reactor's outlet piping; Duration time was determined so that the thermal reaction from the explosive would be finished before impact. Test conditions are summarized in Table 1.

The test facility and configuration are shown in Figure 1 for the 2-in whip test. Pressure history measured by a pressure transducer mounted inside the rocket motor's combustion chamber, was converted into thrust force history, using a calibrated relation between pressure and thrust force. Acceleration history, obtained from an accelerometer mounted on the pipe near the free end, was time-integrated to generate the pipe's velocity history. A high speed motion picture (1500 frames per second) was also used for evaluating överall velocity distribution of the pipe. Dynamic and residual pipe strain on the outer surface was. also measured with strain gauges.

122

Local strain near the fixed end of the pipe was also measured after sectioning the pipe. The same facility slightly modified in the configuration was used for the pipe-to-pipe impact test.

3 NUMERICAL ANALYSIS

Pipe impact behavior was analyzed using a nonlinear finite element code ADINA with a beam element after adding the following capabilities to the original version: (i) Follower load capability, necessary for the large rotation analysis in which force direction is built in the deforming body. (ii) Capability to take account of rigitity-decrease relation caused by pipe ovalization. The relation was obtained from a static three point bending test of a 2-inch pipe, shown in Figure 2. The elastic stiffness matrix used in the analysis was multiplied by the rigitity-decrease ratio, which was given as a function of bending angle. (iii) Explicit computation of kinetic energy, elastic strain energy, and dissipation due to plastic deformation energy, to be used in the evaluation of the results.

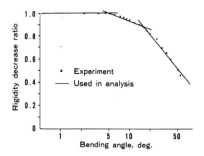

Figure 2. Rigidity decrease effect obtained from static 3-point bending test of a 2-in. pipe.

4 TEST RESULTS COMPARED WITH CALCULATED RESULTS

4.1 Whip test

The overall pipe deformation is shown in Figure 3 at a 40 ms interval compared with the calculated configuration. The pipe was crushed against the floor at 184 ms; In the calculation, time passing the floor line was 190 ms. From the comparison it can be said that overall deformation of the pipe is well simulated by the calculation.

Velocity history, obtained from time-integration of the measured acceleration in the rotating direction of the pipe, is shown in Figure 4. Calculated velocity, corresponding to the same component as experimental one, is also shown in the figure.

Measured and calculated acceleration histories are shown in Figure 5. Component in the rotating direction, same as the velocity history, is plotted in the figure.

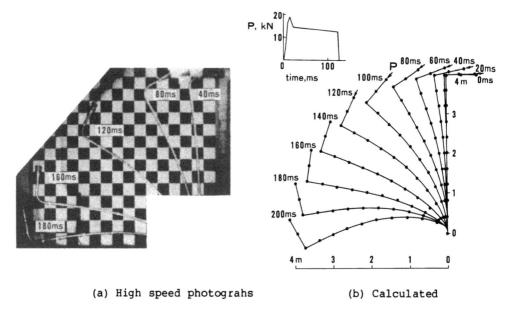

(a) High speed photograhs   (b) Calculated

Figure 3.   Deformation of 2-in. pipe accelerated by detonation at free
end of the pipe.

(a) Measured   (b) Calculated

Figure 4.   Velocity history of 2-in. whipping pipe measured at free end.

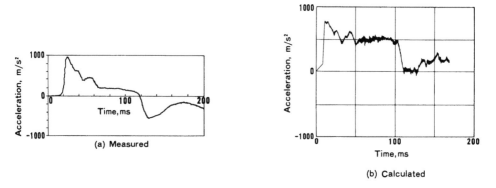

(a) Measured   (b) Calculated

Figure 5.   Acceleration history of 2-in. whipping pipe measured at free
end.

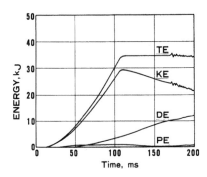

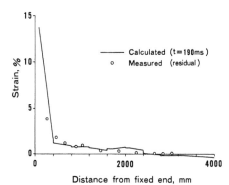

Figure 6. Calculated energy history for 2-in. whip test. TE,KE,DE and PE mean total, kinetic, dissipation, and elastic strain energies, respectively.

Figure 7. Residual strain measured at outer surface in axial direction of 2-in pipe.

Figure 8. Movement of impact and target pipes of 3"-3" impact test. (a) 0 ms (left), (b) 29 ms (center), and (c) 40 ms (right).

Table 2. Summary of calculated results for 2-inch whip test.

	Calculation	Experiment
Displacement	Passed floor line at 190ms	Crushed against floor at 184ms
Max. velocity	47.2m/s at 108ms	43.2m/s at 130ms
Max. acceleration	781m/s² at 11ms	951m/s² at 11ms
Max. residual strain	13.4% at 10cm from fixed end	3.89% at 25cm from fixed end
Bending angle at fixed end	70.6°	71°
Max. kinetic energy	29kJ at 110ms	———

Table 3. Summary of calculated results for 3"-3" impact test.

		Calculation	Experiment
Displacement	Impact pipe at free end	1600mm (100ms)	1141mm (residual)
	Target pipe at impacted zone	1000mm (100ms)	700mm (residual)
Max. Velocity	Impact pipe at free end	39m/s at 29ms	37m/s at 29ms
	Target pipe at impacted zone	40m/s at 33ms	17m/s at 35ms
Kinetic energy of impact pipe		30kJ	31.4kJ
Bending angle	Target pipe at impacted zone	13°	12°

Calculated energy history is shown in Figure 6, where TE, KE, DE and PE stand for total, kinetic, dissipation, and elastic strain energies, respectively. Total energy reaches its maximum of 34 kJ at 124 ms. After completion of explosion, no energy is supplied from outside the system, and energy is maintaind at the same level. Kinetic energy reaches its maximum of 29 kJ at 110 ms, and subsequently decreases monotonically. Dissipated energy due to plastic deformation increases with time, indicating 34% of total energy has been consumed in plastic deformation at 190 ms (when the pipe was crushed against the floor). Elastic strain energy accouts for only 3% of total energy.

Residual strain distribution in the axial direction both measured and calculated at the outer surface of the pipe is shown in Figure 7. It can be said that agreement between them is fairly good execpt for strain near the fixed end, where local constraint effect could not be ignored. The minimum cross-section of the pipe near the fixed end was found 83% of the original cross-section. Calculated and experimental results are summarized in Table 2. Agreement between them is satisfactory except for local deformation of the pipe.

4.2 Pipe-to-pipe impact test

Movement of impact and target pipes, near the impact zone, of 3in.-to-3in. test is shown in Figure 8. Impact velocity was 37 m/s, remaining no cracks in the target pipe. The minimum cross-section of the target pipe at the impacted zone was found 90% of the original cross-section. Numerical analysis was carried out with the same code as in the whip test. The gap between the impact and target pipes was simulated by a nonlinear truss element. Calculated and experimental results are summarized in Table 3. Agreement between them is good.

5 SUMMARY AND CONCLUSION

Pipe-whip and pipe-to-pipe impact experiments and their numerical analyses were performed in order to assess the integrity of the primary piping system of a pressure-tube-type heavy-water-reactor in a postulated guillotine pipe break. Test pipes, whose size and material were the same as those used in the actual plant, were accelerated by means of an explosion caused by detonating powder installed at one end of the pipe. Pipe impact behavior was analyzed using a finite element code ADINA with beam elements for overall pipe behavior. Calculated results were compared with experimental results in terms of deflection, velocity, acceleration and strain, showing good agreement between them. Therefore, overall pipe behavior upon impact could be well simulated by numerical analysis.

Forces on structures by two-phase jet impingement
Experiments and modelling

P.Gruber, W.Kastner, R.Rippel & R.Altpeter
Kraftwerk Union AG, Erlangen, FR Germany

SUMMARY

Studies of the jet impingement forces caused by perpendicular flow
against flat plates – mostly from circular nozzles – have been performed at KWU and by a number of other experimenters. Measurement
data for flow onto other structures (such as H beams and pipes) from
a non-circular nozzle (0.1-A break simulated by a rectangular cross
section) are presented for the first time in this paper. On the basis
of the information obtained, under quasi-steady-state experimental
conditions, on discharge rates, pressure distributions, central dynamic
pressure on the structure concerned etc., a method of calculation
that is simple to use has been developed and provides quite a good
description of the experimental results, as a comparison with measurements shows. The method of calculation can be applied to different
structures by using the drag coefficients known from single-phase
fluid dynamics.

1 INTRODUCTION

For the design and safety analysis of nuclear power plants, the failure
of pressure-retaining piping is postulated. In this connection, it is
necessary to investigate the effects of discharging jets of fluid on
systems and/or structures adjacent to the leak. In most of the cases
considered, the fluid discharging is pressurized water that expands
and thus flashes partially into steam. Calculation of such two-phase
jets is not possible with analytical methods. One way to solve the
problem is to use complex computer codes that have to be verified
experimentally.

Based on many years of experimental studies and their systematic
evaluation, an empirical method of calculation that is simple to use
has been developed at KWU. It provides a cheap alternative, enabling
conservative estimates of essential parameters to be made with little
more than a pocket calculator. Thus the maximum pressure at the stagnation point, the pressure distribution, and the jet impingement
force, on various structures can be determined for steady-state discharge of pressurized water from nozzles of different shape.

2 EXPERIMENTAL WORK

The experimental arrangement used is shown in Figure 1. It consists in the main of the pressure vessel, a discharge nozzle, the structure on which the jet impinges and a low-pressure vessel.

The pressure vessel has a volume of 8.3 m³. Demineralized and de-aerated water is used that can be heated electrically (292 kW). The maximum permissible operating points are 100 bar pressure and 310°C. Subcooled initial conditions are achieved by increasing the pressure still further with nitrogen at an appropriate temperature (up to 100 bar).

The discharge pipe has an inside diameter of 133 mm and is fitted with a wedge-type gate valve as the isolating valve. Opening this gate valve which frees the entire cross section of the pipe, initiates the discharge operation. The discharge nozzle to be tested is attached at the end of the discharge pipe.

The low-pressure vessel, with a volume of about 50 m³, serves as a protection and for separation of the phases; the water arising is drained off through the bottom, while the wet steam produced is vented to atmosphere through an opening with a nominal diameter of 1000 mm.

Various shapes of discharge nozzle were used. Two examples are shown in Figure 2. In the nozzle with the circular opening, the care-fully shaped inlet, and the length of 400 mm, allow the flow and thermodynamic equilibrium to become more or less fully developed. The rectangular nozzle (132 x 10.4 mm) is a simulation of a 0.1-A break of a pipe. Here, the nozzle acts as a flow restrictor, which encourages thermodynamic disequilibria.

The target structures used were a flat plate (900 x 1200 mm), two H beams, "T 53" (53 x 53 mm) and "T 75" (75 x 75 mm), and a pipe "R 75" (outside diameter of 75 mm), as shown in Figure 3.

In addition, the following parameters were varied:
- degree of subcooling in the pressure vessel
- nozzle-structure distance
- orientation of the rectangular nozzle to the beam and pipe structure
    (parallel or perpendicular to the major axis)

The following list provides an overview:

Vessel pressure:		5	$\leq p_B \leq$ 102 bar
Vessel temperature:		20	$\leq T_B \leq$ 300°C
Nozzle geometry:	Circular form:	10	$\leq D \leq$ 65 mm
	Rectangular form:	10.4	x 132 mm²

Hydraulic coefficient
of resistance: $\qquad 0.15 \leq \zeta \leq 2.69$
Nozzle-structure distance: $\qquad 2.5 \leq z \leq 330$ mm

The aim of the experiments performed under quasi-steady-state con-ditions was to investigate
- the thermohydraulic state of flow in the nozzle
- discharge rate
- pressure distribution on the target structures
- jet impingement force on the structures.

3 CALCULATION PROCEDURE

3.1 Discharge rate

The calculation of discharge rates is performed essentially on the basis of known formulae; only for pronounced thermodynamic disequilib-

ria is a correction introduced. The methods are outlined in the following:
  The initial state of the fluid in the pressure vessel is described
by the pressure p_B, and the degree of subcooling $\varphi = p_{sätt}$ $(T_B)/p_B$.
For the range $0 < \varphi \leq 0.8$, a satisfactory agreement with Pana's method
of calculation /1,2/ is obtained. For the range $0.8 < \varphi < 1.0$, the model
according to Moody /3/, in conjunction with Pana's equation, provides
satisfactory results if there is a thermodynamic equilibrium in the
nozzle.
  In the case of the rectangular nozzle used, which leads to pronounced
states of disequilibria because it acts as a flow restrictor, a correction factor

(1) $$\lambda = 0.76 - 0.1875 \cdot \varphi + 0.1563 \cdot \varphi^2 - 0.4688 \cdot \varphi^3$$

was derived additionally, so that the mass flow density can be determined from

(2) $$\dot{m}/A = (\dot{m}/A)_{PANA} / \lambda$$

The agreement between measurements and calculation can be seen in
Figure 4.

3.2 Pressure distribution over structures

The pressure distribution for perpendicular impingement on flat plates
can generally be described by a Gaussian distribution curve (where
"r" is the radial coordinate):

(3) $$[p(r) - p_\infty] = [p_{r=0} - p_\infty] \cdot \exp(-K_r \cdot r^2)$$

with K_r being a factor dependent upon the state of the fluid and the
nozzle, and the nozzle-to-plate distance

(4) $$K_r = C/D^2 \qquad\qquad C = f(z/D, \varphi)$$

The non-dimensional pressure distribution ψ is thus obtained in the
following form:

(5) $$\psi = [p(r) - p_\infty]/[p_{r=0} - p_\infty] = \exp(-K_r \cdot r^2) = \exp[-C \cdot (r/D)^2]$$

For saturated pressurized water:

(6) $$C = 1.938 - 0.99426 \cdot (z/D - 0.25) - 0.73559 \cdot (z/D - 0.25)^2 + 0.85303 \cdot (z/D - 0.25)^3 -$$
$$0.31809 \cdot (z/D - 0.25)^4 + 0.055679 \cdot (z/D - 0.25)^5 - 0.00458193 \cdot (z/D - 0.25)^6 +$$
$$0.000141356 \cdot (z/D - 0.25)^7$$

For subcooled pressurized water, i.e. for $0 \leq \varphi < 1$,

(7) $$C = 1.4 - 0.4 \cdot \varphi$$

must be used. For non-circular nozzles, an equivalent diameter

(8) $$D_{eq} = \sqrt{4 \cdot A/\pi}$$

must be assumed for the nozzle. Furthermore, for oval or rectangular

129

nozzles, the pressure distribution is not in radial symmetry. It can, however, be described by an equation using two coordinate axes:

(9) $$\psi_x = \exp(-K_x \cdot x^2) \qquad\qquad \psi_y = \exp(-K_y \cdot y^2)$$

For a small nozzle-to-plate distance

(10) $$K_x / K_y = b/a$$

holds as an approximation (where "a" is the length of the major axis, and "b" is the length of the minor axis).
    For larger distances,

(11) $$K_x \approx K_y \approx C / D_{eq}^2$$

In addition,

(12) $$\sqrt{K_x \cdot K_y} = C / D_{eq}^2$$

applies on principle.
    The agreement with the experimental data is shown in Figure 5.

3.3 Central dynamic pressure

To be able to quantify the pressure distribution hitherto represented in a non-dimensional form, knowledge of the central dynamic pressure is needed. On the basis of experiments with circular nozzles, the following formula has been developed as a function of φ, ζ, z and D:
    The dynamic pressure is normalized with the pressure in the vessel

(13) $$\eta = [p_{r=0} - p_\infty] / [p_B - p_\infty]$$

According to /4/, the following has been determined empirically

(14) $$\eta = \exp\left[-N \cdot \varphi + (2.54 \cdot N + 3.31 \cdot M) \cdot \varphi^3 - (1.54 \cdot N + 2.31 \cdot M) \cdot \varphi^{4.3}\right]$$

where

(15) $$N = th\left[0.2 \cdot (z/D_{eq} - 1)\right]^2$$

and

(16) $$M = \ln\left[\varepsilon \cdot 0.65 \cdot (z/D_{eq})^{-2.21}\right]$$

with

(17) $$\varepsilon = \exp(-0.1193 \cdot \zeta^{0.603})$$

With a considerable jet constriction (flow restrictor effect), a large degree of subcooling, and small distances z, the equivalent diameter must be corrected.

(18) $$D_{eq}^* = D_{eq} \cdot 3.55 \cdot \exp(-0.0058 \cdot z)$$

In these cases, the occurrence of a pronounced water core which will reach as far the structure must be assumed in the jet. Figure 6 shows

130

the readings obtained for discharge from a rectangular nozzle onto
different structures. The computed curves are also plotted for comparison.

In order to understand the information provided in Figure 6, the
course of the experiment with the jet discharging from the rectangular
nozzle must be dealt with in greater detail. During the fluid discharge,
the pressure in the pressure vessel changes continually (quasi-steady-
state), while the temperature remains constant. Thus, the degree
of subcooling decreases, i.e. the pressure ratio φ approaches more
and more the value 1. If φ is less than 0.7, a pronounced thermodynamic
disequilibrium prevails in the nozzle. The vaporization of the depres-
surized water takes place essentially after the nozzle. The constricted
jet has a water core over a considerable length. If the distances are
small ($z = 80$ mm), the water core impinges on the structure, and large
values are obtained for the non-dimensional dynamic pressure η. If
the degree of subcooling decreases (φ greater than 0.7), the point
of onset of vaporization shifts into the nozzle, the jet expansion
increases greatly, and the central dynamic pressure decreases considerably. This explains the discontinuous transition of some of the measurements. The change in flow behavior must be taken into account in the
method of calculation.

3.4 Jet impingement force

The jet impingement force exerted on a flat plate can be calculated
from the integration of the pressure distribution. As a result, we
obtain:

(19)
$$F_R = \pi \cdot D_{eq}^2 \cdot (\eta / C) \cdot (p_B - p_\infty)$$

Calculated and measured jet impingement forces are represented in
Figure 7.

If the two-phase jet is directed onto a body of finite extent, such
as an H beam, then a simple way to calculate a conservative value
for the jet impingement force is to perform a geometry-related integration of the spatial pressure profile for a flat plate, according
to Figure 8. A simple numerical method of integration produces the
required result as Figure 9 shows.

It has been found in the experiments with a jet from a rectangular
nozzle impinging on different beam structures that the results of
jet impingement force calculations can be applied to other structures
having the same projected area. The forces are in the same ratio as
the drag coefficients known from single-phase fluid dynamics.

4 REFERENCES

/1/ Pana, P.: A modified Bernoulli equation for the critical flow
    discharge in the subcooled region, CSNI paper, Paris, 16-17 May 1974
/2/ Pana, P. and M. Müller: Subcooled and two-phase critical flow
    states and comparison with data. Nuclear Engineering and Design 45
    (1978), pp. 117-125
/3/ Moody, F.J.: Maximum two-phase vessel blowdown from pipes. General
    Electric report APED-4827, 1965.
/4/ Mohammadian, S.: Strahlaufweitungsmodell (Model of jet expansion),
    KWU report, unpublished, Offenbach, FRG, 1984 (in German)

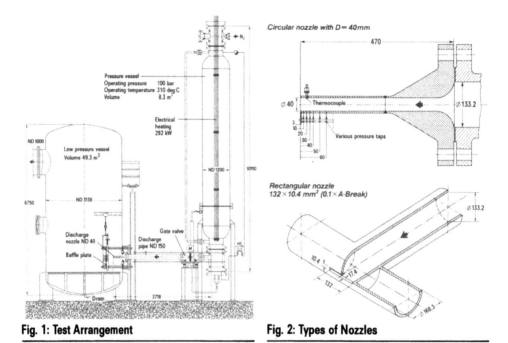

Fig. 1: Test Arrangement

Circular nozzle with D = 40mm

Rectangular nozzle
132 × 10.4 mm² (0.1 × A-Break)

Fig. 2: Types of Nozzles

Test structures

Design of supporting components
for the test structures

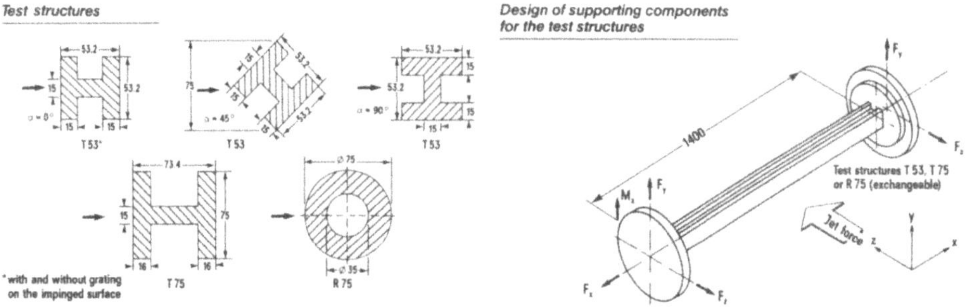

Fig. 3: Design Features

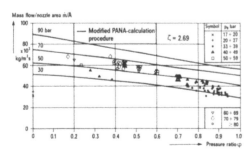

Fig. 4: Critical Mass Flow
Test with Rectangular Nozzle

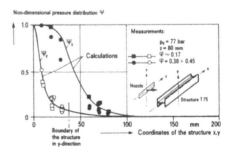

Fig. 5: Pressure Distribution on a Flat Target

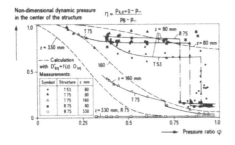

Agreement between measurement and calculation
for two-phase jets with a water core ($\varphi < 0.7$)

Fig. 6: Central Pressure
on Test Structures

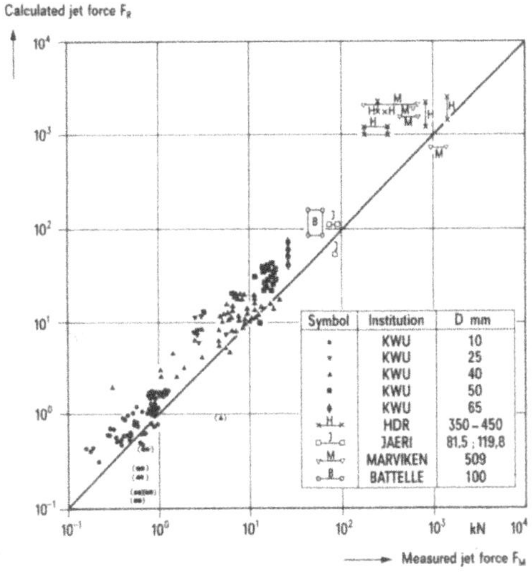

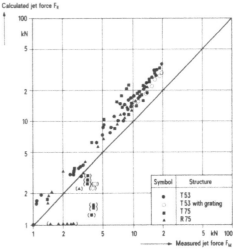

Fig. 7: Comparison of Measurements
and Calculations (Circular Nozzles)

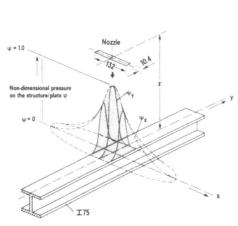

Fig. 8: Calculation Procedure
for Small Targets

Fig. 9: Comparison of Measurements
and Calculations
(Rectangular Nozzle)

133

Environmentally assisted cracking I: Data evaluation

Quantitative measures of environmental enhancement for fatigue crack growth in pressure vessel steels

W.H.Bamford

Westinghouse Generation Technology Systems Division, Pittsburgh, Pa., USA

I.L.Wilson

Westinghouse Research and Development, Pittsburgh, Pa., USA

In 1981 the ASME Code Section XI reference crack growth curves were revised to reflect the state of knowledge of crack growth in light water reactor environments[1]. These curves, shown in figure 1 were developed to allow consideration of the key variables then known, including mean stress (R ratio), BWR vs. PWR environment, and even cyclic frequency effects. The latter two considerations were not directly accounted for but were included through "worst case" arguments, as detailed in reference [1]. This revision of the ASME Code curves remains in effect today, although the variables which affect environmental acceleration in water environments are much better known. We now know that both the microstructure of the material, the chemistry of the water environment, and even the flow rate of the water can have very important impacts on the level of crack growth observed in controlled laboratory tests. The key to further improvement of the reference crack growth curves lies in finding ways to quantify these effects in a way which will allow prediction of growth in structural applications.

1 MATERIALS CHARACTERISATION

Shortly after the revision of the ASME curves it was recognized that the curves were very conservative for some steels, and upon further study it was found that the level of sulfur in the steel had a very important effect [2]. High levels of sulfur in pressure vessel steels, present in manganese sulfide inclusions, resulted in growth rates predicted by the code curves, but lower levels of sulfur (<0.10 wt. percent) produced significantly less enhancement. It seemed that a quantitative measure might be readily available, from the material chemistry, but further study showed that the effect is much more complex, being related to the size, shape and distribution of the inclusions. Plate materials contain a wide range of sizes and shapes of manganese sulfide inclusions, while they are generally much smaller and more spherical in forgings, and even smaller in welds. Therefore the threshold sulfur for significant enhancement appears to vary according to the product form. For plates, as shown in figure 2 for example, even a low sulfur content can result in enhanced growth. On the other hand, figure 3 shows that forgings with low bulk sulfur contents can show very little enhancement, and similar results have been found for welds [3].

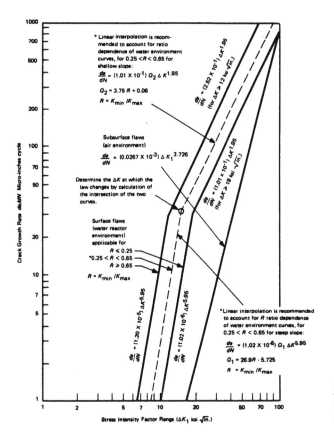

Figure 1. ASME Section XI Reference Fatigue Crack Growth Curves for Pressure Vessel Steels

Recently another attempt has been made to quantify the inclusion content, through use of a standardized method, ASTM E45-76 (method D). The method relies on visual observation of the inclusions on metalographically polished sections.

The inclusion content of selected specimens was recently determined based on surfaces approximately 0.15 ins. from, and parallel to the fracture surfaces after corrosion fatigue testing [3]. The first three steels tested were all low sulfur (0.004%). Two were relatively modern steels and one was an older steel, and all were clean with low inclusion contents. Interestingly the more modern steels exhibited slightly higher inclusion ratings on the ASTM counts, but the older steels had areas where larger inclusions are observed. Sulfur prints from the same surfaces also demonstrated the same finer dispersion in the modern steels, and showed that the sulfides in the older steel were coarser. If the sulfur content and distribution in the steels significantly affects the environmental response, then variations in crack growth rate could be expected as the crack grows through the regions of sulphide segregation in the higher sulfur steels.

The high sulfur (0.025) forging examined using the ASTM method exhibited a higher inclusion count and, although the sulfur content was six times that of the previously described plate steels, the sulfur print for this material displayed a dispersed, relatively homogeneous sulfide distribution. The high sulfur plates exhibited higher inclusion ratings, particularly of the classical sulfide type, and there are more of the heavier type inclusions. Although the weldments have sulfur

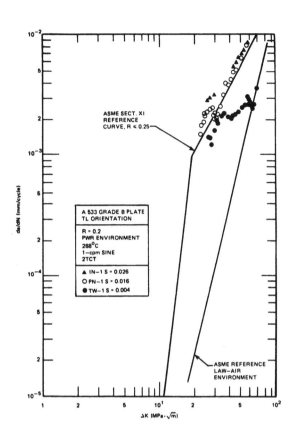

Figure 2. Effect of Sulfur Content on Fatigue Crack Growth in PWR Environment at R=0.2, A533B plates.

contents ranging from 0.008 to 0.014 wt pct., the inclusion sizes were below the scale, and the sulfur prints did not reveal any sulfides.

2 ENVIRONMENTAL CHARACTERISATION, AND MODEL DEVELOPMENT

A great deal of progress has been made in relating the character of the water environment to the level of crack growth enhancement. It is evident that the water in the immediate vicinity of the crack tip is of most importance in determining the rate of crack advance and the corrosion potential and pH are the key characteristics. A number of studies [4,5,6] have confirmed that under controlled oxygen conditions the potential and pH conditions at the crack tip are essentially the same as those in the exposed crack mouth. These values are measurable in operating power systems, and offer some hope of quantification of these effects.

Another important finding, which complicates the environmental quantification, is that the flow of various species in and out of the crack tip vicinity can have an important impact on the crack growth rate. This flow is directly related to the bulk flow, which is also a measurable quantity in an operating system, but the flow rate varies considerably, from nearly static to over one foot per second (.305 m/sec). Attempts to characterize the effects of flow on crack growth enhancement are only in their early stages.

The development of global models to characterize the enhancement of crack growth from water environments is complicated because there

139

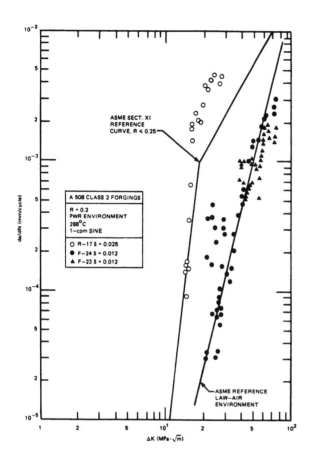

Figure 3. Effect of Sulfur Content on Fatigue Crack Growth in PWR Environment at R=0.2, A508 C2 Forgings

appears to be a strong interaction between the water itself and the material in which the crack is progressing. This is because manganese sulfide inclusions in the steel dissolve readily, and produce sulfur rich anions in the crack tip region, which clearly affect crack growth rates. Of course the amount of manganese sulfide exposed to the environment is dependent on the size and distribution of the inclusions in the steel, as mentioned earlier.

Two different global models have been developed to quantify these effects, one based on slip dissolution (or film rupture) and the other based on a hydrogen embrittlement mechanism. The slip dissolution model is based on the hypothesis that the crack propagation rate is governed by the frequency of oxide rupture at the crack tip and its relation with the passivation rate of the newly exposed material, and has been described in detail [7]. The hydrogen embrittlement model is based on the theory that hydrogen is trapped in the material ahead of the crack, making its advance easier [8]. Since manganese sulfide inclusions are known for trapping hydrogen, this theory relates more strongly to the material affects on crack growth rate. It has also been suggested recently [9] that crack advance may occur by a combination of the two models. An extensive amount of experimental work has been done to develop both of these theories, and although neither are considered mature at this time, the slip dissolution model is more fully quantified. An attempt has been made to develop a model for crack growth based on the available data, but its application to an actual cracking incident [11] indicated that a more refined approach was needed.

3 CONCLUSIONS

Thus, it may be concluded that significant progress has been made in quantifying environmental enhancement of crack growth in the years since publication of the present code model. In addition to more advanced models of cracking, a multitude of additional data has been obtained. These new developments have called attention to the complexity of the actual process, and offer the hope of significant improvements in prediction of environmental cracking. Ford [10] in a recent review of the available theories, found that the crack growth rates predicted by the slip dissolution model could exceed those of the present code reference curves [1], but these growth rates cannot be sustained, and soon fall back under the code predictions. Ford also suggested that a more realistic crack growth prediction should be based on a time dependent analysis, as opposed to the cyclic basis of the current code curves. Such an approach clearly has merit, but will require more work before revision of the ASME Code reference curves are contemplated.

REFERENCES

Bamford, W.H., "Technical Basis for Revised Crack Growth Rate Curves for Pressure Boundary Steels in LWR Environment" ASME Trans. Journal of Pressure Vessel Technology, Vol 102, Nov 1980, pp. 433–442.

Bamford, W. H., "Environmental Cracking of Pressure Boundary Materials, and the Importance of Metallurgical Considerations," in Proceedings ASME Pressure Vessel and Piping Conference, June 1982, PVP Vol 58, p.209.

Bamford, W. H., "Environmentally Assisted Crack Growth in Light Water Reactors," in Heavy Section Steel Technology Program Semi-Annual Progress Report, April September 1986, Report ORNL/TM/9593/V3, November 1986.

Combrade, P., "Prediction of Environmental Crack Growth in Nuclear Power Plant Components," Semi-annual Progress Report No. 4 on EPRI Contracts RP 2006-1 and 2006-8, June 1984.

Gabetta, G., and Buzzanca, G., "Measurement of Corrosion Potential Inside and Outside a Growing Crack During Environmental Fatigue Tests," in Proceedings, IAEA. Second IAEA Specialists Meeting on Subcritical Crack Growth, Sandai, Japan, May 1985, NUREG/CP-0067, April 1986.

Hanninan, H., Illli, H., Torronen, K., and Vulli, M., "On the Electrochemical and Chemical Conditions in Corrosion Fatigue Cracks of Low Alloy Steels in High Temperature Water," in Proceedings, Second IAEA Specialists Meeting on Subcritical Crack Growth, Sandai, Japan, May 1985, NUREG/CP-0067, April 1986.

Ford, F. P., "Mechanisms of Environmental Cracking in Systems Peciliar to the Power Generation Industry," Electric Power Research Institute report EPRI-NP-2589, September 1982.

Hanninan, H., Torronen, K., Kempainen, M., and Salonen, S., in Corrosion Science, Vol 23, 1983.

Sieradski, K., and Newman, R. C., in Philosophical Magazine, Vol 51, pp. 95-132, 1985.

Ford, F. P., "Status of Research on Environmentally-assisted Cracking in LWR Pressure Vessel Steels," in Proceedings ASME Pressure Vessel and Piping Conference, San Diego, CA, June 1987.

Riccardella, P. C., Copeland, J. F., and Gilman, J., "Evaluation of Flaws or Service Induced Cracks in Pressure Vessels," ASME Pressure Vessel and Piping Conference, San Diego, CA, June 1987.

The influence of environment and frequency on near threshold and spectrum loading fatigue behaviour of SA508 RPV steel

R.D.Achilles & J.H.Bulloch

Department of Metallurgy and Materials Engineering, University of the Witwatersrand, Johannesburg, Republic of South Africa

1 INTRODUCTION

Since the discovery of an environmentally assisted fatigue crack growth phenomenon in Light Water Reactor (LWR) environments, numerous studies including Bamford (1980) and Kondo et al (1972), have been conducted over the past fifteen years.

The present paper describes a series of fatigue tests in pure argon and a Pressurised Water Reactor (PWR) environment which have a number of differing loading sequences.

2 EXPERIMENTAL CONDITIONS

An idealised spectrum loading sequence which simulates certain working transients prevalent in a PWR reactor is shown in Figure 1. This complicated loading sequence has been simplified to a series of discrete loading sequences illustrated in Figure 2. Fatigue tests on SA508 Cl III steel in pure argon and PWR water have been conducted utilising the loading sequences shown in Figures 1 and 2. Test frequencies of 3Hz and 0.3Hz and 25mm thick Compact Tension specimens were adopted. All the tests were subjected to 500 blocks except the ripple loading sequence PWR water tests which were allowed to continue in order to obtain more data.

3 RESULTS

The crack growth data for the various loading sequences were characterised by the root mean square stress intensity factor range,

ΔK_{rms} i.e.

$$\Delta K_{rms} = \sqrt{\frac{\Sigma(\Delta K_i)^2 n_i}{\Sigma n_i}}$$

where n_i is the number of load amplitudes corresponding to ΔK_i.

(a) Argon Tests
Crack growth rates of the various loading sequences are shown in Figure 3 and growth rates at the end of the 500 blocks are indicated by an

arrow. The crack growth rate obtained for the R = 0.9 test compares
favourably with relevant argon data reported by Bulloch and Buchanan
(1986). The results also show that:

(i)   trapezoidal loading sequence (TLS) gave fatigue crack growth
rates which were twice as fast as the ripples loading sequence (RLS),

(ii)   peaks loading sequence (PLS) gave fatigue crack growth rates
which were 2,5 times faster than the RLS,

(iii)   spectrum loading sequence (SLS) gave fatigue crack growth rates
which were 4 times faster than the RLS.

Such results indicate that loading sequence affects the fatigue crack
growth behaviour, hence loading sequence effects must be evaluated on the
basis of the effect of overloads and underloads on fatigue crack growth
behaviour.

Scanning Electron Microscope examination of the RLS, Figure 4, and PLS
fracture surfaces reveal a typical stage I fracture mode. This results
from the strong influence of microstructure on stage I crack growth and
occurs when the cyclic plastic zone size is less than the grain size,
i.e. the local stress concentrations are only sufficient to activate one
or a few slip planes. Grain boundaries therefore exert more influence on
the spread of slip to neighbouring grains.

The fracture surfaces of the TLS and SLS specimens however, show stage
II ductile striated growth which suggests that the crack growth behaviour
of the TLS and SLS are no longer influenced by microstructure.

(b) PWR Water Tests

These tests at a frequency of 3Hz were designed to evaluate the
environmental component of fatigue crack growth behaviour. Indeed the
results show that the RLS, PLS, TLS and SLS gave fatigue crack growth
rates that were between 3 and 3.5 times faster than companion tests in an
argon environment, Figure 5. Hence such tests exhibited both mechanical
and environmental components to the fatigue crack extension.

It is well known that, in aqueous environments, reducing the frequency
results in an increase in the fatigue crack growth rate. The 0.3Hz RLS
showed a five fold increase in the fatigue crack growth rate over the 3Hz
test and a sixteen fold increase over the argon test, Figure 6. The
0.3Hz SLS test, however, exhibited growth data that were 2 and 6.5 times
faster than the 3Hz test in PWR water and argon respectively.

The fracture mode of both the RLS and PLS, see Figures 7 & 8, was
again typical stage I cracking; however isolated intergranular facets
were also observed. For the TLS and SLS the fracture mode was again
stage II ductile striated growth; in this case however, the incidence of
intergranular failure (IGF) had increased significantly.

For the 0.3Hz RLS test, Figure 9, the incidence of intergranular
failure increased even further. Area fraction measurements of the
percentage intergranular failure showed that the percentage intergranular
failure decreased from a maximum of 18% at the start of the test to
<1% at a ΔK of 7 MPa$\sqrt{}$m. The fracture surface of the SLS specimen
exhibited several large brittle fan-shaped facets, Figure 10.

4   DISCUSSION

Jones (1983) and Matsuoka et al (1978) have shown that a single overload
of sufficient magnitude (generally 50% of the baseline range) results in
retarded crack growth rates which occur over a crack length increment of
the order of the overload plastic zone size. Indeed cracks overloaded in

this fashion may totally arrest when the effective stress intensity range at the crack tip is reduced below the threshold stress intensity, ΔK_{th}. However, the onset of the retardation or crack arrest often does not coincide with the actual overload cycle but is delayed until the crack has grown some distance into the overload plastic zone (delayed retardation). Further the application of an underload prior to a single overload does not appear to have much effect on the subsequent retardation, whereas an underload following an overload tends to mitigate the retardation effect.

In the case of the argon tests no retardation or crack arrest was observed; indeed crack growth was increased as a result of the overloads and underloads. This lack of retardation of the fatigue crack growth is due to a number of reasons;

(i)   retardation can be explained in terms of crack closure where the larger plastic zone created by the overload cycle results in an increase in the crack opening load giving rise to a lower effective stress intensity range, ΔK_{eff}. However at high R-ratios crack closure does not occur.

(ii)   The degree of retardation is related to the specimen thickness with larger retardations being experienced for thin specimens.

(iii) Decreasing the number of cycles between overloads further reduced the retardation in fatigue crack growth.

The increase in the fatigue crack growth rate is due to the increased stress intensity at the crack tip with each change in R-ratio. The lack of crack closure due to the high R-ratio prevented any decrease in the crack growth rate.

The fractographic observations from the PWR water tests can best be related to the crack growth data in terms of the hydrogen assisted cracking model. In the case of the pure argon tests however, crack growth was purely the result of the mechanical effects.

For the 3Hz PWR water tests, the relatively fast frequency did not allow for much hydrogen diffusion ahead of the crack tip and limited the driving force that causes an environmentally assisted fracture mode. The R = 0.2 cycles of the TLS and SLS allows more diffusion of hydrogen ahead of the crack tip hence the incidence of intergranular failure was more prevalent for the loading sequences containing this low R-ratio cycle.

The reduction in frequency allowed more time for hydrogen diffusion leading to a larger build up of hydrogen ahead of the crack tip giving rise to greater environmental contribution to fatigue crack growth. Depending on the hydrogen concentration and the crack tip stress intensity the environmentally assisted fracture mode was either intergranular failure or brittle transgranular fan-shaped failure.

A significant amount of environmentally assisted fracture IGF was evident at ΔK levels below 7 MPa√m which corresponds to a cyclic plastic zone size of 10 µm. Such an observation agrees well with the work of Cooke and Irving (1975) who suggested that environmentally assisted cracking only occurred when the cyclic plastic zone size was smaller than the material grain size. The observed grain size of SA508 Cl III steel was 10 - 20 µm.

5   CONCLUSIONS and SUMMARY

In the near threshold region fatigue crack growth behaviour of the various loading sequences exhibit good commonality with the R = 0.9, constant amplitude data when plotted as a function of ΔK_{rms}. This holds for both the purely mechanical and the combined mechanical and environmental fatigue situations.

In pure argon no evidence of retardation or delayed retardation due to the overload cycles was found. However, compressive overloads did tend to accelerate crack growth due to the lower crack opening stress intensity; this resulted in the TLS exhibiting a faster fatigue crack growth rate than that of the RLS.

In PWR water the fatigue crack growth rates of all the loading sequences were faster than those recorded in pure argon; these differences were accentuated by decreasing the frequency.

Significant IGF was evident for all the various loading sequences, when environmentally assisted fatigue crack growth was observed in the near threshold region except for the low frequency SLS test which showed evidence of isolated flat, brittle-like fan-shaped regions.

REFERENCES

Bamford, W.H. 1980. Technical basis for revised reference crack growth rate curve for pressure boundary steels in LWR environments. Jnl. Press. Vess. Techn. 102 p443.

Bulloch, J.H. & L.W. Buchanan, 1986. Fatigue crack growth behaviour of A533-B steel in simulated PWR water. Corr. Science 24 p661.

Cooke, R.J. & Irving 1975. The slow fatigue crack growth and threshold behaviour of a medium carbon alloy steel in air and vacuum. Eng. Frac. Mech. 7 p69.

Jones, R.E. 1973. Fatigue crack growth retardation after single cycle peak overload in Ti-6Al-4V Titanium alloys. Eng. Frac. Mech. 5 p585.

Kondo, T., H. Kikuyama, M. Shind, & R. Nagasaki 1972. Corrosion Fatigue of ASTM A-302B steel in high temperature water, in simulated nuclear reactor environment. Corrosion Fatigue (NACE-2) p539.

Matsuoka, S. & K. Tanaka 1978. Delayed retardation phenomenon of fatigue crack growth resulting from a single application of overload. Eng. Frac. Mech. 10 p515.

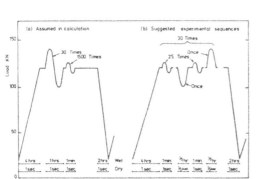

Figure 1: Idealised spectrum loading sequence simulating working transients in a PWR reactor.

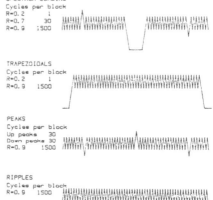

Figure 2: Schematic of various loading sequences adopted in the present study

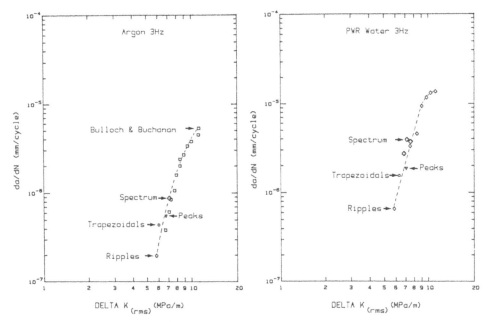

Figure 3: Fatigue crack growth data in argon for 3Hz variable amplitude loading.

Figure 5: Fatigue crack growth data in PWR water for 3Hz variable amplitude loading.

Figure 4: Ripples loading sequence Argon 3Hz.

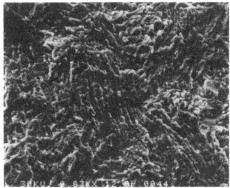

Figure 6: Ripples loading sequence PWR water 3Hz.

147

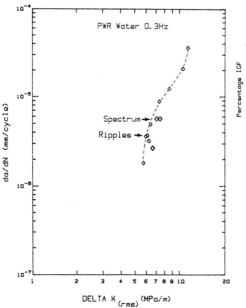

PWR Water 0.3Hz

Spectrum→
Ripples→

da/dN (mm/cycle)

DELTA K $_{(rms)}$ (MPa/m)

Figure 7: Fatigue crack growth data in PWR
water for 0.3Hz variable amplitude loading.

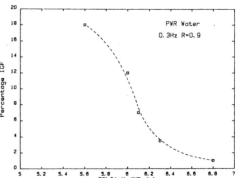

PWR Water
0.3Hz R=0.9

Percentage IGF

DELTA K (MPa/m)

Figure 9: Change in percentage intergranular facets
with increase in stress intensity range.

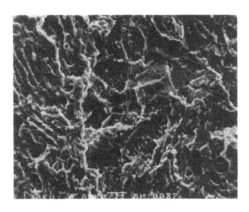

Figure 8: Peaks loading sequence
PWR water 3Hz.

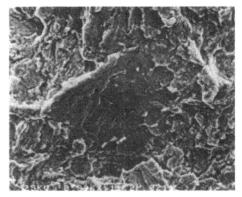

Figure 10: Spectrum loading
sequence PWR water 0.3Hz.

148

The effects of reactor-typical loading on fatigue crack growth of pressure vessel steel in PWR environments

W.H.Cullen
Materials Engineering Associates, Inc., Lanham, Md., USA
D.R.Broek
FractuREsearch, Inc., Galena, Ohio, USA

1. INTRODUCTION

Seven tasks comprise the Characterization of Environmentally-Assisted Crack Growth in LWR Materials program at MEA. Current results on one of them, Development of a Cumulative Damage Factor for Environmentally-Assisted Fatigue Crack Growth, are presented in this summary. Other tasks in the program address aspects of environmental effects on (a) stress-life curves, (b) effect of part-through crack geometry, (c) nucleation and growth, and (d) micromechanisms and models for the calculation and prediction of environmentally-assisted cracking. Completion of these tasks will provide, for the first time, a comprehensive picture of the development and growth to critical size of an environmentally-assisted flaw in a nuclear reactor component.

This improved awareness of flaw growth characteristics and increased accuracy of the calculations of crack extension under reactor-typical conditions of loading and environment will benefit the quality of crack extension calculations which are performed for plant aging studies and in-service inspection evaluations. In the context of this research effort, a variety of pressure vessel and piping steels used in primary boundary applications are used as test specimens. Low-carbon piping steels are used in the stress-life testing and much of the mechanisms research; pressure vessel steels are used in the cumulative damage factor and nucleation and growth studies, and both types of steels are used in the part-through crack growth studies. In many cases, unclad pressure vessel or piping steels are used for studies in which it is conservative to assume that the cladding has been breached or offers no mechanical strengthening for the phenomenon being investigated.

The cumulative usage studies are centered on the testing of compact specimens using a variety of at first simple, and then progressively more and more complex overload and underload cyclic amplitude schemes. Along with the implementation of a data base defining some of the crack acceleration and retardation effects, a calculational model is being developed and calibrated using the baseline and simple spectra data. The hope for this model is that it will be capable of predicting environmentally-assisted crack extension under variable amplitude loading waveforms. Moreover, environmental effects

combined with overload/underload effects have not been carefully
determined for any material/environment system. The question of
whether overloads would cause the same degree of crack growth retardation in the presence of environmental effects remains an open
issue, and there is no basis for assuming any benefits because of
these effects. The development of a calculational model will entail
evaluation of how individual cycles contribute to fatigue crack
growth, and how to integrate these effects to achieve a reasonable
prediction of crack extension under reactor-typical loadings.

2. CUMULATIVE DAMAGE FACTOR FOR ENVIRONMENTALLY-ASSISTED FATIGUE CRACK GROWTH

The specimens for this study came from a heat of steel produced by
Lukens (Heat No. D2819); the chemical and mechanical properties of
this steel are given in Table 1. 2T-CT specimens were cut in the T-L
orientation, using the designations described (ASTM, 1984a). All
specimens were machined to an initial notch depth of 33 mm (1.3 in.)
and precracked an additional 4 mm (0.1 in.). The final applied
stress intensity factor at the conclusion of precracking was
16.2 MPa$\sqrt{m}$ (14.7 ksi$\sqrt{in.}$).

2.1 Experimental Results

The MEA tests were conducted in either a high temperature helium gas
or high-temperature PWR water environment, and in either case, were
carried out in a single specimen autoclave. Each specimen was
instrumented with an LVDT in order to measure the crack mouth opening
of the specimen. Crack mouth opening and load readings were used in
an experimentally-determined formula to calculate the crack lengths
in the specimen while the test were underway. In general, the procedures of ASTM E 647 (ASTM 1984b) were incorporated into the MEA
procedure, but precise procedures for the compliance technique method
of crack extension determination, and for the post-test correction
were not formally part of E 647 at the time this research was carried
out. The initial phase of the testing was to develop crack extension
rates at 288°C for tests in both helium gas and PWR water with
regularly-spaced (in terms of crack length) applications of single
underload or overload cycles. Schematic examples of some of the
waveforms used in this study are shown in Figure 1. Figure 2 shows
results of two tests in PWR water: one a constant amplitude test,
the other with identical loads, but with one full underload/overload
(UL/OL) cycle applied every 1.25 mm (0.050 in.). There are several
interesting observations based on this figure.

(1)   The initial crack growth rate is very high in the test of
        W8R-11, and much lower in the case of W8R-16. The reason for
        this is unknown, but may be a reflection of the strong role of
        local microstructure in the determination of environmentally-
        assisted crack growth rates.

(2)   Complete arrest was obtained after the first UL/OL application,
        and the loads had to be increased in order to extend the crack,
        but crack extension was achieved after successive applications
        of the UL/OL combination.

150

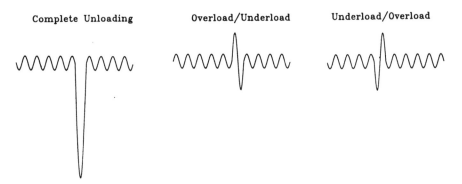

Complete Unloading   Overload/Underload   Underload/Overload

Figure 1. Schematic examples of the waveforms used in the variable
amplitude testing program.

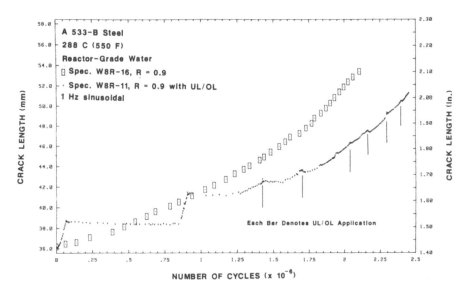

Figure 2. Crack length vs. cyclic counts for specimen W8R-16, tested
at R = 0.9, and specimen W8R-11, interrupted for underload/overload
cycles as shown.

(3)   The crack length and cycle count values for the second UL/OL
       application reside, completely by coincidence, very near the
       values for the constant amplitude test. Therefore it is easy
       to see that, from that point on, the net effect of the UL/OL
       applications is to retard crack extension.

     Several tests have been conducted with a reactor-typical, variable
amplitude waveform consisting of three components with load ratios of
0.2, 0.7, and 0.9. This spectrum, shown in Figure 3, is the same as
that utilized by the International Cyclic Crack Growth Rate (ICCGR)
Group in an ongoing round robin test program. The results of tests
with the ICCGR spectrum waveform are shown in Figure 4. Specimen
W8R-13 was a baseline test in helium gas, while specimen W8R-17 is
the analogue test in PWR water, and the differences between the two
are less than normal experimental differences for such tests even

151

when conducted in identical environments. Similarly, specimen W8R-18
was a second test in PWR water, but the frequency was reduced by a
factor of five. While the crack extension is a little faster, the
difference is not at all proportional to the change in cyclic period.

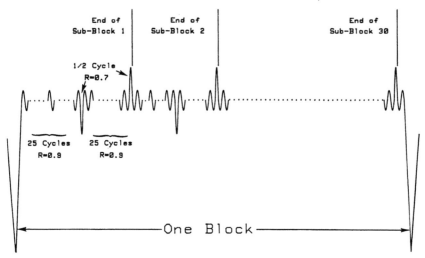

Figure 3. A schematic of the variable amplitude spectrum used by the
International Cyclic Crack Growth Rate (ICCGR) Group in a round robin
test program. This spectrum was also used in this study.

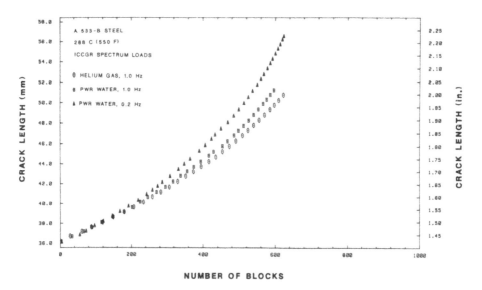

NUMBER OF BLOCKS

Figure 4. Crack length vs. cyclic count for three tests conducted
using the ICCGR variable amplitude waveform. Specimen W8R-13 was
tested in 288 helium gas, while Specimens W8R-17 and W8R-18 were
tested in PWR water using cyclic frequency components of 1 Hz and
0.2 Hz, respectively.

2.2 Analytical Modeling of Variable Amplitude Test Results

The second aspect of this type of work is the development of a calculational model which would utilize the baseline test results as
calibration parameters, and which could be used to predict crack
extensions under variable amplitude loading conditions. The development of this model is in the preliminary stages, but the need for
development of a sophisticated model is clear if a calculation is
made based on the current ASME reference lines using a simple superposition (or linear) method. Figure 5 shows the "prediction" for the
helium gas test at 550°F. Both the prediction based on MEA data
acquired over the appropriate load range, and the prediction based on
ASME subsurface flaw reference line, which is not based accurately on
data obtained in the ΔK range, are low. Similarly, Figure 6 shows
the "prediction" for a PWR environment test using actual MEA data for
PWR environments, and the ASME reference lines for the appropriate
load ratio. In this case the ASME prediction is very far off, over-
conservative by a factor of about ten. The prediction based on MEA
data is closer, but is still off by a significant percentage and is
unconservative.

At the present time, additional spectrum load tests are being
carried out, as well as a few constant amplitude tests to provide the
appropriate calibration data to the calculational model which is
being developed. Late in 1987, a verification test using reactor-
typical cyclic periods and load spectrum will be conducted to
determine if the model can predict in advance the amount of crack
extension to be measured in such a test.

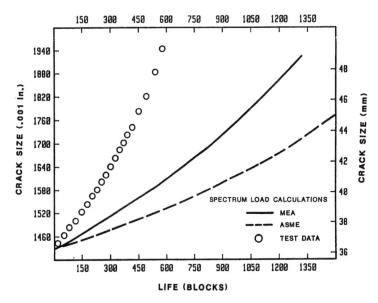

Figure 5. Calculation of crack growth rates for the helium gas
environment tests using a linear superposition model and baseline
growth rates obtained from either the ASME reference line for
subsurface flaws or the MEA-obtained test data in helium gas
environment for the load and ΔK ranges needed.

153

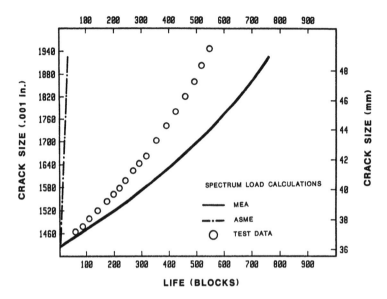

Figure 6. Calculation of crack growth rates for the PWR environment tests using a linear superposition model and baseline growth rates obtained from either the ASME reference lines for surface flaws for the appropriate load ratio, or the MEA-obtained test data in PWR water for the load and ΔK ranges needed.

REFERENCES

ASTM 1984. Standard Method of Plane-Strain Fracture Toughness of Metallic Materials, Designation E 399-83. In 1983 Annual Book of ASTM Standards - Metals Test Methods and Analytical Procedures, Section 3, Vol. 03.01, revised annually, pp. 519-554. Philadelphia:American Society for Testing and Materials.

ASTM 1984. Standard Test Method for Constant Load Amplitude Fatigue Crack Growth Rates Above 10^{-8} m/Cycle," Designation E 647-83. In 1983 Annual ASTM Book of Standards - Metals Test Methods and Analytical Procedure, Section 3, Vol. 03.01, revised annually, pp. 711-731. Philadelphia:American Society for Testing and Materials.

Testing round robin on cyclic crack growth of low and medium sulfur A533-B steels in LWR environments

H.Kitagawa
Department of Mechanical Engineering, Yokohama National University, Japan
K.Komai
Department of Mechanical Engineering, Kyoto University, Japan
H.Nakajima
Department of Fuels and Materials Research, Japan Atomic Energy Research Institute
M.Higuchi
Research Institute, Ishikawajima-Harima Heavy Industries, Yokohama, Japan

1 INTRODUCTION

After the facts of environmentally assisted crack growth of low alloy steel was first observed when cyclically loaded in high temperature water(Kondo et al 1972). The subject has been extensively studied in connection with the evaluation of the integrity of LWR pressure boundary materials. In 1977, International Cooperative Group on Cyclic Crack Growth Rate Testing and Evaluation(the ICCGR group) was organized for more systematic and effective solution of the problem. Successful results have been reported on the programs of the ICCGR activity, particularly in the promotion of a couple of programs of testing round robin and the associated research(Jones 1986).

JAERI also organized a domestic group of 15 organizations as the Corrosion Fatigue Subcommittee(JCF) of the LWR Safety Research Committee to carry out the similar test program. The group has been evaluating the behavior of steels representing the range of quality for the existing Japanese LWR plants.

This paper describes the present status of the Japanese domestic testing round robin and related research especially focused on the test methodology.

2 FIRST PHASE TESTING ROUND ROBIN(Kitagawa et al 1986)

The objectives of this testing round robin are as follows;
 a. Development of common methodological basis.
 b. Evaluation of interlaboratory variability.
 c. Crack growth assessment on the grades of steels used in the domestic power plants.

The participant laboratories are listed in Table 1.

The materials used were manufactured by the Japan Steel Works. The heats of A533B class 1 plates prepared were designated in the plan, as the low and medium sulfur grades with the analyzed sulfur contents of 0.004%(heat L) and 0.014%(heat M), respectively(Table 2). Metallography and mechanical tests of both materials were conducted. All of those results obtained as the basal properties were reasonably within the range for the normal A533B class 1 plates in nuclear applications.

Mechanical and environmental test specifications designated as the common standard are summarized in Table 3 and 4, respectively.

155

Loading conditions were selected so as to follow those applied in the ICCGR low R testing round robin for later comparison. The environmental condition for BWR type water was also designated to adopt the same specification as that for ICCGR high R testing round robin, while that for PWR type water having been somewhat different.

The corrosion fatigue crack growth rate(da/dN) in BWR type water are plotted as a function of stress intensity factor range($\triangle$K) in Figure 1. A half of the results on heat L showed essentially no environmental effect, with the rest half having shown a moderate acceleration relative to the ambient air data. On the other hand, results on heat M showed substantially accelerated crack growth rate. The data points were situated just below the surface flaw(wet) line of ASME Code Sec. XI(1980). Results on the heat M in BWR type water confirmed the results obtained in the ICCGR low R testing round robin.

Figure 2 shows experimentally determined da/dN versus $\triangle$K relationships for PWR type water. All the results showed no essential difference from the air data, with one exception obtained in stagnant water on heat M.

The variability in the data values associated with the crack growth rate obtained by different laboratories was evaluated on the basis of the standard deviation in da/dN. Figure 3 illustrates the variability in different $\triangle$K levels for all the data obtained under BWR type water environment. The trend was seen to be more pronounced in heat M. The extent of the deviation in fatigue test in high temperature water was greater relative to that in air obtained by Clark et al(1975).

In summary there was a consistency among the results from different laboratories, and also an apparent difference due to sulfur content of the materials was recognized in the test results with BWR type water. A common methodological basis was established through this testing round robin.

3 FACTORS OF METHOD AFFECTING CYCLIC CRACK GROWTH IN HIGH
  TEMPERATURE WATER AND SEVERAL ATTEMPTS ON TESTING METHOD IN JAPAN
  (Komai et al 1986)

During the course of the testing round robin, the influence of the test method on the resultant crack growth rate was examined as shown in Table 5. Besides, the typical examples of respective features are shown in Figure 4 - 7. All of the those procedures, i.e., simple preimmersion of test specimens, precracking under high frequency loading in test environment and the starting of fatigue tests from low $\triangle$K region, accelerated the crack growth rate. On the contrary, direct feeding of high temperature water at a high flow velocity to the crack tip decreased the accelerating effect. From these phenomena, the importance of the water chemistry within the cracks was recognized. The crack growth acceleration was hypothesized to occur when the water chemistry within the crack reached a certain critical condition. The observations and the associated hypothesis have been reflected to the methodological development in attempting reliable test data under low $\triangle$K conditions.

In addition to the experiments mentioned above, the following several new attempts have been put into practice;

a. Measurement of crack closure using unloading compliance(Figure 8). The crack opening ratio(U) is higher in the heat L with lower crack growth rate.

b. Two methods of crack length measurement, the reversible direct current potential and the visual observation have yielded mutually consistent results.

c. The fracture surface analysis using computer-aided image processing has been successfully applied to the crack growth rate measurement. The technique has been found to be effective for the less ordered fracture surface in which the analysis was impossible by the conventional techniques.

4 PLAN FOR SECOND PHASE TESTING ROUND ROBIN

Referring to the activities not only of the ICCGR but also of the JCF groups within the recent five years, a great deal of progress has been made in all aspects of subcritical crack growth. In the area of test methodology, mentioned above, reliable test techniques are now available in tracking down possible sources of interlaboratory variability. Exploratory studies have been also attempted in moving from laboratory experiments to application stages. It is clearly recognized that the typical example is to estimate the crack growth in real components. In order that the life prediction of nuclear pressure vessel be made with adequate accuracy, extended research works in the field of basic engineering may be necessary. One of the critical experiments to be dealt with in such an application has been pointed out(Takahashi et al 1986), namely clarification and modeling of the three dimensional growth of surface flaws in LWR coolant environment.

As shown in Table 6, a plan of second phase testing round robin has been settled and a substantial level of effort is being continued in providing key informations to improve the predictability of the environmentally assisted three dimensional crack growth.

5 SUMMARY

For implementation of structural integrity assessment on nuclear reactor pressure vessels, the JCF group was organized aiming at promoting a few series of testing round robin and related research program. The activities focused on the testing round robin and test methodology were reviewed and future direction of the group was introduced.

REFERENCES

Kondo, T. et al. 1972. Proc. of 1st Intern. Conf. on "Corrosion
    Fatigue". 539.
Jones, R.L. 1986, NUREG/CP-0067. vol. 1. 93.
Kitagawa, H. et al. 1986. NUREG/CP-0067, vol.1. 135.
Clarke, W.G., Jr. 1975. J. of Testing & Evaluation. vol. 3. 454.
Komai, K. 1986. NUREG/CP-0067. vol. 1. 69.
Takahashi, H. 1986. NUREG/CP-0067, vol. 2. 411.

Table 1 Participants in the Japanese testing round robin of the JCF

Test Environment	Organization	Principle Investigator
Ambient air	JSW	T. Iwadate
	JAERI	T. Oku
BWR type water	NRIM	N. Nagata.
	Hitachi	R. Sasaki
	IHI	M. Higuchi
	Toshiba	M. Arii
	Babcock-Hitachi	Y. Sakaguchi
	Nippon Kokan	M. Kurihara
	JAERI	T. Kondo
PWR type water	CRIEPI	H. Takaku
	MHI	T. Endo
	Sumitomo Metal In.	K. Tokimasa

Table 3 Loading conditions of testing round robin

Test type	Constant load range
Specimen	1 TCT (initial a/w = 0.40)
Orientation	T-L
Wave form	sine
Frequency	1.0 cpm
R Ratio	0.2
Load	P_{max} = 27.4 kN, P_{min} = 5.5 kN
	starting ΔK = 28 MPa$\sqrt{m}$

Table 2 Chemical composition of materials

Material	C	Si	Mn	P	S	Ni	Cr	Mo	Cu
JCF-Low Sulfur	0.17	0.24	1.37	0.003	0.004	0.60	0.07	0.46	0.02
JCF-Medium Sulfur	0.21	0.29	1.45	0.007	0.014	0.65	0.03	0.51	0.03
Specification	0.25 max	0.15-0.30	1.15-1.50	0.035 max	0.40 max	0.40-0.70	-	0.45-0.60	-

Table 4 Environmental conditions of testing round robin

	BWR	PWR
System pressure	8.3 MPa	15.7 MPa
Test temperature	288°C	320°C
Conductivity	<1 μS	<20 μS
pH	6.0 ± 1.0	
Dissolved O_2	200 ± 50ppb	<10 ppb
Cover gas	Ar or N_2 (if necessary)	—
Li*	—	2.0 ppm
B	—	500 ppm
Cl⁻	<0.1 ppm	<0.1 ppm
F⁻	<0.1 ppm	<0.1 ppm

Table 5 Influence of test method on EAC growth rate

Test Method	Steel	Specimen	Result	Labo. / Figure
Pre-immersion of Specimen	A533B LS* MS**	CDCB	Pre-immersion (10-90days) accelerated crack growth rate at the beginning of experiment	JAERI / Fig.4
High Frequency Precracking in Water	Ditto	1TCT	High frequency precracking in water followed by low frequency test led to high rate of crack growth	IHI / Fig.5
Flow Rate at Crack Tip	Ditto	1TCT	The acceleration of environmental crack growth rate was reduced by high flow rate of water at crack tip	NRIM / Fig.6
Test Beginning at Low ΔK	A533B MS	1TCT	The initial steep rise of crack growth rate curve was shifted to lower ΔK with decrease in initial ΔK level	JAERI / Fig.7

at 288°C, 0.2ppm DO   *LS: Low Sulfur (0.004%)   **MS: Medium Sulfur (0.014%)

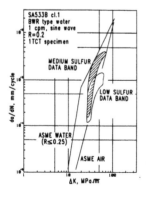

Fig.1 Results of testing round robin in BWR type water

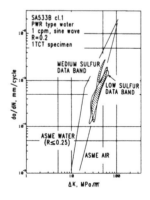

Fig.2 Results of testing round robin in PWR type water

Table 6 Plan of 2nd phase testing round robin(Material: A508 cl.2)

Variables	Core test	Share test	Related test
1)Frequency	1 cpm	1 Hz - 0.1 cpm	
2)Stress ratio	0.2	0.7	
3)Flow rate	low	high, jet	
4)Temperature	288°C/320°C	$\geq$200°C	<200°C
5)Environment	BWR/PWR		
6)D. O.	200ppb/10ppb		
7)Sulfur content	0.014%	0.003%	
8)Specimen size	1T-CT	2T-CT	4T-CT
9)Wave form	Sine		
10)Starting ΔK	25 MPa$\sqrt{m}$	20 MPa$\sqrt{m}$	
			11)Low ΔK
			12)Conductivity (SO_4^{2-})
			13)Fractography
			14)Surface flaw growth

D. O. = Dissolved oxygen

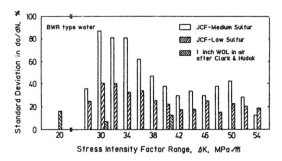

Fig.3 Variation in standard deviation with ΔK range in BWR type water

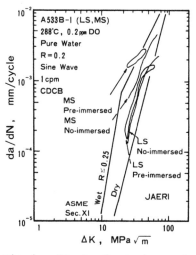

Fig.4 effect of pre-immersion

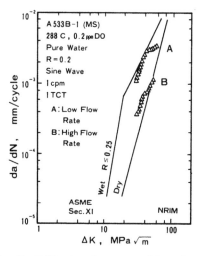

Fig.5 Effect of precracking in water

159

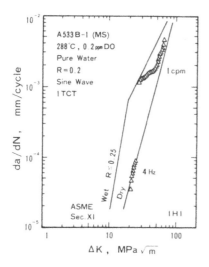

Fig.6 Effect of water flow rate   Fig.7 Effect of starting ΔK

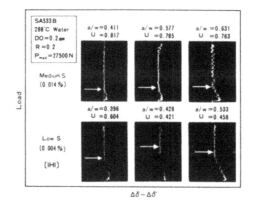

Fig.8 Examples of measurement of P_{op}

Crack growth study on carbon steel in simulated BWR environments

N.Takeda, M.Hishida & M.Kikuchi
Nuclear Energy Group, Toshiba Corp., Yokohama, Japan

K.Hasegawa
Mechanical Engineering Research Laboratory, Hitachi Ltd, Japan

K.Suzuki
Nuclear Power Research and Development Center, TEPCO, Tokyo, Japan

1 INTRODUCTION

For further advancement of BWR structural integrity, life time estimation and prediction of final failure mode for BWR components are
indispensable. These analyses require quantitative characterization
of subcritical flaw growth during plant operations.

This study focuses on crack propagation for BWR piping systems
of carbon steels. Fatigue crack growth data were generated in
simulated BWR water environments using compact tension specimens.
Metallurgical, environmental and stress factor effects on the crack
growth behavior were investigated.

A surface crack study relative to this was described elsewhere
(Hasegawa 1987).

2 EXPERIMENTAL

The 1T CT specimens with side grooves were machined from 20B Sch.
100 pipe of JIS G 3455 STS42 carbon steel and from 26B Sch. 80 pipe
of STS49 carbon steel with the crack extension directions being parallel to the circumference of these pipes. The chemical composition
of these steel pipes are shown in Table 1. A few CT specimens made
of weld metal were fabricated from a butt weld joint of STS42 Sch.
100 pipes.

Prior to the experiments, CT specimens were pre-cracked in air
at room temperature by fatigue at a stress intensity factor not beyond
an initial value for the experiments.

Corrosion fatigue crack growth test as basic data was accomplished
under 2×10^{-2} Hz triangular waveform with stress ratio 0.5 in 288°C
pure water containing 8 ppm dissolved oxygen. Then, the test conditions were changed independently in the basic crack growth test condition in order to examine their effects on the crack growth rates,
from 2×10^{-2} Hz to 2×10^{-3} Hz, from 0.5 stress ratio to 0.2 stress
ratio and from 288°C to 150°C. Trapezoidal waveform tests were also
carried out to identify stress corrosion cracking contribution to
crack growth rates during corrosion fatigue tests. Furthermore,
some specimens were tested in 288°C air-saturated steam environment.
Fatigue crack growth rates in room temperature air were obtained
for each specimen as the reference to those in water and steam environment as above. The test conditions are summarized in Table 2.

Crack extension was monitored by a compliance method and crack
growth rates were determined by seven points incremental polynomial
method in most of the tests. All the rates were plotted against
applied cyclic stress intensity factor range, ΔK, in double logarithm
scales.

3 RESULTS AND DISCUSSION

Fatigue crack growth rates obtained in room temperature air condition
are shown in Fig.1. There was not any noticeable difference in crack
growth rates between STS42 and STS49 pipes. No difference was also
found in crack growth rates between the base material and the weld
metal. These crack growth rates in low ΔK range exceeded the reference
curve for carbon and low alloy steels in air of ASME Code Sec. XI
(ASME 1983) as shown in Fig.1.
   Crack propagation rates obtained in 288°C pure water under triangular
waveform at 2×10^{-2} Hz and 2×10^{-3} Hz are also shown in Fig. 1.
ΔK dependency of the crack growth rate was complicated. It seems
that only the rates in low ΔK range followed Paris law relationship
and the rates in high ΔK range were retarded or remained unchanged
In some cases, the crack growth rate decreased followed by another
increase. All the obtained crack growth rates except those under
2×10^{-3} Hz triangular waveform were located within the ASME code
reference curve in water at stress ratio $R \geq 0.65$. In 2×10^{-3} Hz
triangular waveform test, the rates were several times higher than
the rates at 2×10^{-2} Hz at high stress intensity factor range and
the rates were beyond the ASME code reference curve.
   Crack growth rates in 288°C 8 ppm DO water under 2×10^{-3} Hz and
2×10^{-4} Hz trapezoidal waveform are shown in Fig.2. Note that no
detectable crack growth was obtained under 2×10^{-4} Hz trapezoidal
waveform until stress intensity factor range ΔK exceeding about 30
MPa.m$^{1/2}$ for both pipe steels.
   The 2×10^{-3} Hz trapezoidal waveform test and the 2×10^{-2} Hz
triangular waveform test did not give any noticeable differences
in crack growth rates, however, the 2×10^{-4} Hz trapezoidal waveform
test showed slightly higher crack growth rates at around $\Delta K=30$ MPa.
m$^{1/2}$. In the trapezoidal waveform tests, the waveforms at 2×10^{-3} Hz
and 2×10^{-4} Hz were composed by holding for 450 seconds and 4950
seconds at the top load in the 2×10^{-2} Hz triangular waveform test,
respectively. Therefore, crack growth rate acceleration in the
2×10^{-4} Hz trapezoidal waveform test was due to stress corrosion
cracking for top load holding periods of 4950 seconds. Subtraction
of crack growth rates under 2×10^{-2} Hz triangular waveform from those
under 2×10^{-4} Hz trapezoidal waveform could give a stress corrosion
cracking propagation rate (Kawakubo 1980). The rates from 3×10^{-10}
m/s to 8×10^{-10} m/s are calculated at around $K=60$ MPa.m$^{1/2}$ as stress
corrosion cracking propagation rates of carbon steel in 288°C 8 ppm
DO water.
   As for stress ratio, decrease from 0.5 to 0.2 shifted the crack
growth rate curve to a high ΔK direction as shown in Fig.3. An
effective cyclic stress intensity factor, K_{eff}, instead of ΔK, seemed
to be available for plotting different sets of rates at stress ratios
of 0.5 and 0.2.
   Temperature change from 288°C to 150°C under 2×10^{-2} Hz triangular
waveform apparently lowered the crack growth rate as shown in Fig.4.
temperature air.

Table 1. Chemical Composition of STS 42 and 49 Carbon Steel

(Wt%)

	C	Si	Mn	P	S
STS 42	0.22	0.30	1.20	0.025	0.012
STS 49	0.20	0.33	1.16	0.026	0.012

Table 2. Test Conditions of Corrosion Fatigue Tests

Material	Environment			Temperature(C)			Dissolved Oxgen	Stress Waveform		Stress ratio		Frequency (Hz)		
	Water	Steam	Air	288	150	RT	8(ppm)	Triangular	Trapezoidal	0.5	0.2	2×10^{-2}	2×10^{-3}	2×10^{-4}
STS 42 20B Sch 100	O			O			O	O		O		O		
	O			O			O	O		O			O	
	O			O			O	O			O	O		
	O			O			O		O	O			O	
	O			O			O		O	O				O
	O				O		O	O		O		O		
		O		O			O	O		O		O		
		O		O			O		O	O			O	
STS 49 26B Sch 80			O			O		O						
		O		O			O	O		O		O		
			O			O		O		O				
Weld Metal	O			O			O	O		O		O		
			O			O		O		O				

Environmental change from 8 ppm dissolved oxygen containing water to air-saturated steam lowered the crack growth rate as shown in Fig.5. Changing stress waveform from 10^{-2} Hz triangular to 10^{-3} Hz trapezoidal did not give any noticeable increase in growth rate in steam environment as well as that in 288°C 8 ppm DO water. And the rate in steam environment is generally lower than that in 288°C 8 ppm DO water.

Therefore, crack growth in steam environment can be evaluated conservatively by the rate in 288°C 8 ppm DO water.

Crack growth rate of weld metal in 288°C 8 ppm DO water under 2×10^{-2} Hz triangular waveform are shown in Fig.6. The growth rate is almost equal to or lower than those of the base materials.

4 CONCLUSION

Fatigue crack growth rates of carbon steels are obtained using compact tension specimen in high temperature water and in steam environments and the results are summarized as follows.

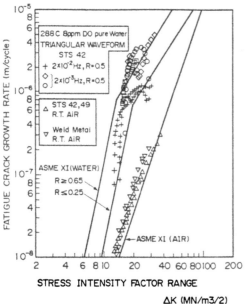

Fig.1 Crack growth rates for STS 42 carbon
steel obtained in triangular waveform tests

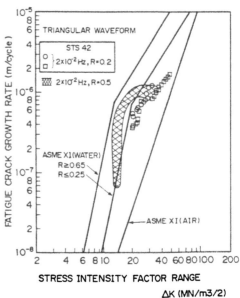

Fig. 3 Crack growth rates for STS 42 carbon
steel under stress ratio R=0.2 in 288°C 8ppm
DO water

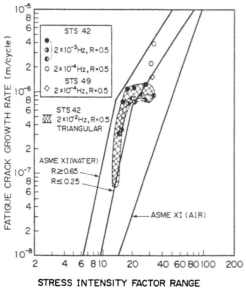

Fig. 2 Crack growth rates for STS 42
and STS 49 carbon steels in trapezoidal
waveform tests (288°C 8ppm DO, pure water)

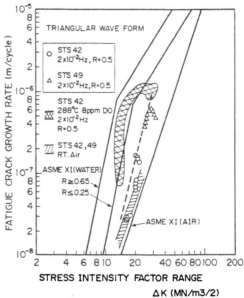

Fig. 4 Crack growth rates for STS 42 and
STS 49 carbon steels in 150°C 8ppm DO water

164

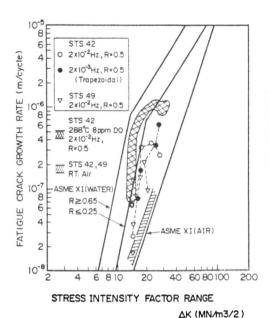

Fig. 5 Crack growth rates for STS 42 and STS 49 carbon steels in 288°C air-satulated steam

Fig. 6 Crack growth rates for weld metal in 288°C 8ppm DO water

(1) Crack growth rates strongly depended on the frequency in triangular waveform. The rates became higher as the frequency lowered from 2×10^{-2} Hz to 2×10^{-3} Hz.

(2) Crack growth rate are almost equal under 2×10^{-3} Hz Trapezoidal waveform and under 2×10^{-2} Hz triangular waveform.

(3) Crack growth rate are lowered as stress ratio decreased from 0.5 to 0.2.

(4) Temperature change from 288°c to 150°C apparently lowered the crack growth rate.

(5) 288°C 8 ppm DO water environment gave much higher acceleration on crack growth rate than 288°C air-saturated steam.

(6) Crack growth rate for weld metal in 288°C pure water are almost equal to or lower than that of base metal.

ACKNOWLEDGEMENTS

This project has been performed under a joint research program of Tokyo Electric Power Co. Inc., Hitachi Ltd, and Toshiba Corp.

REFERENCES

Hasegawa, K., et al. 1987. Surface Crack Growth Behavior for Carbon Steel Piping in BWR Water Environment to be presented at the 9th International Conference on Structural Mechanics in Reactor Technology.

ASME. 1983. ASME Boiler and Pressure Code; Section XI.

Kawakubo, T., et al. 1980. Crack Growth Behavior of Type 304 Stainless Steel in Oxygenated 290°C pure Water under Low Frequency Cyclic Loading. Corrosion, Vol.36, No.11, p.638-647.

Effects of environmental factors on fatigue crack growth behaviours of A533B steel in BWR water

N.Nagata & Y.Katada

National Research Institute for Metals, Tokyo, Japan

1 INTRODUCTION

Much progress has been made in a knowledge of corrosion fatigue behaviours of low alloy pressure vessel steels in LWR environments by
collaborative works, for instance, ICCGR (International cooperative
group on Cyclic Crack Growth Rate) and JCF(Japanese committee on Corrosion Fatigue). Quantitative evaluation of influential factors not
only of mechanical ones such as stress ratio, frequency, wave form but
also of environmental ones such as temperature, dissolved oxygen,flow
rate and material ones such as sulfur content in steels has been
disclosed(Cullen 1983a,1986). Compared to the mechanical factors,
however, the knowledge of the environmental factors seems to be insufficient and there still remain areas to be investigated. Among the environmental factors influencing the fatigue crack growth rate in low
alloy steels in LWR water both temperature and dissolved oxygen(DO) are
considered to be critical variables.

The effect of temperature on crack growth rates in low alloy steels
has so far been investigated mainly in PWR conditions(Atkinson 1986,
Cullen 1983b) after the preceding work in BWR condition(Kondo 1971). It
was found that the results obtained were complicated and inconsistent in
some means. As for the effect of DO concentration, ICCGR round robin
test results have shown no or little dependence in a range from 0ppb to
8ppm(Hale, Jones 1983). However, there were very few data to see the effect and further investigations are required. The present research has
been undertaken to examine the effects of temperature and DO concentration in wide ranges including PWR and BWR conditions on the crack growth
rates in low alloy pressure vessel steel in pressurized pure water.

2 EXPERIMENTAL PROCEDURE

2.1 Apparatus

The experimental apparatus employed consists of an electro servo-
hydraulic fatigue testing machine of 98KN, an autoclave with a capacity
of 15ℓ and high temperature pressurized water loop. The loop was
operated at a pressure of 7.8MPa and a flow rate of 60 ℓ/h for recirculating water. Applied loads were compensated by the friction of pull
rod at pressure seals. Water chemistry was basically simulated to BWR
coolant using deionized and deoxygenated pure water. Measurements

Table1. Test conditions and water chemistry specifications

Material	ASTM A533B cl.1	Temperature	320,288,250,225,200,
Specimens	Compact tension		175,150,100C,RT
Stress ratio	0.1	Dissolved	1,10,100,1000,
Frequency	0.0167Hz	oxygen	8000ppb
Wave form	Sinusoidal	pH	6.3 - 6.8
Pressure	7.8MPa	Conductivity	<0.2μS/cm

of the water chemistry were carried out in terms of DO, conductivity and pH at both the feed water line and the return water line in the loop during operation. DO concentrations were monitored continuously and controlled within an accuracy of 5% of each test condition using a combination of nitrogen gas deaeration and argon-oxygen mixed gas oxygenation. Actual levels of DO concentration in returned water were normally down to 60-80% of the controlled levels. Both the mechanical and environmental test conditions employed in the present study are shown in Table 1.

2.2 Material and specimen

The material used was a low alloy pressure vessel steel plate equivalent to ASTM A533B cl.1(specified as JIS G3120 SQV2A) subjected to the same heat treatment as those for the reactor pressure vessels. Sulfur content considered as a critical element for corrosion fatigue behaviours was 0.007 wt%. Two types of compact tension specimens were employed. One was typical 1TCT of a L-S orientation. The other was 25mm thick and 120mm in width of a T-S orientation. Monitoring of the crack length of the specimen was conducted by a compliance method using a linear variable differential transformer which was mounted on the crack mouth of the specimen in the autoclave. The crack length measurements for the large type specimens were also carried out simultaneously by a direct observation method using an optical microscope through the pressure boundary.

3 RESULTS

3.1 Effect of temperature

For corrosion fatigue tests on the effect of temperature DO concentration was fixed at 100ppb to simulate BWR water condition. Typical examples of the relations between crack length a and number of cycles N for 1TCT specimens at temperatures tested are shown in Figure 1. It is seen that the trends of a-N curves are similar, but the total numbers of cycles vary with the test temperature. It was found that the effect of temperature for the large type CT specimens was the same as those for 1TCT e.g. the number of cycles attained to the same crack length increment was the largest at 175C and the smallest at 100C. Therefore, following descriptions will be made mainly on the typical results for 1TCT specimens.

The fatigue crack growth rates were calculated by using the seven points incremental polynomial method from the a-N data and plotted against stress intensity factor range ΔK as shown in Figure 2. In the figure the reference curves for both water and air in ASME Code

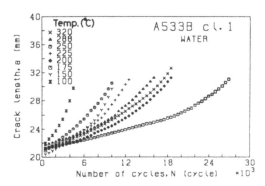

Fig.1 Effect of temperature on crack length vs number of cycles curves in water.

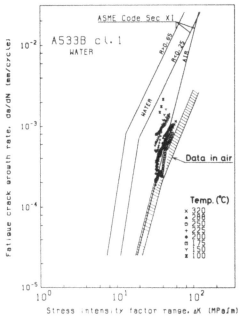

Fig.2 Effect of temperature on crack growth rates vs ΔK data.

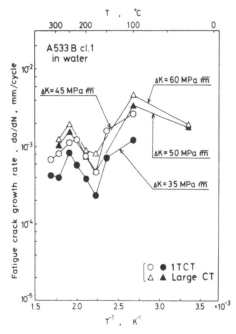

Fig.3 Effects of temperature on crack growth rates at specific values of ΔK.

Sec.XI (ASME 1980) are shown for comparison. It is clearly seen that the crack growth rates vary with test temperature: a maximum is observed at 100C while a minimum at 175C. However, no data points can exceed the reference curve for the lower stress ratio. The results for the BWR operating condition i.e. 288C and 100ppb DO concentration showed the crack growth rates as low as the ASME reference air line. Temperature dependence of the crack growth rates in A533B steel in water was demonstrated as a function of inverse of temperature as shown in Figure 3, in which the crack growth rates for both types of specimens were obtained for specific values of ΔK. Complicated behaviours of crack growth rates were observed: the maxima appeared at 100C and 250C while the minimum at 175C, and the curves are divided into four sectors at each discontinuity. It is apparent that the general trends are independent of the values of ΔK, specimen size and specimen orientation.

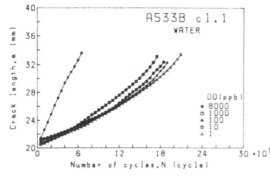

Fig.4 Effect of DO concentration
on crack length vs cycles curves
at 288C.

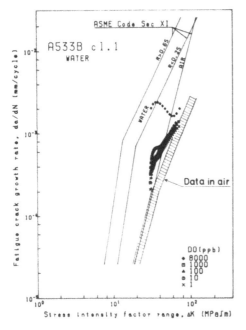

Fig.5 Effect of DO concentration
on crack growth rate vs ΔK data.

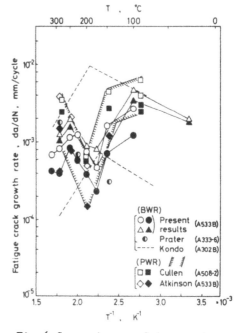

Fig.6 Comparisons of temperature
dependence of crack growth rates
in BWR and PWR environments.

3.2 Effect of dissolved oxygen

The effect of dissolved oxygen on crack growth rates in A533B in water
was examined in a wide range of DO concentration from 1ppb to 8000ppb.
In this case 1TCT specimens were used, and temperature and pressure of
water were fixed at 288C and 7.8MPa respectively in order to simulate
the BWR environment. Figure 4 shows examples of a-N curves at DO concentrations of 1, 10, 100, 1000 and 8000ppb. The number of cycles to
attain to the same crack length increment decreases with increasing DO
concentration, especially at DO above 1000ppb. The relations between
crack growth rate and stress intensity factor range for the data in
Figure 4 are shown in Figure 5. At DO concentrations below 1000ppb the
crack growth rates increase with increasing ΔK and scatter around ASME
Code air line. At DO concentration of 8000ppb, on the other hand, the

crack growth rates tend to decrease with increasing ∆K, although the values of growth rate exceed those for lower DO concentrations and likely to cross to the ASME Code water line at initial portion of the curve. The crack growth rate at this point was the highest value among the data obtained by the present experiments for the low sulfur steel.

3.3 Fractography

Fractography of the cracked surfaces were examined by a low power optical microscope and a scanning electron microscope. It was found that the fracture surfaces of all the specimens tested except those at room temperature were covered with oxide films and stained in black colour. The specimen tested at DO concentration of 8000ppb showed locally greenish gray colour. X-ray diffraction analyses revealed that the black oxide film consisted of magnetite while the greenish gray oxide film hematite. Surface roughness varied with test temperature and DO concentration. The roughest fracture surfaces were observed at temperatures from 175C to 250C accompanying with extended facets and subcracks.

4 DISCUSSION

The present study has shown that the fatigue crack growth rates of A533B steel were accelerated by the water environment and the extent of acceleration was strongly influenced by water temperature and DO concentration. These conclusions were derived from the comparison of the present data to the reference curve for the crack growth rates in air in ASME Code Sec. XI, though some of the present data fell on below the reference curve for air. It is observed, however, that when compared to the crack growth rates of the present material tested in air at room temperature, no data for water environment scatter below the data for air shown as hatched bands in Figures 2 and 5 even for the lowest crack growth rates obtained in water at 175C. Therefore, it can be said that the acceleration by water environment necessarily occurs in corrosion fatigue behaviours of low alloy steels. The temperature dependence of the crack growth rates shown in Figure 3 reflects the extent of acceleration. It is obvious that within the range of temperature tested the maximum crack growth rate appears at 100C and the minimum at 175C, and that the temperature dependence is independent of specimen size and orientation. The present results have been compared to the data appeared in the literatures as shown in Figure 6(Katada 1985). It is apparent that the general trends of the temperature dependence are similar between BWR and PWR conditions except that reported by Kondo et al., in which the test conditions were different from the others in terms of material, specimen geometry and loading condition. There are few direct comparisons of crack growth rates between BWR and PWR water conditions. It seems, however, that the difference between both water conditions might not influence the crack growth rates in water even though some extents of scatter were observed due probably to interlaboratory test conditions. Since the present results showed that the crack growth rates are apparently independent of DO concentration below 1000ppb at BWR temperature,it can be concluded that the crack growth behaviours in low sulfur steels in BWR and PWR water conditions are inherently similar in terms of temperature and dissolved oxygen.

As aforementioned, macroscopic fracture surfaces varied with temperature and rough surfaces were observed in a range from 175C to 250C, especially at 200C. Similar results were also observed under a PWR condition by Atkinson et al.. In addition, zig-zag crack propagations and subcrackings were recognized by the direct observation. Surface roughness of the fatigue cracked planes may induce closure effect so that effective stress intensity factor range reduces during the tests resulting in reduction of actual crack growth rates. Therefore, the low crack growth rates obtained at the temperature range from 150C to 225C may partly be attributed to these closure effect and subcracking. It is interesting to note that the crack growth rates observed at DO concentration of 8000ppb decrease with increasing ΔK as seen in Figure 4. This may imply that in high temperature water of high DO concentration cracked surfaces are oxidized so heavy that the oxide induced closure becomes operative resulting in the reduction of effective ΔK.

5 CONCLUSION

(1) The temperature dependence of the crack growth rates in a range from room temperature to 320C indicated a complicated behaviour:the maximum growth rate was observed at 100C and the minimum at 175C.
(2) The DO concentration dependence of the crack growth rates in a range from 1ppb to 8000ppb showed that the crack growth rates unchanged in a range of DO concentration below 1000ppb, above which, on the other hand, they increased with increasing DO concentration.
(3) The crack growth rates in low sulfur steels in high temperature water are inherently independent of environmental factors in terms of temperature and DO concentration between BWR and PWR conditions.

REFERENCES

ASME 1980. Boiler and Pressure Vessel Code, Section XI, Appendix A.
Atkinson, J.D. & J.E. Forrest 1986. Factors influencing the rate of growth of fatigue cracks in RPV steels exposed to a simulated PWR primary water environment. Corr. Sci., 25: 607-631.
Cullen, W.H.(ed.) 1983a. Proceedings of IAEA specialists' meeting on subcritical crack growth, NUREG/CP-0044.
Cullen, W.H.(ed.) 1986. Proceedings of the second IAEA specialists' meeting on subcritical crack growth, NUREG/CP-0067.
Cullen, W.H., K.Torronen & M. Kemppainen 1983b. Effects of temperature on fatigue crack growth of A508-2 steel in LWR environment, NUREG/CR-3230.
Hale, D.A., C.H. Lange & J.N. Kass 1983. Crack growth resistance of low alloy steel in high temperature oxygenated water. NUREG/CP-0044:141-179.
Jones, R.L. 1983. Cyclic crack growth in high temperature water: Results of a international testing round robin. ibid: 65-88.
Katada, Y. & N. Nagata 1985. The effect of temperature on fatigue crack growth behaviour of a low alloy pressure vessel steel in a simulated BWR environment. Corr. Sci., 25: 693-704.
Kondo, T., T. Kikuyama, H. Nakajima, M. Shindo & R. Nagasaki 1971. Proceedings of a conference on corrosion fatigue on chemistry, mechanics and microstructure, NACE-2: 539.

Fatigue crack growth behaviour of different stainless steels in pressurized water reactor environments

C.Amzallag
UNIREC, Firminy, France
J.L.Maillard
E.C.A.N. INDRET, La Montagne, France

1 INTRODUCTION

Stainless steels are utilized widely in a number of components of the primary coolant of a pressurized water reactor. Such components may be subjected to cyclic loadings in service which could cause flaws or cracks to grow. Linear-Elastic-Fracture-Mechanics (LEFM) concepts offer a very useful tool to evaluate the crack growth of these flaws, and LEFM methods are presently included in Section XI of the ASME Boiler and Pressure Vessel Code. The Section XI curves, however, are presently applicable only to low-alloy ferritic pressure vessel steels.

Since stainless steels are also utilized in pressure vessel and piping applications, there is a need to define reference fatigue crack growth rate (FCGR) curves for these materials.

Over the course of the 1970's, considerable work has been done in characterizing the effect of various parameters (e.g. waveform, temperature, steel composition, water chemistry, etc...) upon the fatigue-crack propagation of low-alloy ferritic pressure vessel steels (Cullen (1981), Scott (1983), Bamford (1982), Scott (1984), Amzallag (1981, 1983, 1985), Dufresne (1982)). Several effects are now well established and research is continuing to explain certain phenomena peculiar to this metal-environment system.

Some studies have shifted to the behaviour of stainless steels (Bamford (1979), Bernard (1982), Amzallag (1981), Hale (1982), Cullen (1985)).

In this context, an experimental programme has been conducted in order to :

1. Determine the fatigue crack growth rate curves of different stainless steels for nuclear pressure vessels and pipings in pressurized water reactor environments.

2. Define reference fatigue crack growth rate curves for these materials.

2 EXPERIMENTAL PROCEDURE

2.1 Test programme

All tests were conducted with a sinusoidal waveform at a frequency of 1 cpm.

The parameters studied are :
1. the steel structure : different stainless steels, cast and forged, with austenitic, austenitic-ferritic and martensitic structures.
2. the load ratio : R = Pmin/Pmax = 0,1 and 0,7.
3. the water chemistry : 2 environments (P1 and P2).
   The P1 environment corresponds to the nominal working conditions.
   The P2 environment corresponds to a pollution of P1 environment by oxygen and chlorides.

2.2 Test environments

The specifications of the 2 environments P1 and P2 are listed below.

Specifications	P1 environment	P2 environment
Temperature	300 °C	300 °C
Pressure	140 bars	140 bars
pH (R.T.)	$9,5 < pH < 10,2$	$9,5 < pH < 10,2$
Cl	$< 0,1$ ppm	0,5 ppm
O_2	$< 0,1$ ppm	1 ppm

The P2 environment corresponds to a pollution of the P1 environment by 1 ppm of oxygene and 0,5 ppm of chlorides.
The specified pH at room temperature is higher than the pH normally specified for PWR tests.

2.3 Materials

4 stainless steels have been studied. The main characteristics are given in the tables hereafter.

Steel	Utilization	Form	Heat treatment	austenitic
1-Z6 CNDNb 17-12	internal bolting	forged bar	quenched from 1100 °C	austenitic
2-Z3 CNMD 23-17	tube support plate	forged plate	quenched from 1100 °C	austenitic
3-Z5 CNDU 21-8	primary piping	cast pipe	quenched from 1100 °C	austenitic-ferritic
4-Z6 CND 17-4-1	internal bolting	forged bar	quenched from 1100 °C and temperated at 550 / 625 °C	martensitic

174

. Chemical composition

Steel	C	Mn	Si	S	P	Ni	Cr	Mo	Cu	Nb	N
1-Z6CNDNb17-12	0.086	1.63	0.41	0.002	0.023	13.3	17.57	2.3	0.052	0.78	
2-Z3CNMD23-17	0.027	5.56	0.04	0.001	0.021	15.3	22.48	2.9	0.054	0.26	0.4
3-Z5CNDU21-8	0.031	0.66	0.8	0.001	0.018	7.83	21.24	2.5	0.084		0.07
4-Z6CND17-4-1	0.049	0.875	0.28	0.019	0.017	4.43	15.14	0.8	0.09		

. Mechanical properties

Steel	Temperatures	Sy MPa	Su MPa	A %	Z %	KCU J/cm^2
1-Z6 CNDNb 17-12	R.T.	278	596	52	70	150
	300 °C	196	478	>39	64	
2-Z3 CNMD 23-17	R.T.	439	792	50	61	124
	300 °C	256	623	49	60	
3-Z5 CNDU 21-8	R.T.	364	600	32	65	
	300 °C	264	526	32	66	
4-Z6 CND 17-4-1	R.T.	655	938	17	67	120
	300 °C	692	854	12	54	

2.4 Test rigs

The FCGR tests have been carried out on servo-electrohydraulic machines, in autoclaves designed to operate at a temperature of 300 °C and a pressure of 140 bars.

The tests in P1 environment ((O$_2$) < 0.1 ppm) are performed in static autoclaves, i.e. without water circulation. In these autoclaves, the oxygene is consumed on the walls and rapidly reaches levels of the order of 10 ppb.

The tests in P2 environment ((O$_2$) = 1 ppm) are performed in low flow rate circulating autoclaves.

The crack propagation is measured with a Linear Variable Differential Transformer (LVDT), made in stainless steel, attached to the lips of the specimen and located inside the autoclave.

3. RESULTS

The figure 1 shows the results obtained on the austenitic-ferritic cast
steel :
- in air, the temperature has no influence on the FCGR.
- the curves obtained in environment are parallel to the curves obtained
  in air, and shifted by a factor of about 3, which indicates a slight
  but clear effect of environment.
- the results obtained in the 2 environments P1 and P2 are very close :
  a pollution of the P1 environment by 1 ppm oxygene and 0.5 ppm
  chlorides does not seem to affect the FCGR.

The figure 2 shows results obtained on the 23 Cr-17 Ni austenitic
steel, in P1 environment at R = 0.1 and 0.7, and in P2 environment at
R = 0.7 :
- as in the previous case, practically same results were obtained in
  environments P1 and P2.
- the FCGR increases when R ratio increases from R = 0.1 to R = 0.7.

In fact, the results obtained on the 4 materials have shown the same
trends concerning the effect of environment and load ratio.

In order to identify parameters which have a significant effect and to
define reference curves, the results have been plotted in function of
the environments P1 and P2 and in function of the R ratio R = 0.1 and
R = 0.7.
The figure 3 presents the results obtained in the nominal environment
P1 at R = 0.1. This figure shows that :
1. on one hand, the 2 austenitic steels have similar FCGR.
2. on the order hand, austenitic-ferritic and martensitic steel give
also similar results, in spite of quite different mechanical
properties.
3. the austenitic steels give FCGR higher than austenitic-ferritic and
martensitic steels.
The same clasification was also obtained in the polluted environment
P2.

The figure 4 shows the results obtained on the 4 materials, in identifying only the environment and the R ratio. It clearly shows :
- that the results obtained in P1 and P2 environments are equivalent.
- a significant effect of R ratio.
In these conditions, in order to define references curves, the results
have been plotted in function of the R ratio only.
In order to take into account this effect, the FCGR da/dN have been
expressed in function of $\Delta Keff. = \Delta K/(1-R/2)$, relation used by
Bernard (1982), for air results.
The figure 5 shows that this expression correctly describes the
results.
The upper bound of air results, proposed by Bernard (1982), is plotted
on this figure.
The fitting of all results by a relation : $da/dN = C \times (\Delta K)^n$ gave a
slope close to 2, rather than 4 proposed in air by Bernard (1982).

The overall results have been adjusted by an equation of the form :
da/dN = C x $(\triangle K)^2$ (figure 6). The scatter of the results varies from 1
to 5, which is relatively small for this type of test.

The figure 7 compares all results with the actual ASME Code curves.
This figure shows that these curves constitute an upper limit of the
results obtained to date. They may be not conservative for results
obtained at high R ratio (R > 0.7) and/or low values of
$\triangle K$($\triangle K$ < 9 MPa $\sqrt{m}$).

The FCGR behaviour of stainless steels has shown a certain number of
differences with that of low alloy steels :
   1. Chemical composition
It has been shown that the sulfur content has an essential role in the
FCGR of low alloy steels. The martensitic steel 17Cr-4Ni-1Mo, which has
a sulfur content of 0.019 % (considered as high sulfur for a low alloy
steel) gave a low FCGR.
   Thus, it seems that the sulfur contained in the steel does not play
the same role in stainless steels and in low alloy steels.
   2. Water chemistry
The figure 8 shows the effect of the pollution of PWR environment
obtained on a low alloy steel in a previous study (Dufresne (1982)) :
- pollution of chlorides and caustic soda without oxygene has no effect
- the presence of 1 to 3 ppm oxygene increases the FCGR by a factor of
  about 10.
- the simultaneous addition of chlorides and oxygene give same results
  as oxygene alone.
   3. Loading
The effect of R ratio is more pronounced in the case of stainless steels
than in low alloy steels.

4. CONCLUSIONS

A slight, but definite effect of environment was observed for the 4
steels studied.
   The effect of the different parameters can be summarized as follows :
- structure of the steel :
  Austenitic steels give FCGR higher than austenitic - ferritic and
  martensitic steels.
- load ratio :
  R ratio has a pronounced effect. The relation : $\triangle Keff = \triangle K/(1-R/2)$
  correctly takes into account the effect of R ratio.
- water chemistry :
  P1 environment (O_2 < 0.1 ppm ; Cl < 0.1 ppm) and P2 environment
  (1 ppm O_2 + 0.5 ppm Cl) give equivalent results : a pollution of P1
  environment by 1 ppm O_2 + 0.5 ppm Cl does not seem to affect the
  FCGR.
   References curves have been derived from all results. The following
relations have been obtained :
Mean curve   : da/dN = 4.25 x 10^{-7} x $(\triangle K)^2$
Lower bound : da/dN = 2.1 x 10^{-7} x $(\triangle K)^2$
Upper bound : da/dN = 10.5 x 10^{-7} x $(\triangle K)^2$
Variation   : 1 to 5
   The actual ASME curves for low alloy steels may be not conservative
for high R ratios (R > 0.7) and/or low values of $\triangle K$ ($\triangle K$ < 9 MPa $\sqrt{m}$).

177

REFERENCES

Cullen, W.H. 1981. IAEA Specialists Meeting on Subcritical Crack Growth Freiburg : NUREG/CP-0044.

Scott, P.M. & A.E. Truswell. 1982. Corrosion Fatigue Crack Growth in Reactor Pressure Vessel Steels in PWR Primary Water. J. Pressure Vessel Tech. Vol. 105 (3).

Bamford, W.H. 1982. Aspects of Fracture Mechanics in Pressure Vessels and Piping. PVP-58, ASME.

Scott, P.M., A.E. Truswell & S.G. Druce. 1984. Corrosion Fatigue of Pressure Vessel Steels in PWR Environments. Influence of Steel Sulfur Content. Corrosion, Vol. 40(7).

Amzallag, C & J.L. Bernard. 1981. IAEA Specialists Meeting on Subcritical Crack Growth Freiburg : NUREG/CP-0044.

Amzallag, C, J.L. Bernard & G. Slama. 1983. International Symposium on Environmental Degradation of Materials in Nuclear Power Systems-Water Reactors. Myrtle Beach.

Amzallag, C., J.L. Bernard & G. Slama. 1985. IAEA Specialists Meeting on Subcritical Crack Growth. Sendai : NUREG/CP-0067.

Amzallag, C & G. Baudry. 1981. Propagation des fissures en milieu PWR de composition anormale. Rapport Creusot-Loire N° 1470.

Dufresne,J. & J.B. Rieunier. 1982. Probabilistic evaluation of fatigue crack growth in a SA508 and SA533B steel. ASME Paper PVP-82-67.

Bamford, W.H.1979. Fatigue Crack Growth of Stainless Steel Piping in a Pressurized Water Reactor Environment. J. Pressure Vessel Tech., Vol. 101

Bernard, J.L. & G. Slama. 1982. Fatigue Crack Growth Curve in Air Environment at 300 °C for Stainless Steels. Nuclear Tech., Vol. 59.

Amzallag, C., G. Baudry & J.L. Bernard. 1981. IAEA Specilists Meeting on Subcritical Crack Growth Freiburg : NUREG/CP-0044.

Hale, D.A. 1982. Materials Performance in a Startup Environment. EPRI Contract RP-1332-2, GE Report NEDC-24392, Vol. 1.

Cullen, W.H. 1985. Fatigue Crack Growth Rates of Low-Carbon and Stainless Piping Steels in PWR environment. NUREG/CR-3945.

ASME Code. 1980 Edition. Revised Appendix A Crack growth reference curves.

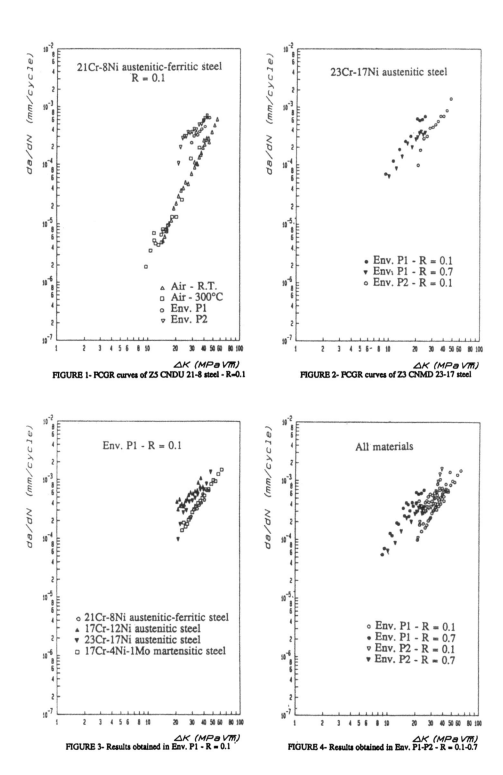

FIGURE 1- FCGR curves of Z5 CNDU 21-8 steel - R=0.1

FIGURE 2- FCGR curves of Z3 CNMD 23-17 steel

FIGURE 3- Results obtained in Env. P1 - R = 0.1

FIGURE 4- Results obtained in Env. P1-P2 - R = 0.1-0.7

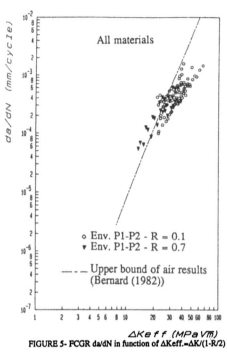

FIGURE 5- FCGR da/dN in function of ΔKeff.=ΔK/(1-R/2)

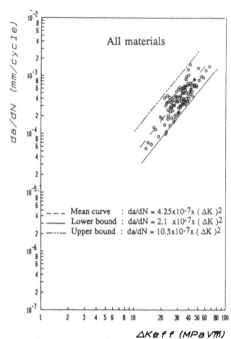

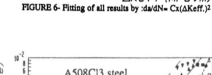

FIGURE 6- Fitting of all results by :da/dN= Cx(ΔKeff.)²

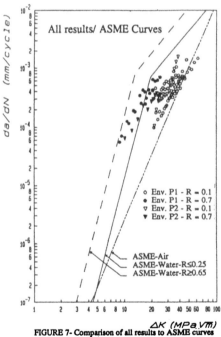

FIGURE 7- Comparison of all results to ASME curves

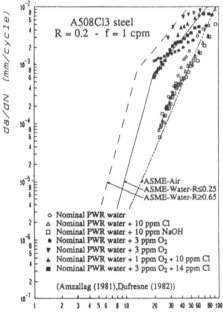

FIGURE 8- A508C13 -Effect of the pollution of PWR water

180

Fatigue and corrosion fatigue data analysis using FATDAC

E.D.Eason, S.P.Andrew & S.B.Warmbrodt
Failure Analysis Associates, Palo Alto, Calif., USA
J.D.Gilman
Electric Power Research Institute, Palo Alto, Calif., USA

ABSTRACT

This paper describes FATDAC, an interactive fatigue and corrosion FATigue Data Analysis Code. FATDAC provides a variety of analysis techniques, including: (1) an optimization process which minimizes the numerical scatter in da/dN that arises from differentiation of (a,N) data, (2) a feature for combining data from many investigators while taking into account the varying amount of crack extension or life represented by each point, (3) weighted least squares techniques for fitting standard or user-defined models, (4) analysis of corrosion fatigue effects by state-of-the-art time domain analysis techniques, and (5) methods for handling outliers, confidence bounds or intervals, printing and plotting.

1 INTRODUCTION

For more complete understanding of corrosion fatigue phenomena it is desirable to combine fatigue data generated by different investigators under similar test conditions and analyze all data together. Before data pooling can be performed, however, differences between investigators in data acquisition and data reduction techniques must be addressed. The Δa or ΔN measurement intervals generally vary by investigator, and may vary for different tests by the same investigator. This problem has been magnified by computerized testing, which allows data to be collected at much smaller increments compared to manual techniques. Different data collection intervals result in different numbers of (a,N) points per specimen, so with standard data reduction techniques the number of (da/dN,ΔK) points will differ from test to test. This can cause problems when modeling, as specimens with many (da/dN,ΔK) points can dominate curve fits to a collection of data which includes many equally-valid specimens with fewer points. Other data reduction differences include the method of (a,N) differentiation (some methods introduce more scatter than others), the type of model fitted to (da/dN,ΔK), and the method of determining model parameters.

The need to analyze data pooled from many sources with minimum numerical error has been recognized by several organizations, including the Electric Power Research Institute (EPRI), the International Cyclic Crack Growth Rate Committee and the Metal Properties Council Task Force

181

on Crack Propagation Technology. EPRI is funding projects to (1) create
a unified data base of nuclear reactor materials, called EDEAC (EPRI
Database for Environmentally Assisted Cracking) (Mindlin et al 1986),
and (2) develop an advanced data analysis code FATDAC (FATigue Data
Analysis Code), the subject of this paper.

The following list outlines the main FATDAC features, which are more
fully described in subsequent sections:

1) $(da/dN, \Delta K)$ data may be calculated from the experimental (a,N) data
   using an optimization routine to minimize scatter, or through the
   conventional secant and seven-point polynomial methods.

2) Corrosion fatigue effects can be assessed by considering crack
   growth on a time domain basis, in either $\dot{a}_e$ vs. $\dot{a}_b$ or $\dot{a}_e$ vs. $\dot{\varepsilon}$
   format.

3) A variety of models, including user-defined models, can be fit to
   the $(da/dN, \Delta K)$, $(\dot{a}_e, \dot{a}_b)$ or $(\dot{a}_e, \dot{\varepsilon})$ data.

4) Confidence intervals or bounds can be developed for the $(da/dN, \Delta K)$
   data or the associated models.

5) $(da/dN, \Delta K)$ points, or models fitted to the points, may be integrated
   for comparison with the original (a,N) data.

6) Outliers in (a,N), da/dN, or da/dt data may be identified either
   manually or through statistical or ASTM criteria, and they can be
   included or excluded in modeling as desired.

7) $(da/dN, \Delta K)$, $(\dot{a}_e, \dot{a}_b)$, and $(\dot{a}_e, \dot{\varepsilon})$ data from many different tests can
   be merged into a single set. The data can be weighted by the amount
   of crack growth or life (cycles) represented by each point.

8) Many data sets may be automatically processed through the same path.

9) Output may be printed or plotted or saved in a permanent file.

2 DESCRIPTION OF FATDAC

FATDAC is an interactive, modular, menu-driven computer program
developed for the analysis of fatigue and corrosion fatigue data. The
structure of the code is shown in Figure 1. The source code is written
in FORTRAN 77, and it is currently running under VMS 4.3 on a VAX 11/785
computer. FATDAC is specifically designed to interface with the EDEAC
database, using the EDEAC accession numbers to identify each data set.
However, the code also allows the user to analyze (a,N) or $(da/dN, \Delta K)$
data that are not included in this database.

2.1 (da/dN) Calculation

FATDAC calculates da/dN from (a,N) data developed in either constant
load range or constant ΔK tests. If the test is of the constant ΔK
type, line segments are fit over Δa or ΔK intervals in the (a,N) data by
least squares, determining da/dN for each interval. For constant load

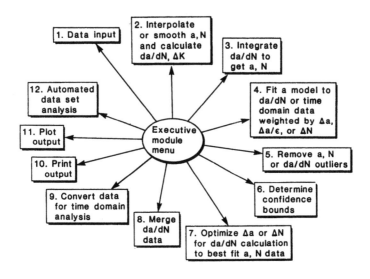

Figure 1. FATDAC modular program structure.

range tests, the user has three options. Conventional secant or seven-point polynomial methods as prescribed by the ASTM E647 standard (ASTM 1984) can be used, or an optimized method can be utilized. This latter technique allows the user to fit linear or exponential functions over optimally-sized Δa or ΔN intervals, calculating (da/dN,ΔK) by evaluating the derivatives at the points of average slope in each interval. The optimized technique generally produces superior results compared to the ASTM methods.

The size of the Δa or ΔN differentiation interval is a critical issue when attempting to reduce the numerical data scatter (Clark and Hudak 1974, 1979, Wei et al 1979, Ostergaard et al 1981). The differentiation interval must be large relative to the experimental error in crack length measurement, or the da/dN points will have excessive numerical scatter. Hence, these intervals are key variables for data analysis. Within FATDAC, the differentiation interval may be:

(a) defined by the (a,N) point spacing (the ASTM approach);
(b) chosen by the user; or
(c) optimized to minimize the deviations between the original (a,N) data and the curve produced by integrating the final (da/dN,ΔK) points.

2.2 Optimized Differentiation Intervals

The optimization technique is shown schematically in Figure 2. The user selects the type of differentiation interval (Δa or ΔN) and its allowable range (defaults are provided). FATDAC then selects a trial differentiation interval Δa or ΔN within the allowable range, and divides the (a,N) data into intervals of this size. Smoothing functions

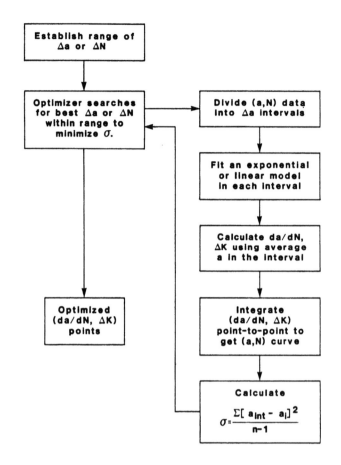

```
┌─────────────────────┐
│  Establish range of │
│      Δa or ΔN       │
└─────────────────────┘
           │
           ▼
┌─────────────────────┐        ┌─────────────────────┐
│  Optimizer searches │        │   Divide (a,N) data │
│  for best Δa or ΔN  │───────▶│   into Δa intervals │
│  within range to    │◀──┐    └─────────────────────┘
│    minimize σ.      │   │               │
└─────────────────────┘   │               ▼
           │              │    ┌─────────────────────┐
           │              │    │  Fit an exponential │
           │              │    │   or linear model   │
           │              │    │    in each interval │
           │              │    └─────────────────────┘
           │              │               │
           │              │               ▼
           │              │    ┌─────────────────────┐
           │              │    │  Calculate da/dN,   │
           │              │    │  ΔK using average   │
           │              │    │   a in the interval │
           │              │    └─────────────────────┘
           ▼              │               │
┌─────────────────────┐   │               ▼
│     Optimized       │   │    ┌─────────────────────┐
│   (da/dN, ΔK)       │   │    │     Integrate       │
│      points         │   │    │    (da/dN, ΔK)      │
└─────────────────────┘   │    │  point-to-point to  │
                          │    │   get (a,N) curve   │
                          │    └─────────────────────┘
                          │               │
                          │               ▼
                          │    ┌─────────────────────┐
                          │    │     Calculate       │
                          └────│                     │
```

$$\sigma = \frac{\Sigma[\,a_{int} - a_i\,]^2}{n-1}$$

Figure 2. Logic of interval optimization for minimum deviation
from (a,N) data

(linear or exponential) are fitted over each interval, and a (da/dN,ΔK)
point is computed at the average slope in each interval. These
(da/dN,ΔK) pairs are then integrated point-to-point to produce an (a,N)
curve. A straight line on log-log coordinates is assumed between
(da/dN,ΔK) points (in sequence) to allow integration steps to be
independent of (da/dN,ΔK) point spacing. The sum of the squared
deviations between the original (a,N) points and the integrated (a,N)
curve is computed for the trial differentiation interval. This process
is then repeated with new differentiation intervals chosen by the
optimizer until the value of Δa or ΔN that minimizes the sum of squared
deviations has been found.

Use of the optimized technique with Δa intervals on hundreds of
pressure vessel steel specimens tested in air, Boiling Water Reactor
(BWR), and Pressurized Water Reactor (PWR) environments, has shown that
most constant load range data sets can be accurately represented by 5 to
10 (da/dN,ΔK) points. The resulting improvement in da/dN scatter,
compared to secant or 7-point polynomial differentiation of the (a,N)
data, can be significant, as illustrated in Figure 3 for a single
pressure vessel steel specimen tested in air. The (a,N) data and the

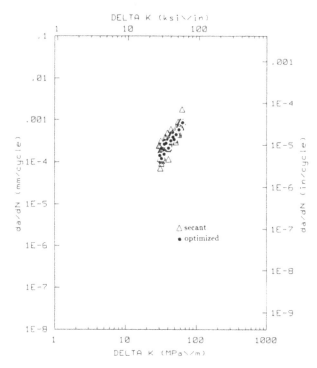

Figure 3. Comparison of scatter from secant and optimized differentiation.

integrated curve resulting from the optimized differentiation intervals are shown in Figure 4. It is clear in Figure 4 that the optimized technique is capable of smoothing through minor perturbations in the (a,N) data while accurately following the major changes in shape. As the experimental error in the (a,N) data increases, more scatter is observed in da/dN values from the secant and polynomial methods, increasing the advantage of optimizing the differentiation interval.

2.3 Time Domain Analysis

FATDAC allows for the analysis of corrosion fatigue data by both frequency domain (da/dN) and time domain (da/dt) methods. The particular time domain methods have been presented in recent work (Shoji and Takahashi 1983, Shoji et al 1981, Gilman 1985, Ford 1986 and Ford et al 1985). The basic approach is to compare the crack growth rate in the environment, $\dot{a}_e$, with a baseline rate $\dot{a}_b$. FATDAC calculates both of these time domain growth rates and allows the user to fit models to the $(\dot{a}_e, \dot{a}_b)$ data.

The baseline fatigue crack growth rate $(da/dN)_b$ is calculated from applied ΔK values using an appropriate crack growth model, such as a Paris fatigue model fitted to air or inert environment data. The time domain rate is then calculated from:

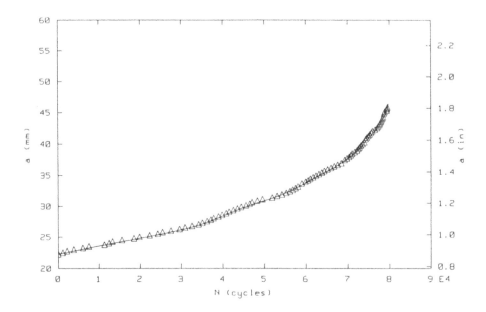

Figure 4. Original data and curve integrated from optimized points.

(1) $\dot{a}_b = (da/dN)_b (\frac{1}{t_r})$

where t_r is the period of the rising-load portion of the loading cycle. The environmental crack growth rate is determined by calculating $(da/dN)_e$ from (a,N) data taken in the corrosive environment, then converting to time domain $\dot{a}_e$ by using $(da/dN)_e$ in place of $(da/dN)_b$ in equation 1.

An example of this type of environmental rate calculation is shown in Figure 5. The bilinear shape is characteristic of tests exhibiting an environmental enhancement in crack growth rate relative to the base rate (which was crack growth rate in air in this case).

The strain rate approach developed by Ford (1985, 1986) and others shows promise for fundamentally understanding corrosion fatigue effects. This approach is based on theoretical relationships between crack tip strain rate, $\dot{\varepsilon}$, and the environmental crack growth rate. The strain rate approach is the underlying theory behind the $\dot{a}_e$ vs. $\dot{a}_b$ approach; the connection between the two approaches is $\dot{\varepsilon} = f(\dot{a}_b)$. In the strain rate correlations, the environmental (time-domain) crack growth rate is calculated as previously described, then FATDAC calculates $\dot{\varepsilon}$ using the models proposed by Ford (1986). An option is available for users to input their own crack tip strain rate models. The resulting strain rate and crack growth rate values can be modeled through a power law equation:

(2) $\dot{a}_e = A\dot{\varepsilon}^n$

186

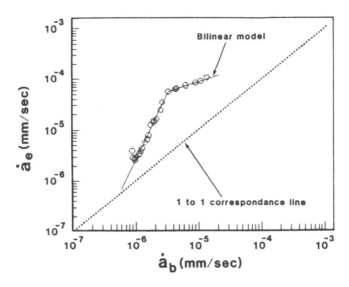

Figure 5. Total environmental rate vs base rate plot.

and plotted on $\dot{a}_e$ vs $\dot{\epsilon}$ plot, as shown in Figure 6. The slope of such curves is typically less than unity, as it is here.

2.4 Modeling Capabilities

FATDAC currently is able to fit Paris, piecewise Paris (up to 5 segments), Walker, or user-defined models to (da/dN,ΔK), and additional models are under development. Time domain $\dot{a}_e$ vs $\dot{a}_b$ correlations can be fit with a linear or piecewise linear model, and $\dot{a}_e$ vs ϵ correlations can be fit with equation 2. The piecewise linear model is especially useful for modeling corrosion fatigue data, which can exhibit bi- or tri-linear behavior, and for modeling apparent ΔK thresholds.

Paris models are fit by linear least squares procedures, whereas the other models use a nonlinear least squares algorithm. All methods allow the minimization of either vertical or horizontal residuals, and the piecewise model also allows minimization of perpendicular residuals.

Fits in both the horizontal and vertical direction are statistically valid in that they both minimize residuals in a particular direction. Neither approach is fundamentally "correct" in fracture mechanics correlations, because both the abscissa and ordinate contain a combination of experimental and analytical error. In addition, when dealing with large collections of data, the numerical effects of scatter are more important to the result than any theoretical consideration of the proper direction for fitting. The authors have standardized on the use of horizontal residuals for (da/dN,ΔK) correlations where the slope is greater than two, and vertical residuals for $(\dot{a}_e,\dot{a}_b)$ or $(\dot{a}_e,\epsilon)$ correlations where the slope is less than or equal to one.

When fitting models, FATDAC allows four data weighting options. The purpose of this weighting is to give equal weight to equal crack growth or number of cycles when fitting (da/dN,ΔK) or time domain models, regardless of the number of points determined experimentally or analytically. The weighting options available are:

187

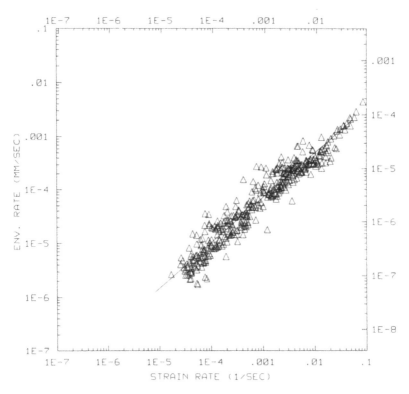

Figure 6. Environmental rate vs cyclic strain rate.

(1) uniform (each point has equal weight);
(2) by crack extension (Δa) corresponding to each (da/dN,ΔK) point;
(3) by life expended (ΔN) per point; and
(4) by the ratio of crack extension to measurement precision (Δa/ε) per
     point.

2.5 Merging and Automated Path Capabilities

FATDAC facilitates combining and analyzing (da/dN,ΔK) and time domain
data from multiple data sets. Through this option, as many as 2000
points or 400 data sets may be combined, stored, analyzed, and saved as
a permanent file. When storing these merged sets, the differentiation
intervals (Δa) used in calculating da/dN, rise time for calculating
da/dt, R ratio, frequency, temperature and other test variables are also
stored along with the appropriate outlier status information.
    When the user wishes to process many data sets by the same path
through the code, the process of answering the same menu-driven
questions over and over again becomes tedious and time consuming.
Hence, the code can be programmed to automatically process a list of
many data sets through the same path. As each set is analyzed, the
results can be added to the merge file. Any logical combination of
modules can be placed within this path, and the list of data sets to be
processed can either be created while inside FATDAC or separately, using
the system editor. Each defined path and data set list can be stored
and retrieved whenever the automated analysis option is chosen. After
all data sets have been processed, a list of any processing errors and

188

the associated data sets are displayed at the terminal and stored in a permanent file. This allows automated processing of all but the exceptional specimens, which can then be analyzed in detail by choosing options from the menus.

2.6 Outlier Identification and Removal

In any data set, the user may question the validity of certain points and decide to ignore them in a particular phase of analysis. FATDAC has three options available for outlier identification, manual selection of points, Chauvenet's criterion, and ASTM criteria. Once a point is declared an outlier, it is not actually deleted from the data set, but rather is flagged as an outlier in FATDAC. When other modules are entered with the same data set, the user has the option of including or excluding outliers in the analysis.

Manual outlier identification can be applied to (a,N), $(da/dN,\Delta K)$, $(\dot{a}_e,\dot{a}_b)$ and $(\dot{a}_e,\varepsilon)$ data. It allows one or more individual points to be identified as outliers. Chauvenet's criterion (Young 1962:78) is a statistical method that identifies all points with low probabilities of occurrence as outliers. Since this method identifies points that are a specified number of standard deviations away from a fitted model, it is only available for $(da/dN,\Delta K)$ points after fitting a model. The ASTM criteria within FATDAC will identify points that violate the linear elastic fracture mechanics criterion, or points with less than a 10^{-5} mm/cycle crack growth rate, according to ASTM E647 (1983).

2.7 Other Features

Three types of confidence intervals or bounds, with confidence levels of 90, 95, 99, and 99.9 percent, are available:

(1) bounds on the $(da/dN,\Delta K)$ data (assuming a lognormal distribution in da/dN);
(2) bounds on the values of the slope and intercept (Paris model only); and
(3) hyperbolic bounds on the mean model line.

This last type of bound includes the effect of variations in both slope and intercept, but all three bounds are based on the usual simplifying assumption that all errors are in the vertical direction. None of these bounds account for the correlation between the fitted Paris model slope and intercept that is evident in low alloy steel and other materials. Because of the above assumptions, the confidence levels are approximate.

Two types of integration are available in FATDAC, point-to-point integration of $(da/dN,\Delta K)$ data, and integration of any $(da/dN,\Delta K)$ model. Point-to-point integration is performed on a, using a modified trapezoidal integration method, assuming the $(da/dN,\Delta K)$ points are connected in sequence by piecewise Paris models. The integration interval, Δa, can be either uniform or successively larger multiples of Δa_0 (i.e.: $\Delta a_2 = 1.01\Delta a_1$, $\Delta a_3 = 1.01\Delta a_2...$). The starting point for the integration can be the minimum value of crack length from the experimental data or any user-defined point. Model integration is performed by directly integrating the analytic model using a trapezoidal rule.

The printing and plotting routines allow real-time terminal display of the FATDAC analysis results. The plot option offers a full complement of plots (complete with symbol control) including optimization output, (a,N), (da/dN, ΔK), and time domain (a_e, a_b) and (a_e, ε) points, models and bounds. Similar information is available through the print routine, whose output may also be sent to a permanent file for future recall. In addition to printing and plotting, FATDAC also can automatically document the path followed through the program.

3 EXAMPLE

An example illustrating several FATDAC capabilities is shown in Figure 7. This plot contains about 1150 da/dN data points from over 240 specimens of pressure vessel steel tested in air at various frequencies, R ratios, and temperatures, by investigators from all around the world. A 3-parameter model of the form used by Yuen et al (1974) was fitted to a four-fifths subset of the data,

$$(3) \quad da/dN = 7.87 \times 10^{-8} \left(\frac{\Delta K}{2.88 - R}\right)^{3.07} \text{ mm/cycle}$$

where ΔK is in MPa $\sqrt{m}$. The points have been normalized to the same condition as the model and confidence interval (R = 0.1) to display the degree of agreement. Certain points are solid to indicate that they are outliers because of linear elastic fracture mechanics criteria, manual selection or Chauvenet's criterion.

The ASME Section XI code line (1983) for evaluating subsurface flaws in pressure vessels is also shown in Figure 7. It is apparent that the slope of this code line,

$$(4) \quad da/dN = 4.77 \times 10^{-10} (\Delta K)^{3.276} \text{ mm/cycle}$$

where ΔK is in MPa $\sqrt{m}$, is steeper than the best-fit air data curve. Equation 3 was fitted using horizontal (ΔK) residuals, which gives it the steepest slope that can be justified by the data. It is apparent that some refinement in the code line is now possible. The development of equation 3 is documented in detail in Eason et al (1987) and described in a companion paper in this conference.

4 CONCLUSIONS

The FATDAC software has proven to be a convenient method for analyzing fatigue and corrosion fatigue data. It is capable of handling massive amounts of data with careful, consistent analysis of each individual specimen. Based on synthetic benchmarks and actual applications to hundreds of data sets, the optimized differentiation provided within FATDAC is generally more accurate in reproducing the (a,N) data and produces less scatter in calculated da/dN than the ASTM secant and 7-point polynomial methods.

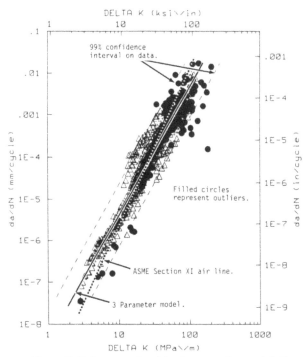

Figure 7. Data normalized to 3 parameter model at R = 0.1 with 99% confidence interval.

5 ACKNOWLEDGEMENTS

The authors gratefully acknowledge the encouragement and support of R. Jones of the Electric Power Research Institute, and of many other members of the International Cyclic Crack Growth Rate Committee and the Metal Properties Council Committee on Crack Propagation Technology. This project is funded under EPRI Contract RP2006-4.

6 REFERENCES

ASME 1983. Boiler and Pressure Vessel Code, Section XI, Article A-4000, Fig. A-4300-1.

ASTM 1984. Standard Test Method for Constant-Load-Amplitude Fatigue Crack Growth Rates Above 10^{-8}m/cycle. In Annual Book of ASTM Standards. ASTM E647-83:710-730.

Besuner et al 1981. BIGIF User's Manual 1, Introduction and Theoretical Background. EPRI Report NP-1830-CCM:3.1-3.4.

Clark, W. G., Jr. and Hudak, S. J., Jr. 1979. The Analysis of Fatigue Crack Growth Rate Data. In Application of Fracture Mechanics to Design. New York:Plenum Press:67-81.

Clark, W. G., Jr. and Hudak, S. J., Jr. 1974. Variability in Fatigue Crack Growth Rate Testing. ASTM Task Group E24.04.01 Report.

Eason, E. D., Andrew, S. P., Warmbrodt, S. B. and Nelson, E. E. 1987. Analysis of Pressure Vessel Steel Fatigue Tests in Air. Topical Report, EPRI Contract RP2006-4.

Ford, F. P. 1986. Mechanistic Interpretation of Design and Evaluation Codes for Environmentally Assisted Cracking. Paper 327, NACE Corrosion '86 Conference. Houston, Texas.

Ford, F. P., Taylor, D. F., Anderson, P. L., and Ballinger, R. G. 1985. Environmentally-Controlled Cracking of Stainless and Low-Alloy Steels in Light-Water Reactor Environments. Supplementary Report, EPRI Contract RP2006-6.

Gilman, J. D. 1985. Application of a Model for Predicting Corrosion Fatigue Crack Growth in Reactor Pressure Vessel Steels in LWR Environments. In Predictive Capabilities in EAC. ASME Winter Annual Meeting.

Mindlin, H., Rungta, R., Koehl, K. and Gubiotti, R. 1986. EPRI Database for Environmentally Assisted Cracking (EDEAC). EPRI Research Project 2006-2, Interim Report. Palo Alto:EPRI.

Ostergaard, D. F., Thomas, J. R., and Hillberry, B. M. 1981. Effect of Δa-increment on Calculating da/dN from a versus N Data. In Fatigue Crack Growth Measurement and Data Analysis. S. J. Hudak, Jr. and R. J. Bucci (eds). ASTM STP 738:194-204.

Shoji, T. and Takahashi, J. 1983. Role of Loading Variables in Environment-Enhanced Crack Growth for Water-Cooled Nuclear Reactor Pressure Vessel Steels. May 1981 Proceedings of the IAEA Specialists Meeting on Subcritical Crack Growth. USNRC Report NUREG/CP-0044.

Shoji, T., Takahashi, H., Suzuki, M., and Kondo, T. 1981. A New Parameter for Characterizing Corrosion Fatigue Crack Growth. Journal of Engineering Materials and Technology. 103:298.

Wei, R. P., Wei, W., and Miller, G. A. 1979. Effect of Measurement Precision and Data Processing Procedures on Variability in Fatigue Crack Growth Rate Data. Journal of Testing and Evaluation. 7(2):90-95.

Young, H. D. 1962. Statistical Treatment of Experimental Data. New York:McGraw-Hill.

Yuen, A., Hopkins, S. W., Leverant, G. R. and Rau, C. A. 1974. Correlations Between Fracture Surface Appearance and Fracture Mechanics Parameters for Stage II Fatigue Crack Propagation in Ti-6A1-4V. Metallurgical Transactions. 5:1833-1842.

Vibration and piping dynamics

Fluid-elastic instability of heat exchanger tube arrays in potential cross-flow

P.J.M.van der Hoogt

Twente University of Technology, Enschede, Netherlands

1 INTRODUCTION

The fluid-elastic excitation mechanism in tube banks is a flutter-type of instability, caused by a change of the mutual position of the tubes within a bundle, hence excerting asymmetric fluid forces. This self-excited mechanism, which is distinct from vortex-induced vibration, since the unsteady fluid forces are generated by the motion of the tubes, manifests itself by large amplitude vibrations and often by so-called "oval" or "whirling" motions.

Connors (1970) performed wind tunnel experiments on a single row of flexibly mounted cylinders (tubes) in cross flow and noticed an abrupt increase of amplitude when the flow velocity exceeded a critical value. Adopting a quasi-static analysis, Connors developed a stability criterion for predicting the critical flow velocity for a row of identical cylinders, yielding

$$(1) \qquad W_g/f_n D = K(m\delta/\rho D^2)^{\frac{1}{2}}$$

where W_g is the critical mean flow velocity through the minimum gap between adjacent cylinders in the same row, f_n is the lowest natural frequency of the cylinders, D is the diameter of the cylinder, m is the cylinder mass per unit length (including the added mass), ρ is the fluid density, δ is the logarithmic decrement of damping of the cylinders in still fluid (viscous and structural damping) and K is a constant, its magnitude depending on geometry parameters. The left-hand side of (1) is called "reduced velocity", whereas $m\delta/\rho D^2$ is referred to as "mass-damping parameter". For the single row with a pitch-to-diameter ratio of 1.41, Connors found a value of K=9.9.

In order to study the fluid-elastic instability effect of tube arrays in a uniform cross-flow, Van der Hoogt and Van Campen (1984) adopted a two-dimensional complex velocity potential approach and calculations were performed on two tubes out of a tube array. A potential flow approach was used, because, for closely packed staggered tube arrays, the wakes behind the tubes are considerably suppressed and the flow distribution around the tubes shows a "potential-like" character. In the present paper the model is extended to account for more complex configurations, followed by an examination of their dynamical features. For a specific staggered configuration the results of the critical flow velocities will be compared with Païdoussis' (1984) data, who dealt with a comparable approach.

2 TWO-DIMENSIONAL VELOCITY POTENTIAL FORMULATION

In the potential flow approximation, the fluid velocity field related to a solitary cylindrical tube k vibrating in unbounded cross-flow can be described by Laplace's equation

(2) $\qquad \Delta \chi_k = 0$

where χ_k is the complex velocity potential.

Using complex moving coordinates z_k to define an arbitrary point in the fluid with respect to the axis of symmetry of tube k, as indicated in Fig. 1, the general solution of (2) can be written as

(3) $\qquad \chi_k = -\bar{W}_0 z_k + \sum_{n=1}^{N} A_{nk} z_k^{-n}$

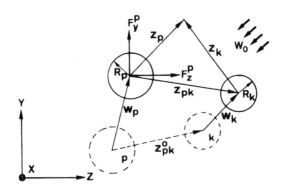

Fig. 1. Coordinate notations for two tubes p and k in cross-flow.

where W_0 is the undisturbed complex fluid velocity, A_{nk} are complex constants and N is the number of terms, used in the calculations. A bar over a complex quantity stands for its complex conjugate value. For an array of K parallel tubes vibrating in unbounded cross-flow the influence of each separate body k should be expressed in terms of each body p. This can be achieved by means of the following coordinate transformation

(4) $\qquad z_k = z_p - z_{pk} \qquad p=1,2,..,K; \quad k=1,2,..,K$

Substituting (4) into (3), the total complex velocity potential of the tubes, expressed in the coordinates of tube p, can be written as

(5) $\qquad \chi^P = -\bar{W}_0 z_p + \sum_{k=1}^{K} \sum_{n=1}^{N} A_{nk} (z_p - z_{pk})^{-n}$

The unknown complex constants A_{nk} of (5) will be established by means of the impermeability condition for each tube p, i.e.

(6) $\qquad [z_p \partial \chi^P / \partial z_p + \bar{z}_p \partial \bar{\chi}^P / \partial \bar{z}_p] = [\dot{w}_p \bar{z}_p + \dot{\bar{w}}_p z_p] \quad \text{for } |z_p| = R_p$

where $\dot{w}_p$ is the complex tube velocity.

Substituting (5) into (6) yields a set of equations for the constants A_{nk} (for detailed information the reader is referred to Van der Hoogt (1986)).

196

(7) $$-nA_{np} R_p^{-n-1} + \sum_{k=1}^{K} {}^{*} \sum_{m=1}^{N} \bar{A}_{mk} R_k^{-m+1} \bar{T}_{nmpk} = \delta_n^1 (W_0 + \dot{w}_p)$$

where

$$T_{nmpk} = \frac{(-1)^m (m+n-1)!}{(n-1)!(m-1)!} R_p^{n-1} R_k^{m-1} z_{pk}^{-n-m}$$

The asterisk * in (7) indicates that $k=p$ will be excluded during summation and δ_n^1 is the Kronecker delta.

Once the constants of integration have been solved from (7), the velocity potential (5) can be written in terms of the velocities of the tubes, $\dot{w}_p$, the undisturbed fluid velocity W_0 and the geometry parameters R_p and z_{pk}. Since the tubes vibrate at specific modes, they change their mutual positions periodically, hence altering their original positions. By defining the relative complex displacement between two arbitrary tubes p and k by

(8) $$w_{pk} = w_k - w_p$$

the following simple relation can be derived from Fig. 1.

(9) $$w_{pk} = z_{pk} - z_{pk}^o$$

where z_{pk}^o is the mutual position between two tubes p and k in the fixed coordinate system.

The first order effect of a change of position between the bodies and its influence on the fluid forces acting on the tubes will be established by linearizing χ^p with respect to the original state of the system, leading to

(10) $$\chi_{(1)}^p = \chi^p - \partial\chi^p/\partial z_p w_p + \sum_{\ell=1}^{K} {}^{*} \{\partial\chi^p/\partial z_{p\ell} w_{p\ell} + \partial\chi^p/\partial \bar{z}_{p\ell} \bar{w}_{p\ell}\}$$

The interaction between all the vibrating bodies due to the presence of the fluid causes coupled motions of these bodies. This interaction effect is represented by the pressure distribution p^p at the fluid-tube interface, yielding

(11) $$p^p = -\tfrac{1}{2}\rho \left[\partial\chi_{(1)}^p/\partial t + \partial\bar{\chi}_{(1)}^p/\partial t + \partial\chi_{(1)}^p/\partial z_p {}^{*} \partial\bar{\chi}_{(1)}^p/\partial \bar{z}_p \right]$$

ρ being the fluid density.

Once the pressure distribution has been established, the fluid forces (per unit tube-length) acting on each tube p are obtained by integrating p^p along the tube periphery, hence

(12) $$F^p = i \int p^p \, dz_p$$

where the real and imaginary part of (12) are the force components in z-direction (F_z^p) and y-direction (F_y^p), respectively (see Fig. 1).

The fluid forces (up to first order) are added to the equations of motion of the tubes as a set of generalized forces.

The displacement vector $\underset{\sim}{w}^T = (w_1, w_2, \ldots, w_k)$ is written as a series of so-called beamfunctions, X_r, hence

(13) $$\underset{\sim}{w} = \sum_{r=1}^{R} \left[\underset{\sim}{U}_{1r} X_r(x) \exp(i\omega t) + \underset{\sim}{U}_{2r} X_r(x) \exp(i\bar{\omega}t) \right]$$

197

Where $\underset{\sim}{U}_{1r}$ and $\underset{\sim}{U}_{2r}$ are the complex amplitude vectors and ω is a complex angular frequency of the system.

Assuming pure bending theory to be valid for the tubes, the expression (13) is substituted into the equations of motions of the tubes, while using the orthogonality properties of the beamfunctions, yielding

$$(14) \quad \left[\omega^2 \, (\underline{M} - \underline{M}^*) - i\omega \, (\underline{D} - \underline{D}^* - \underline{D}^{**}) - (\lambda_r^4 \, L^{-4} \, \underline{S} - \underline{S}^*) \right] \left| \begin{matrix} \underset{\sim}{U}_{1r} \\ \overline{\underset{\sim}{U}}_{2r} \end{matrix} \right| = \left| \begin{matrix} \underset{\sim}{F}_0 \\ \underset{\sim}{F}_0 \end{matrix} \right|$$

where λ_r depends on the supporting conditions of the tubes, L is the length of the tubes, and $\underline{M}$, $\underline{D}$ and $\underline{S}$ are the structural mass matrix, the structural damping matrix and the structural stiffness matrix, respectively. $\underline{M}^*$ is the added mass matrix, $\underline{D}^*$ and $\underline{D}^{**}$ are the hydrodynamic damping matrices, $\underline{S}^*$ is the added stiffness matrix, $\underset{\sim}{F}_0$ is a vector, containing the static fluid forces.

Solution of the eigenfrequencies ω and the associated vibrational mode shapes in any cross-sectional plane of the tubes can be obtained from (14) by using standard complex eigenvalue routines.

2.1 Adjustment of the velocity potential approximation.

Until so far it was assumed that the fluid-elastic forces follow the motions of the tubes instantaneously, without producing any phase lagging. In studying aeroelastic vibrations in a cascade of circular cylinders, Roberts (1966) discussed a jet switch phenomenon, interacting with the motions of the cylinders and the time delay involved. Tanaka & Takahara (1981) measured the fluid dynamic forces acting on a square array and found a positive phase difference between cylinder vibration and the fluid force. In order to account for a phase lagging effect in the equations of motion, let us consider the (complex) displacement of a tube p in an array of K tubes

$$(15) \quad w_p = w \exp(i\omega t)$$

where w is the complex amplitude coefficient and ω is an angular eigenfrequency of the system. The fluid forces acting on each tube are a function of w_p and they can formally be written as

$$(16) \quad F_{\ell p} = C_{\ell p} w_p \exp(i\eta_{\ell p})$$

where $F_{\ell p}$ is the fluid force acting on tube ℓ due to the displacement of tube p, $\ell^p C_{\ell p}$ is a function of the system geometry, in fact representing added stiffness terms and $\eta_{\ell p}$ is an angle indicating the phase lagging of the force $F_{\ell p}$ with respect to the displacement w_p of tube p. The equations of motion can easily be modified by multiplying each term of the added stiffness matrix with the corresponding value of $\exp(i\eta_{\ell p})$. From literature insufficient experimental data are available to assign a specific set of values of the phase lagging to a particular configuration, so, according to Païdoussis et al. (1984), who dealt with a similar approach, all phase angles are chosen identical and are put equal to η.

At this stage it is worthwhile to compare Païdoussis' approach with the one outlined in this section. Due to his derivation of the equations of motion in real quantities, the displacement v_p of tube p in, say, the y-direction was written as

$$(17) \quad v_p = v \sin(\omega t + \phi_p)$$

where ϕ_p is a reference phase angle of the tube displacement. The fluid force acting on tube ℓ due to the displacement v_p will also vary harmonically, but lagging by a phase angle η, so

(18) $F_{\ell p} = C_{\ell p} v \sin(\omega t + \phi_p + \eta)$

In order to modify the equations of motion properly, Païdoussis had to rewrite (18) to the form

(19) $F_{\ell p} = C_{\ell p}\left[v_p \cos(\eta) + \dot{v}_p \sin(\eta)/\omega\right]$

From (19) it can be seen that the second part of the right-hand side yields velocity dependent forces and because of the explicit appearance of ω in these additional terms, calculations require the use of iterative eigenvalue procedures. To avoid complications arising from the latter solution technique, Païdoussis put ω in (19) equal to the natural frequency of the tubes. In the present model this restriction can be dropped, thanks to the application of a complex formulation, which automatically accounts for the total contribution of (16) in the added stiffness matrix.

3 RESULTS

In order to examine the sensitivity of the critical velocity and of the number of tubes K with respect to changes of η, calculations were performed on three, six and ten identical tubes placed in an equilateral arrangement. Fig. 2 presents the variation of the reduced critical velocity $W_c = W_g/f_n D$ (cf. (1)) as a function of the phase angle η for the three tube arrangements. $W_g = W_0 P/(P-D)$, where W_0 is the undisturbed flow velocity and P is the pitch (i.e. the center-to-center distance between a tube and an adjacent one). As can be seen from Fig. 2 the size of the system strongly affects the critical flow velocity, whereas this velocity becomes rather insensible for changes in the phase lagging η for values of $\eta > 30°$. The latter effect is confirmed by the results of Païdoussis et al (1984).

A typical stability diagram is shown in Fig. 3 for the three configurations under consideration. In this figure the reduced flow velocity is given as a function of the mass-damping parameter $m\delta/\rho D^2$ for a phase angle of $\eta = 30°$.

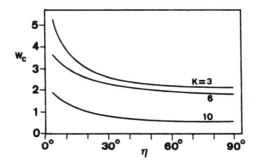

Fig. 2   Critical flow velocities for three configurations as a function of the phase angle η (P/D=1.5; R_p=ρ=1; δ=0.01; $m/\rho D^2$=0.25).

The configuration consisting of three tubes reveals two kinds of instabilities: for small values of $m\delta/\rho D^2$, dynamical instability is found,
whereas at higher values of this parameter a buckling (divergence) instability dominates (the straight line in the figure). For the remaining two
systems, dynamical instability is found over the whole range of the parameter $m\delta/\rho D^2$. It should be noticed that, apart from the small $m\delta/\rho D^2$ -
region, Connors' power-law formulation, given by (1), is fairly predicted
by the model. The critical velocities obtained by the present model were
compared with the results from Païdoussis' approach for a system of seven
identical tubes with $\delta=0.01$. Two phase angle values were used: $\eta=10^\circ$ and
$\eta=30^\circ$. From Fig. 4 it can be seen that, although qualitative agreement
between both models is achieved, considerable discrepancies in the magnitudes of the critical flutter velocities appear, which, as outlined before, might be due to the different way of incorporating the phase lagging
effect in both models.

Some experimental data from various investigators have been reproduced
from Païdoussis' paper in Fig. 4. As can be observed both theoretical models overpredict these experimental data. It is believed that this over-
prediction might be caused by limitations of the potential flow approximation. At low values of $m\delta/\rho D^2$ a qualitative prediction of the slight

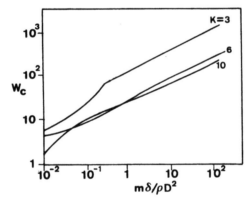

Fig. 3   Critical flow velocities as a function of $m\delta/\rho D^2$ for three, six
and ten tubes (P/D=1.5; $\delta=0.01$; $\eta=30^\circ$).

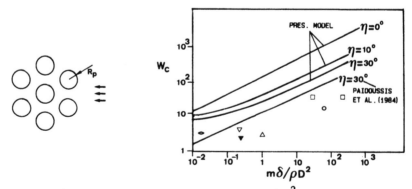

Fig. 4   Critical velocities as a function of $m\delta/\rho D^2$ for seven tubes.
(P/D=1.5; $\delta=0.01$).

200

curvature of the experimental data towards the W_c-axis by the present model can be observed. For $\eta=0$, so in the absence of a phase lagging, no dynamical instability was found, which is in agreement with Païdoussis' observations. The upper curve in Fig. 4 represents the divergence (buckling) instability. It was found that the critical buckling velocities (which are independent of the damping of the system) for $\eta=0^o$ coincide with Païdoussis' values.

4 CONCLUSIONS

Possibly occuring instabilities may be classified into buckling instability (divergence) and dynamical instability (flutter). Calculations revealed only instabilities of the divergence type, whenever a phase lagging angle equal to zero between tube and fluid motion was used. The associated critical flow velocity appeared to be strongly dependent on the system geometry, thus on the number of tubes and their mutual distance.

In order to account for the influence of a phase lagging effect between the motions of the tubes and the fluid forces, a modification of the model was performed by introducing a phase lag angle η. Dynamical (flutter) instability was found to occur for specific configurations of tubes whenever a non-zero phase lagging effect was incorporated in the calculations. For a specific configuration critical velocities were compared with the results obtained by Païdoussis et al. (1984), dealing with a similar approach. Although agreement was found for the critical buckling velocities using $\eta=0^o$, discrepancies occurred for the critical flutter velocities, using $\eta = 10^o$ and $\eta = 30^o$ possibly due to the different way of incorporating the phase lag effect in the equations of motion. Although the present model predicts that the reduced critical velocity depends on the square root of the mass-damping parameter $m\delta/\rho D^2$ in a similar way as found by Connors, the actual values are larger than the experimental values from literature.

REFERENCES

Connors, H.J. 1970, Fluidelastic Vibration of Tube Arrays Excited by
    Cross Flow. Flow-Induced Vibration in Heat Exchangers, ed., D.D. Reiff,
    ASME, New York 1970, pp. 42-56.
Hoogt, P.J.M. van der, Campen, D.H. van, 1984, Self-Induced Instabilities
    of Parallel Tubes in Potential Cross-Flow. Symp. on Flow Induced-Vibra-
    tion, ASME Winter Annual Meeting, New Orleans 1984, Vol. 2, pp. 53-66.
Hoogt, P.J.M. van der, 1986, Vibrations of Co-Axial Cylindrical Bodies in
    Axial or Cross Flow. PhD-Thesis, Twente University of Technology, En
    schede, The Netherlands, ISBN 90 71382 05 2.
Païdoussis, M.P., Mavriplis, D., Price, S.J., 1984, A Potential-Flow
    Theory For the Dynamics of Cylinder Arrays in Cross Flow. Journal of
    Fluid Mechanics, Vol. 146, 1984, pp. 227-252.
Roberts, B.W., 1966, Low Frequency Aeroelastic Vibrations in a Cascade of
    Circular Cylinders. Mechanical Engineering Science Monograph No. 4.
    London, The Institution of Mechanical Engineers.
Tanaka, H., Takahara, S., 1981, Fluid Elastic Vibration of Tube Array in
    Cross Flow. Journal of Sound and Vibration (1981) 77 (1), pp. 19-37.

Dynamic behaviour of a safety injection pump with barrel

H.Guesnon & D.Pehau
Framatome, Paris la Défense, France

1. INTRODUCTION

In the nuclear field, safety systems reliability is a prime necessity, mainly for the safety injection pumps. So, these pumps have to be designed and qualified for severe environment, particularly for seismic conditions, in order to certify their ability to operate surely in the different configurations.

Most of the manufacturers must verify it by the following ways :
. calculations, of the natural structural frequencies by the means of computer codes not always very precise about the boundary conditions and above all rarely qualified by on site tests,
. testing on vibrating table,
. on site experimental measurements with a great deal of instrumentation : accelerometers, transducers...

But these methods are very expensive and need new measurements for each minor modification.

These are the main reasons why FRAMATOME has decided to develop mathematical tool (see fig. 1) to allow a reliable analysis of the dynamical behaviour for the PWR low had safety injection pump (LHSI) (see fig. 2).

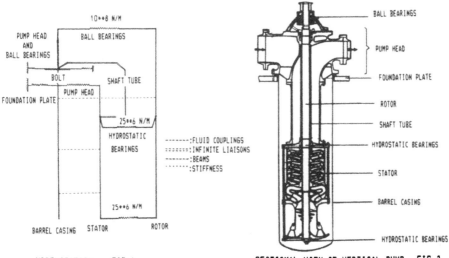

MODELISATION . FIG 1   SECTIONAL VIEW OF VERTICAL PUMP . FIG 2

Tests have been performed at the manufacturer workshop and on site
(with LHSI pump - see hereafter pump description and technical informations). The results of these experimental studies were used to develop
the code and to establish qualification and validation of the model.

- PUMP SERVICE: SAFEGUARD PUMP,LOW HEAD SAFETY INJECTION
  PUMP USED ON THE 1300 MW P'4 (4LOOPS PLANT
  PROGRAM.)

- PUMP TYPE   : VERTICAL CENTRIFUGAL PUMP
  SUCTION IN A VERTICAL TANK

- MANUFACTURER : POMPES GUINARD

- NUMBER OF STAGES : FOUR STAGES
  1RST STAGE WITH INDUCER

- BEARINGS,SUPPORTING THE PUMP SHAFT,OF HYDROSTATIC TYPE

- AXIAL THRUST BEARING,OF BALL BEARING TYPE(DOUBLE)

- COUPLING BETWEEN PUMP AND MOTOR OF UNIVERSAL JOINT

  TYPE (CARDAN TYPE,GLAENZER UNIVERSAL JOINT)

- HYDRAULIC AND MECHANICAL DATA
  N:1485 RPM
  NOMINAL FLOW:700 M3/HR (MAXI FLOW=1100M3/HR)
  DISCHARGE HEAD:170 M COL.WATER
  INPUT POWER:600 KW

- GENERAL ARRANGEMENT(SEE FIG 2)
  PUMP SUPPORTED BY A METAL FRAME,GROUNDED IN CONCRETE
  THE PUMP IS DRIVEN BY A VERTICALLY MOUNTED MOTOR;ON
  A SEPARATED FLOOR ABOVE THE PUMP.

2 AIMS OF THE STUDY

As previously said, the main objective of this study is to produce a
refined and improved calculation code, which allows us to properly
design and calculate a centrifugal pump with vertical shafting, in
regard to dynamic behaviour under exceptionnal or accidental seismic
conditions, and also to pay attention to the natural frequencies against
the excitating frequencies.

Two characteristics were given, as main objectives at the beginning of
the study :

- the calculation code and model, must be of high performance, and give
extended and refined informations about the structure,

- the calculation code and model, must be qualified (experimental tested) and give good accuracy between calculation results and reality.

Against the first point above mentionned, the choice of a developped
calculation code, well elaborated and improved, is considered as a
necessity, because of the mechanical properties of the vertical pumps
structures. These equipments are generally flexible equipments, i.e, with
low natural rotor and stator frequencies, in the same range of speed as
the rotating speed, with important mechanical interaction between rotoric
and statoric parts, and having high hydrodynamic effects in operation.

For this purpose, the SYSTUS code, created and developed by FRAMATOME,
is used. The numerical results produced, are following :

- calculation of the natural resonance frequencies of the mechanical
pump structure (rotor and stator),

- definition of the stiffness and damping values of the whole structure, and also of the different components,

- parametric study and influence of different design configurations,

- verification and importance of the hydrodynamic effects, related to
pump operating conditions.

- Seismic behaviour of the pump : calculation of dynamic stresses and
deformations.

The second point is the need to obtain an accurate and qualified calculation model. This, because of the proximity of natural structural
frequencies and excitating frequencies (rotating speed). Two particular
(or specific) steps must be carefully considered for this accuracy.

- Study of boundary conditions (consequences of the different assumptions on the results : sensibility - Study of the interface or components

Assembly conditions...).
- Impact of the on site arrangement, compared with test-bed conditions grounding of baseplate, concrete floor stiffness, connection with suction and discharge piping....
Following a parametric calculation, which takes in account the variation of these input boundary conditions, a comparison and correlation with experimental results, has to be made, to recognize and select actual site boundary conditions. This is a prime condition to obtain a qualified calculation model.

3 GENERAL PROGRAM PRESENTATION

The development of this calculation model, results in a combination of several theoretical calculation steps and experimental tests :
   A. Initial development of a simplified mathematical model (according manufacturer drawing) : "First" structure modelisation.
   B. First experimental study, on the pump, at pump manufacturer test bed facility :
   - shock test in air and water, and accelerometric measurements,
   - dynamic behaviour under hydraulic excitation, by nominal operating conditions,
   - use of the natural rotor unbalance effect, by deceleration and trip-out,
   - excitation of the structure, at variable frequency, by use of an unbalance excitation system (sine excitation from 2 to 100 Hz).
   C. Improvement of the "First" modelisation :
   - Based on the first test results, at manufacturer workshop, a parametric study of different influence factors and boundary conditions assumptions, was realized.
   - This step is going parallel with a development and improvement work on the mathematical model.
   - Calculation by Manufacturer test loop conditions.
   - Calculation by on site conditions.
   D. Experimental testing, on site, in actual site arrangement, on a nuclear plant :
   . shock test in water on the pump, and accelerometric measurements ; comparison with shop tests and calculations,
   . check of dynamic behaviour of the pump, by running conditions (use of hydraulic excitation).
   This step D, allows us to check and validate the mathematical model previously developed, and to estimate the actual impact of site boundary conditions.
   E. Application of the same analytical model, on another vertical pump structure (other manufacturer, and new technological design for which all dynamic and structural characteristics are already experimentaly known (containment spray pump 1300 - P'4)

4 DESCRIPTION OF MATHEMATICAL TOOL, AND MODELISATION

4.1. The FRAMATOME code "SYSTUS" allows to get a reliable analysis of the dynamic behaviour of vertical pump, i.e., to obtain the natural resonance frequencies of the structure. The pump structure is modelised with beams, each of them being introduced in the calculation by its inertia, mass and cross section values. So going from the geometry, stiffness and inertia of the different pump components, and after computation, the dynamic response of the structure can be produced.

205

For solving this problem, we need to use the SYSTUS dynamical option, according the GIVEN'S Method.

This allows us to obtain :
. The n natural structure frequencies.
. The n proper vectors.
. The displacement (relative to each node of the structure (see fig. 4-5-6-7-8).
. Forces and stresses on each pump component.

4.2. Modelisation

The calculation process is performed in two steps :
- Modelisation, and calculation of the stiffness Matrix of baseplate, pump head, and foundations.
- Modelisation of the total pump structure, included jonction between the different pump components.
In the program, the pump structure is modelised by 5 beams :
. barrel casing,
. pump head,
. rotor parts,
. stator parts,
. shaft tube.
The "SYSTUS" option (called "PORTIQUE") allows us to describe the linear behaviour in one plane xy (or zy, at 90° from xy plane). In each plane the pump structure has 3 degrees of freedom :
UX : displacement on x axis
UY : displacement on y axis
OZ : rotation around the z axis.
- Water in the pump :
The model must be representative of the whole structure included the water inside the pump ; the mass of water is modelised by hydrodynamic coupling effect, (symetric mass matrice, according Fritz Method), or by mass modifications of the parts considered (distributed masses).
. Axial displacement of water is neglected.

- Metal components :
The mass of metal, is taken in account like :
. distributed masses along the beams,
. or, concentrated masses at differents points of the structure (nodes)
- Modelisation of bearings, pump head, and pump baseplate :
The pump journal bearings are of hydrostatic type, designed and calculated by the pump manufacturer. Because of this design, the bearing is modelised in the model by a diagonal Matrix, having a K value of $25 \cdot 10^6$ N/m.

The axial thrust bearing in the pump head, of ball bearing type, becomes a K value of 10^8 N/m.

For the pump head, and due to the complicated volume shape of this component, a specific calculation has to be made, prior and independently of the general calculation. The same occurs for the bedplate and foundation plate between the pump and the concrete.

For all these specific calculations, the use of SYSTUS code, option "coques spatiales" is needed (see example of pump head model - fig. 3).
- General pump structure computation :
After preliminary matrix calculation, and going from two kinds of assumptions for the boundary conditions (clamped or hinged), the second step of the study, is to compute and produce the dynamic behaviour (displacements, rotations, natural frequencies) of the total pump structure

206

For this, all the elements (or beams) of the pump, are connected
to each other by stiffness Matrices which can be described as follows :
- Selection between two types of boundary conditions either clamped
(three displacements and three rotations) or hinged (only three displacements).
- All the five beams (representative of pump) are joined according
to the following scheme (see on fig. 1)   :
. stator connected to the suction barrel both by fluid coupling and by
stiffness pump-head matrix,
. discharge tube connected to the stator by a hydrodynamic coupling
matrix and by an infinitely stiff connection,
. rotor connected to the stator by stiffness bearing matrices,
. the pump head is connected to the concrete by mean of the stiffness
foundation or baseplate matrix, previously calculated.

5 RESULTS OF PARAMETRIC CALCULATIONS

In order to check the influence, on the natural frequencies, of different
parameters assumed to be the most significant, a parametric study has
been systematically performed :
Main results :
- The parameters which appeared to be the most important in the calcu-
lation are :
- Water inside the pump or not (hydrodynamic effect).
- Presence of a baseplate under the pump.
- Boundary conditions between pump and baseplate.
- Suction and discharge piping connections.
- The main results of the analytical simulation are shown in the follo-
wing table :

Pump with water	Plan of modélisation	Pump head and foundation plate	Foundation plate and bed plate	Bed plate and concrete	Liaison with concrete	Value of frist mode (HZ) (*)
no	x,y	"	"	"	"	21
no	x,y	"	"	"	"	26
Yes	x,z	"	"	"	"	22
Yes	x,y	"	hinged	"	"	21.7
Yes	x,y	"	clamped	"	bed plate	18
Yes	x,y	clamped	clamped	clamped	foundation plate	22

(*) All frequencies are bending frequency

COMMENTS AND EXPLANATIONS : We give the influence, on the natural
frequencies, of principal different parameters.
BASE PLATE BOUNDARY CONDITIONS : Fixed conditions doubles the vertical
stiffness ; the horizontal stiffness is multiplying by a factor 1.5

PLAN OF MODELISATION : - little weight on the first mode (frequency < 1 Hz)
                       - pump head non axisymetric (more flexible
                         in x z plane)

WATER IN THE PUMP   : - important weight on the first mode frequency
                         (20 % on the 1rst mode).

- Water lowers the 1rst mode of 4 Hz
  Damping values of 1 to 4 %, by pump operating

SUCTION AND DISCHARGE : - Important weight about all the modes
PIPING CONNECTIONS   (stiffening effect in the xy plane).

MASS OF FOUNDATION PLATE : - No influence about the modes 1 - 3 - 4 - 5 -
                            A reduction of 60 % modifies only the
                            vertical mode n° 2

VALIDATION OF THE MATHEMATICAL TOOL :

The development of the code needs preliminary calculations ($\S D_1 - D_1$)
and experimental validations (comparison to workshop test and on site
test).

RESULTS :

		AT MANUFACTURER WORK SHOP		ON SITE CONFIGURATION	
		TESTS	COMPUTATION RESULTS	TESTS	COMPUTATION RESULTS
X, Y		15,3 HZ	13,8 HZ	19,6 HZ	20,4 HZ
		21,6	23,2	34	39
			32 ∗		43,5
			38,1 ∗		
			45,1 ∗		49,6
X, Z		16,1 HZ	14 HZ	19,9 HZ	19 HZ
		22,2	23,7	34	38,5
			30,4 ∗		43,2
			31,5 ∗		49,5
			42,6 ∗		

∗ frequencies (rotor mode) not been detected during tests because of the
  lack of instrumentation on the rotor

- Optimization of the pump design :
  The use of the developped model and computation code, soon at the
design stage, allows us to optimize the structure through different ways,
in regard to natural resonance frequencies and stresses :
    - Number of rotor bearings (2 or 3 ?)
    - Different bearing stiffness factors
    - Position of the bearings
    - Pumps colum thickness
    - Use of ribs to reinforce the structure

6 CONCLUSION

This action completes a Research and Development program by combining both
the theoretical mathematical analysis and the experimental dynamic study,
in the case of a vertical pump.
This study has permitted us to obtain a refined calculation model,
validated and qualified by experimental tests (influence of each component of the pump and influence of different boundary conditions).
After that, we can use this model for knowledge of dynamic behaviour of
any vertical pump. For the new french N4 Program, this calculation has
been used for the study of the LHSI pump (to optimize and improve his
design). This is the first result and immediate application. The second
objective is the use of this model in the development of a survey and
monitoring system (simulation of structural defects).
At least, we think that this work can be used not only for Nuclear
applications, but also for non nuclear fields, like pumping stations,
offshore.

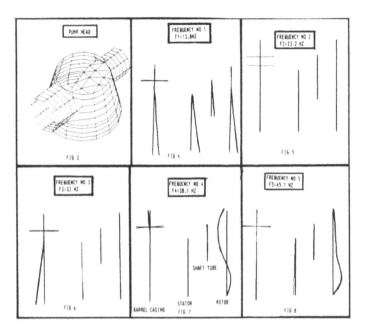

EXPERIMENTAL ANALYSIS
ON SITE TEST

- ## INSTRUMENTATIONS

  TWO ACCELEROMETERS ON THE BARREL. [1,2]
  TWO ACCELEROMETERS ON THE STATOR. [3,4]
  FOUR ACCELEROMETERS ON THE BOLTS OF PUMP HEAD. [5,6,7,8]
  TWO PROXIMITY PROBES BETWEEN STATOR AND ROTOR. [9,10]
  TWO PROXIMITY PROBES BETWEEN STATOR AND ROTOR. [11,12]
  TWO STRAIN GAUGES ON THE BARREL CASING. [13,14]
  TWO STRAIN GAUGES ON THE STATOR. [15,16]

-## TEST PROGRAMME

  -SHOCK TESTS IN WATER
  -TEST DURING THE ACCELERATION
  -TEST AT NOMINAL SPEED
  -DECELERATION TEST

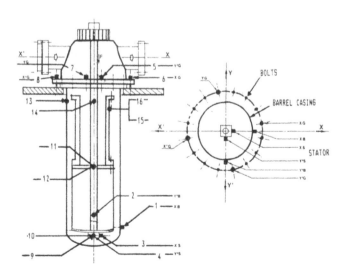

Impact test of components

L.Borsoi
Framatome, Paris la Défense, France

P.Buland
Commissariat à l'Energie Atomique, C.E.N. Saclay, DEMT, Gif-sur-Yvette, France

P.Labbe
Electricité de France, SEPTEN, Villeurbanne

ABSTRACT

Stops with gaps are currently used to support components and piping: it is simple, low cost, efficient and permits free thermal expansion. In order to keep the nonlinear nature of stops, such design is often modeled by beam elements (for the component) and nonlinear springs (for the stops). This paper deals with the validity and the limits of these models through the comparison of computational and experimental results. The experimental results come from impact laboratory tests on a simplified mockup.

1. TEST OBJECTIVE

Piping and components of Nuclear Power Plants are often supported by stops with gaps. This support design has many advantages: it is simple and thus low cost, efficient for high energy loadings such as earthquake or pipe break, reliable without significant maintenance, at least it permits free thermal expansion if gaps are large enough.
However such design involves two kinds of problems:
- adjustement of gaps. Generally the gaps size must be controlled what implies a set of operations: measurement on site, shims fabrication, periodic size verifications,...
- structural analysis. The nonlinear nature of stops makes more difficult the dynamic analysis work.
Regarding this point an elegant way is to perform nonlinear time history calculations. Nonlinear calculations can be made for R&D tasks to improve for example linear equivalent models, but also for engineering purpose since the computer costs are now reduced even for seismic analyses.
Dynamic analyses of components are mainly conducted to provide global results such as loads in supports, in structure load set or displacements. Consequently the used mathematical models, although nonlinear, are unsophisticated: the components are modeled by beam elements and the stops by nonlinear springs (see example on figure 1). It is clear that such simplified models can give correct results only under certain conditions and in a limited range which must be specified as pointed out in a previous paper (SMIRT 7 - paper F4/7).
For this, a french nuclear cooperation EDF,CEA and FRAMATOME has erected a restricted R&D program based on simple experiments involving component

211

and stops. Experimental results are then compared to computational ones
in order to assess the limits of impact beam models.
The experiments comprise two parts:
- elementary tests of rebounds of a component on its stops,
- seismic shaking table tests.
This paper deals only with the first part.

2. EXPERIMENTAL FACILITY

2.1 Vessel Mockup

PWR components are often pressure vessels. The modeling problem is thus
essentially due to the difficulty of impact beam models for simulating
the dynamic behaviour of the impacted shell. In order to respect this
feature but also to conduct simple experiments, the tested component is
simply an unpressurized tank without internals whose sides impinge
stops.
In fact total simplicity view has been a little distorted since this
tank is a very rough approximation of a real component, a PWR primary
Steam Generator (SG), 1/6 scaled. It is mainly formed by a common steel
cylinder, 3. meters high, 0.65 m diameter and 0.014 m thick, which
models the SG external shell. This tank is closed at both ends, at the
bottom by a squared plate which represents the solid tube plate and at
the top by a simple disk. The spatial mass distribution of real SG is
kept with additional lead masses equivalent to the internals masses. The
masses are screwed on the shell and can be removed or transfered. As in
reality, stops are located at two different levels, at the bottom and
about at the midheight of the tank. But the stops design is quite different from reality and is very simplified. Moreover, spherical high tensile steel pins are added for reducing the contact surfaces. Nevertheless the general behaviour of stops is respected: at the bottom the
total stiffness (vessel+stop) is entirely due to the stop stiffness
(bending of a plate) ; at the opposite, the middle stop stiffness is
essentially due to the vessel shell stiffness.
The total mass is about 1.75 tons. The scaling process is the classical
one used in dynamic which keeps velocity unchanged, but on the other
hand which mutiplies acceleration and frequency by the scale factor.

2.2 Pendular Device

The first part of experiments, which have been conducted at the CEA facility, are rebound tests. For this purpose, the vessel is horizontally
placed and supported by 4 cables. Then a pendular device permits to
project by gravity the vessel on its stops which are anchored in a reaction wall (figure 2). The initial deflected position is maintained with
the help of an electromagnet which is cut afterward.
In a first step, elementary tests consist in simple rebounds of the
mass balanced model on its single midheight stop. For this, additional
lead masses are transfered from the bottom to the top in such way that
the vessel mass center coincides with the impact force. Global rotations
are so prevented after impact in order to simplify the determination of
restitution coefficients.
In a second step, rebounds of the mockup on its two stops are performed.
A connected pipe is thus added at the vessel base. This pipe may be
considered as a model of the actual hot leg although its nozzle on the

212

Reactor Coolant Vessel (as a fixed point) is not represented.

In reality, the PWR Steam Generator is vertically maintained on support legs which present twin ball bushing joints which ensure free horizontal displacements. For simplicity and since this is not a key point (gravity scaling has no influence) the legs presence has been dropped and the "SG" horizontally placed.

The impact velocities, unchanged in the scaling process, correspond to the velocity range found for seismic responses: 0.05 to 0.30 m/s. The behaviour of the whole structure (vessel + stops + reaction wall) remains in the elastic domain for this impact range (excluding the local behaviour of stops pins).

Impact forces, accelerations at different locations (including on the pipe attached to the SG), shell deformations, global vessel displacement are transduced. The velocities just before and after impact are deduced from displacements which are measured with the help of Zymmer cameras focused on optical targets placed on the vessel.

3. TEST ANALYSIS

3.1 Mathematical Models

Rebound tests are simulated with both sophisticated shell model and simple beam model (see figure 2). The shell model, which in fact comprises basic 3 nodes thin plates, is developed according a R&D view, but the beam model is similar to production models which are currently used (8 nodes for the straight line).

Both models give first modes in good agreement with actual ones. The following table shows the first frequencies found, rigid modes excluded. Obviously shell modes are not represented through beam models.

Mode description	Experimental	Shell model	Beam model
shell 1*2	98.7 Hz	100.5	-
beam 1	187.4	194.9	198.4
shell 2*2	204.7	214.9	-

Rebound calculations are made by nonlinear modal superposition. Non-linear elements (see figure 1) restore impact forces which are developed at stops locations. Each stop is modeled by a single nonlinear element. This is sufficient due to presence of spherical pins.

The gravity effect, which governs the pendular motion, is transcribed by "gravitational" springs which are horizontally added at the 2 ends of the vessel. The duration before impact is about 0.8 second (1/4 fundamental pendular period). The initial static deflection is imposed by concentrated forces which are then removed.

3.2 Impact force shape

The impact force shape is related to the stop design. At the vessel bottom, a quite rigid mass structure impacts a massless and low stiffness stop anchored in a mass reaction wall. As this system can be reduced to a Single Degree Of Freedom (SDOF), the corresponding impact force shape is simply an half damped Sinus.

At the vessel midheight, the impact kinetics is more complicated since a deformable mass structure - the vessel - impacts a high stiffness stop. In first approximation this can also be reduced to a SDOF on condition

213

that the spring stiffness takes into account the static shell deflection. In reality vessel modes are excited and destroy the half sinus
regularity. The same impact force shape - like a stretched finger - is
found in all experiments (with 1 or 2 stops). Such shape is quite well
recovered by shell model as shown figure 3 (1 stop rebound).
Of course the beam model gives only half sinus impact shape since vessel
shell modes are missing. Nevertheless this shape may be correct on condition that the nonlinear element, which models the stop, includes the
local shell stiffness. The shell stiffness inclusion in the beam model
permits to find accurate impact durations. Consequently the impact
maximum, although not rigorous - peak level and occurrence time -, is
not so far from reality since the areas limited by impact shapes must be
equal for any models since they represent the same momentum variation
of the component due to impact (see figure 4).
Incidentally it can be asked if the impact force maximum value is really
significant. Indeed this value, which is very sensitive to local dynamic
behaviours, is often extracted for design verifications based on static
criteria. In this case, it could be of interest to connect design
"maximum" values to the impact durations.
Concerning the bottom stop without shell deformation, it is clear that
beam models are satisfactory.

3.3 Impact Damping

Impact damping is a general concept which recovers a lot of physical
phenomena. It is related to energy loss or transfer due to impact. Thus
its value closely depends on mathematical models. In the tests where the
whole structure remains in the elastic domain, the main part of impact
damping is due to the energy transfer which occurs between modes during
impact. Figure 5 shows the kinetic energy, versus time, of the
fundamental pendular mode. It is derived from a shell model run simulating a rebound on the middle stop. The energy starts from zero and grows
according to a sinus law up to its maximum value E_0 just before impact,
falls to zero for the reverse motion, then reaches a second maximum E_1
just after impact and finally decreases to zero. If the system is
assimilated to a SDOF, the quantity E_1/E_0 represents the well known
restitution (or rebound) coefficient. The missing energy $(E_0 - E_1)$ is not
lost but transfered to other modes which were not excited before impact
as seen on figure 5_a. This energy is quickly dissipated by structural
damping, but even if it was not the case (figure 5_b), it would not change the restitution coefficient. The accurate value of vessel structural
damping (which is low without heat insulation and supports - below 1% -)
is thus not essential.
The SDOF assimilation is correct in these simulations since the mockup
mass balancing prevents significant rotations after impact.
Typical numerical results are :

.Energy lost by rigid mode E_1/E_0 = 0.87
.Energy transfered to 1st mode (shell mode) E_{S1}/E_0 = 0.09
.Energy transfered to 2nd mode (beam mode) E_{B1}/E_0 = 0.005
.Energy transfered to higher shell modes E_{S*}/E_0 = 0.035

It must be pointed out that the elastic energy is mainly transferred to
the first shell mode and not to the beam mode. For a SDOF, the found
restitution coefficient C_R corresponds to a critical damping percentage
ξ =2% (for low damping : C_R= 1-2$\xi\pi$).

214

In order to maintain such energy balance in beam models, the dashpot value c of the nonlinear element has to be increased by $\Delta c = 2\xi(k/\omega)$ according to SDOF analogy (the dashpot value c provides a viscous force $f = c\,dx/dt$, k is the stop stiffness including shell deformation, and ω is the pulsation related to the impact duration).

Measured impact damping (through restitution coefficient) are found between 2.5 and 4.2% with a 3.3% average for low impact velocities (less than 0.15 m/s). This dispersion may be explained by uncertainties on the determination of restitution coefficients (velocities, presence of small parasite rotations,...) . The results are nevertheless consistent with the previous calculated 2% due to elastic energy tranfer and show that other damping causes are less influent.

The impact damping range (about 3%) found in the experiments can appear very weak. But this impact damping is present even for low impact velocities since the energy transfer process is linear. In addition, energy transfer is larger for real structures which are more complex (presence of fluid, internals, concrete walls,...). Moreover impact damping is greatly increased by local plasticity and friction which are always taking place.(For example an impact test with medium velocity (0.30 m/s) leads to a 5.5% impact damping).

3.4 High acceleration

It is well known that impacts create high accelerations in structures. In fact these accelerations are high level but low energy as their dura-tions are short (it is particularly true for single impact). Therefore they are not destructive in most cases more especially as they generally correspond to high frequencies i.e. to small displacements.

However, in structure response spectra, computed from nonlinear seismic time histories, exhibit high acceleration peaks and high zero period acceleration (see ref.1).

Although response spectra are commonly related to seimic analyses and not to impact excitation ,it has seem interesting to compare experimen-tal and analytical response spectra of the present study.

Figure 6 presents such a comparison. From these response spectra (calcu-lated for 2% damping) and other ones not shown here, it can be noted:

. shell model spectra are very close to experimental spectra at two vessel ends where only the 200 hz beam mode appears. This mode is not excited enough in the beam model.

. At middle stop location, very near to contact, the agreement is less correct. Nevertheless the shell model gives better results than the beam model due to the excitation of shell modes.

. At the end of the connected pipe, the beam model gives correct results in the impact direction, but the 3 spectra are relatively close. In the perpendicular direction, the pipe is undirectely excited by shell modes. It is quite well transduced by the shell model but of course the beam model does not give anything.

General conclusions cannot be drawn from these tests. However it can be mentioned the following points:

. Beam model response spectra are unsignificant at locations where shell modes are influent.

. Beam modes of beam models seem less excited than beam modes of shell models.

. Response spectra on connected pipes which are not directly impacted are correctly transcribed by beam models if the kinetics of the supporting vessel is well represented.

215

4. CONCLUSION

Stops with gaps are an efficient and simple way to support mass components like pressure vessels. Time history calculations are often
performed to analyse such nonlinear structures. For this purpose,
mathematical models combine beam elements for the vessel and nonlinear
springs for the stops. Simple experiments have been carried out in order
to assess the validity and the limits of these models.
The first step rebound tests show the following points.
. Impact damping is low (about 3%) for simplified structures which remain in the elastic domain. The main part of impact damping is due to an
energy transfer from rigid body modes (or basic modes) to shell modes.
. Sophisticated Finite Element shell models permit to recover most of
experimental results: local stiffness, impact force shape, coefficient
of restitution,... But beam models give sufficient results for
engineering purpose providing that the internal parameters of nonlinear
springs are correctly adjusted. The spring stiffness must include the
shell deformation and the dashpot damper value must be increased for
taking into account the neglected energy transfer.
Doing this, the global kinetics of the vessel is well transduced. The
calculated impact force, although not rigorous, is sufficiently accurate
more especially as the maximum value is often extracted for static
criteria verifications.
The creation of high level acceleration by impact is one of the limits
of beam models. However it appears that this phenomena is not so badly
exhausted for locations where beam inertial effects are preponderant.
Evidently in a general way beam models are inadequate when shell modes
have significant effects.

REFERENCE

(1) L.BORSOI, J.P.THOMAS   :" Influence of gaps on seismic behaviour of
the Primary System of a PWR".
SMIRT 7 - paper F4/7.

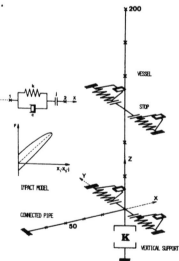

Figure 1 : Example of

Impact Beam Model

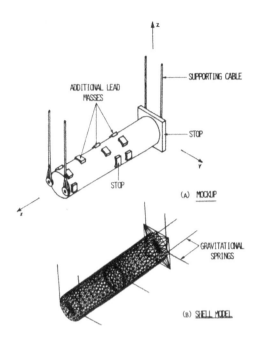

ADDITIONAL LEAD
MASSES

SUPPORTING CABLE

STOP

STOP

(A) MOCKUP

GRAVITATIONAL
SPRINGS

(B) SHELL MODEL

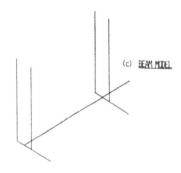

(c) BEAM MODEL

Fig. 2 : Mockup and
Mathematical Models

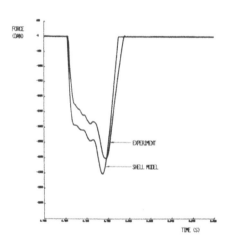

FORCE
(DAN)

EXPERIMENT

SHELL MODEL

TIME (S)

Fig. 3 : Impact Force (IF)

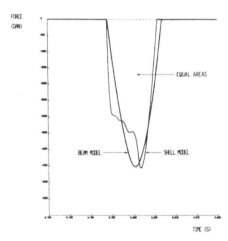

FORCE
(DAN)

EQUAL AREAS

BEAM MODEL

SHELL MODEL

TIME (S)

Fig. 4 : Real and Simplified IF

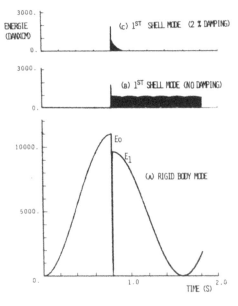

Fig. 5 : Energy Transfer
due to Impact

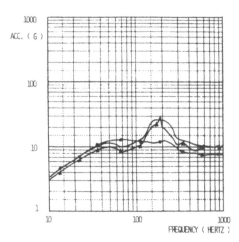

VESSEL BASE (ONE STOP REBOUND)

Fig. 6 : Response Spectra

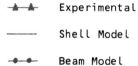

▲▲▲▲   Experimental

————   Shell Model

●●●●   Beam Model

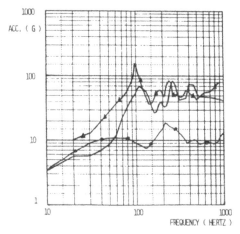

MIDDLE STOP (ONE STOP REBOUND)

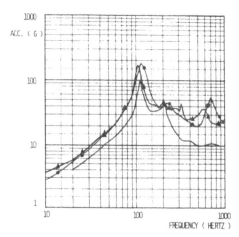

END OF CONNECTED PIPE (IMPACT DIRECTION)
TWO STOPS REBOUNDS

On the influence of ductility on the computation of the response of 1 dof system with superimposed static loads

L.Lazzeri

Ansaldo S.p.A., Italy

1 INTRODUCTION

Structure do yield even under normal loading conditions , in particular
stress concentration areas, (e.g. elbows in piping) and a large energy
dissipation takes place in such spots. This energy dissipation is instrumental in the reduction of the maximum amplitude of the displacement under load. Hence in reality the strong peaks as predicted using
purely elastic methods are not present due to the energy dissipation
during hysteretic cycles. Besides the structural softening due to
yielding causes changes in the frequency characteristics which have a
large beneficial effect on the response in the peak region. This is
particularly true in the case of the strongly peaked floor response
spectra (FRS), where a slight change in the frequency may lead to a
substantial reduction in the amplification. On the other hand ratcheting
may take place as a consequence of the superposition of static and dynamic loads leading to excessive strains.
  While the usual tecniques of Time History analysis have been used
extensively in order to compute such response under partial yielding
conditions /1/, it is of obvious advantage to have an analytical method
for the computation of the three factors:
  1. Sh frequency shifting
  2. τ equivalent damping
  3. Dt ratcheting coefficient
as a function of the two parameters
  Do   ductility
  Ds   static coefficient (ratio of static load to yield load)
  Such a method has been presented by the author /2,3,4/ and it will be
revisited herein; design considerations appropriate for the use of such
method can be found in the companion paper /5/.

2 KRYLOV-BOGOLYUBOFF TECNIQUE

Consider a non linear system, whose equilibrium conditions are:

(1) $\ddot{x} + g(x) + w^{*2}x = 0$

  x displacement w * elastic eigenfrequency
  $\dot{}$ time derivative $g(x,\dot{x})$ a function
  Such equation can be changed into the equivalent one

(2) $\ddot{x} + \tilde{\omega}^2 x + 2\tilde{\omega}\zeta\dot{x} = 0$

$\tilde{\omega}$ = Sh. ω^* elastoplastic eigenfrequency

ζ =   equivalent damping

by means of Krylov-Bogolyuboff tecnique /6/, letting

(3a) $Sh = 1 + [1/(2\pi x_{max} \omega^{*2})] \int_0^{2\pi} g(x_{max} \sin\lambda) \sin\lambda \, d\lambda$

(3b) and $\zeta = [1/(2\omega^{*2} Sh)] \int_0^{2\pi} g(x_{max} \cdot \sin\lambda) \cdot \cos\lambda \, d\lambda$

x max = maximum displacement

In the particular case of a plastic-perfectly elastic system equations (3a and 3b) can be put in closed form as follows:

(4a) $Sh = 1 + 1/2\pi$. $h1 (D,\theta^*)/D$

(4b) $\zeta = (1/2\pi).(1/Sh).[h2(D,\theta^*)/D + h3(D,\theta^*)]$

(4c) $\theta^* = a \cos (1-2/D)$

(4d) $h1(\theta^*,D) = 2 \sin 2\theta^* (2-D) - D .(\pi - \theta^* - 0.5 \bullet \sin 2 \theta^*)$

(4e) $h2(\theta^*,D) = 2.D + 2.\cos \theta^* .(2-D)$

(4f) $h3(\theta^*) = - \sin^2 \theta^*$

where $D = x max/x^*$, where x^* is the yield displacement

3   THE ENERGY APPROACH AND SADE MODEL

Consider the moment in which the velocity is zero and let $(D x^*)$ designate the displacement, measured from the corresponding point. Then the energy equilibrium yields the following equations (where the mass has been assumed unitary):

(5) $\dot{x}^2/2 = \omega^{*2} S^{*2}[\varphi(D) + \psi^* . \psi(D) + Ds . D]$

where the potential energy of the system is given by:

(6a) $t^* = \omega^{*2} S^{*2} \varphi(D)$

and the work of the external forces is :

(6b) $t_e = \omega^{*2} S^{*2} \psi^* . \psi (D)$

where $\varphi(D)$ and $\psi (D)$ are functions of the normalized displacement D and where the work of the stationary force is :

(6c) $t_s = \omega^{*2} S^{*2} D . Ds$

note that the sign of Ds shall be reversed in the reverse half cycle.

If eq.(5) is integrated one can compute the time for the half cycle as given by:

(7) $t = 1/(\sqrt{2} \omega^*) \int_0^D dD/\sqrt{\varphi(D)+\psi^* \psi(D)+D_s D}$

The function $\varphi(D)$ is given by:

(8a) $\varphi (D) = D-D^2/2$ $D \leq 2$; $\varphi (D) = 2-D$ $D > 2$

The maximum value of D will be then $2\bar{D}$, for which the balance of energy as in eq (5) is such that

(9) $\varphi(2\bar{D}) + \psi^* . \psi (2\bar{D}) + Ds . 2\bar{D} = 0$

The function $\psi (D)$ cannot be specified exactly without knowing the forcing function, then one may assume the simplest function obeying the logical boundary conditions :

$\psi(0) = (\partial \psi/\partial D)_0 = (\partial\psi/\partial D)_{2\bar{D}} = 0$ $\psi(2\bar{D}) = 1$

so that:

(8b) $\psi (D) = -2 (D/2\bar{D})^3 + 3 (D/2\bar{D})^2$

In the evaluation of the ψ^* one may take advantage of eq (9,8b), then if Ds = 0 , one has :

(8c) $\psi^* = - \varphi(2\bar{D})$

A first approximation model has been devised by Lazzeri /3/, who has supposed that the amount of energy supplied by the external force t_e does not depend from the value of Ds, hence letting Do the ductility value achieved with Ds=0,equation (9) is rewritten as :

(9') $\psi(2\,\bar{D}s) - \psi(2\,\bar{D}o) + Ds \cdot 2\,\bar{D}s = 0$

From eq (9') $\bar{D}s$ can be computed as a function of Ds, $\bar{D}o$ considering the ψ function as given in eq (8a). Note that $\bar{D}s$ will be larger or smaller than $\bar{D}o$ according to the sign of Ds; this effect will result in a ratcheting, i.e. progressive increment of the deformation during each cycle.

In general then this approach will yield a ratcheting value as given by:

(10) $Dr = /2\,D^* - 2\,D^{**}/$

where D^*, D^{**} are the solutions of the implicit equations :

(11) $\psi(2D^*) - \psi(2\bar{D}o) + Ds.2D^* = \psi(2D^{**}) - \psi(2Do) - Ds.2D^{**} = 0$

In general (excluding small amplitude cycles) Dr is given by:

(10') $Dr = 4\,\bar{D}o \cdot Ds/\,(1-Ds^2)$

Subsequently eq (7) can be used to evaluate the duration of each half cycle and consequently of the frequency and of the shift factor. The computation of the frequency shift factor Sh has been made acccrding to the Krylov-Bogolyuboff approach, to the present one and to the secant modulus method; the results are shown in table 3.1.

TABLE 3.1

Frequency shift coefficient, with no static force

Ductility	Kry-Bog	Ener Appr	Secant	SADE
1.0	1.0	1.0	1.0	1.0
1.25	0.93	0.93	0.89	0.93
1.50	0.85	0.81	0.82	0.85
1.75	0.79	0.76	0.76	0.78
2.00	0.75	0.70	0.71	0.73
2.50	0.70	0.60	0.63	0.65
3.00	0.64	0.53	0.58	0.58
4.00	0.60	0.40	0.50	0.52

The SADE model results from a weighted average cf the models by Krylov-Bogolyuboff and the Energy Approach. In Table 3.2 one can find the ratcheting coefficient Dr as computed by the two parameters Do,Ds. In Table 3.3 one can find the frequency shift coefficient as used in the SADE model as a function of Do, Ds coefficients

TABLE 3.2

Ratcheting coefficient, with static force

Ductility no static force	Static coefficient					
	0.1	0.2	0.3	0.4	0.5	0.6
1.0	0	0	0	0	0	0
1.25	0.50	0.90	1.40	1.70	2.0	2.5
1.50	0.60	1.20	1.90	2.80	3.8	5.0
1.75	0.70	1.45	2.30	3.30	4.7	6.5
2.00	0.80	1.65	2.60	4.10	5.3	8.0
2.50	1.05	2.10	3.30	4.80	6.7	9.4
3.00	1.20	2.5	3.90	5.80	8.0	11.2
4.00	1.65	3.4	5.30	7.60	10.7	15.0

221

TABLE 3.3
SADE model, Frequency shift

Ductility
no static force

Static Coefficient

	0.0	0.1	0.2	0.3	0.4	0.5	0.6
1.0	1.0	1.0	1.0	1.0	1.0	1.0	1.0
1.25	0.93	0.92	0.90	0.88	0.87	0.86	0.84
1.50	0.85	0.84	0.83	0.81	0.76	0.72	0.65
1.75	0.78	0.77	0.76	0.74	0.71	0.65	0.58
2.00	0.72	0.73	0.72	0.69	0.65	0.60	0.53
2.50	0.65	0.64	0.63	0.60	0.57	0.52	0.43
3.00	0.58	0.58	0.57	0.55	0.51	0.45	0.33
4.00	0.52	0.50	0.49	0.46	0.40	0.29	-

TABLE 3.4
Equivalent Damping Coefficient

Ductility	Kryl Bog	Ener Equiv	SADE
1.0	0.0	0.0	0.
1.25	0.11	0.12	0.11
1.50	0.16	0.18	0.16
1.75	0.18	0.25	0.18
2.00	0.21	0.30	0.21
2.50	0.25	0.38	0.25
3.00	0.22	0.42	0.25
4.00	0.20	0.48	0.25

The equivalent damping coefficient is evaluated according to the theory by Krylov and Bogolyuboff and according to the energy equivalence formulation /7,8,9,2,3/, where the damping is :

(11) $\Upsilon = (2/\pi \, (D-1)/D^2) \cdot (1/Sh^2)$

The SADE model is obtained computing the damping basically according to Krylov Bogolyuboff method.

4   RESPONSE COMPUTATION

The SADE model presents basically an analog for he computation of the three parameters Sh, Dr, Υ as a function of Do, Ds; such a method has been used in the DANILS code to compute the response of a system to a given loads, specified in terms of FRS (functions of the damping).

The total ductility has been computed according to

(12)   Dtot = Do + n . Dr (Do,Ds)

where   n, total number of "equivalent cycles" is set equal to one for shock loads and, based on experimental comparisons, in the case of seismic :

(13)   n = 6   Do $\leqslant$ 2   n = 3 + 1.5 Do   Do $>$ 2

However till now no mention has been made of the load factor, as all the model has been based on the ductility factor, which then must be related to the load. In the present model it is taken consideration by means of the "modified β' factor":

(14) $\beta' = a_o / (x * w*^2) = D + Sh^2 /A$

where a_o = reference acceleration, in general ZPA

          A = acceleration amplification

A particular problem has been identified in ref /1/, consisting in the larger transitory response (with respect to the stationary one) of the system, when in resonance conditions. This effect is due to the fact the first cycle is incomplete and the corresponding ductility is lower than the one achieved during the stationary situation. A correction has been then added, whereas for a given stationary ductility a lesser value is postulated for the first cycle. The effect has been found such as to better explain experimental results.

5 LOADS DUCTILITY

Typical cases have been examined for the computation of the response of a typical system to the dynamic excitation; a few data are presented in figs.1. In these cases total ductility vs load factor β' plots are shown for the two cases of shock load and a typical seismic FRS; the frequencies chosen are nearly in correspondence with the peaks of the FRS itself. A comparison is drawn with typical TH data generated by means of a modified version of DYNRES program /10/.

The comparison is in general satisfactory and confirms the engineering significance of the data obtained by means of DANILS code, which makes use of the present model together with FRS analyses tecniques.

A discussion of the data as well as an analysis of their engineering significance can be found in ref./5/.

6 CONCLUSIONS

A method has been presented to compute the response of a one dof system to dynamic excitations under the contemporary action of a static force, in the plastic range. In this way the non linear equation is reduced to an equivalent linear equation, computing:

- the frequency shift
- the equivalent damping
- the ratcheting coefficient

as a funtion of the ductility parameter and static coefficient.

The basic parameters are presented, in numerical form in this paper.

Comparisons have been performed with time history results, with good agreement , confirming the engineering validity of the present approach.

It should be mentioned that a large reduction of the response peaks is produced by even a small amount of plasticity and that the present method is instrumental in such a calculation.

7 REFERENCES

/1/ Giuliano V., Lazzeri L., Scala M. "3D analysis of piping in the linear field under extreme loading conditions" K10/11, 6 SMiRT Conf, Paris, Aug 1981.

/2/ Lazzeri L., Scala M. "Equivalent dampings in piping due to local yielding" ASME PVP-CED meeting, Chicago, July 1986 PVP, Vol.108 pg.5 .

/3/ Lazzeri L."On the analysis of plasticity in the dynamic analysis of structures with reference to piping" ASME Winter Meeting , Anaheim, Dec. 1986.

/4/ Lazzeri L., Scala M., Viti G., Agrone M. "Design considerations for the analysis of piping in dynamic conditions with moderate ductility values" ASME PVP Meeting, San Diego, July 1987.

/5/ Lazzeri L. "On the design of piping in non linear conditions under dynamic loads" K15/10, 9 SMiRT Conf, Lausanne, Aug 1987.

/6/ Krylov N., Bogolyuboff N. "Non linear oscillations" transl.English

Princeton University Press, Princeton 1943
/7/ Shibata H.et al "A study of the damping characteristics of equip-
    ment and piping systems for nuclear reactor facilities" K13/4,
    6 SMiRT Conf, Paris 1981.
/8/ Masri S.F., Hadjan A.H. "On the problem of equivalent damping
    estimation from experimental data" K18/5, 8 SMiRT Conf, Brussels
    1985.
/9/ Hadjan A.H. "A reevaluation of equivalent linear models for simple
    yielding systems" Earthquake Eng and Struct Dynamics Vol.10,759–
    767 (1982).
/10/ Viti G., Olivieri M., Travi S. "Development of non linear floor
    response spectra" Nucl Eng Design, 64, march 1981.

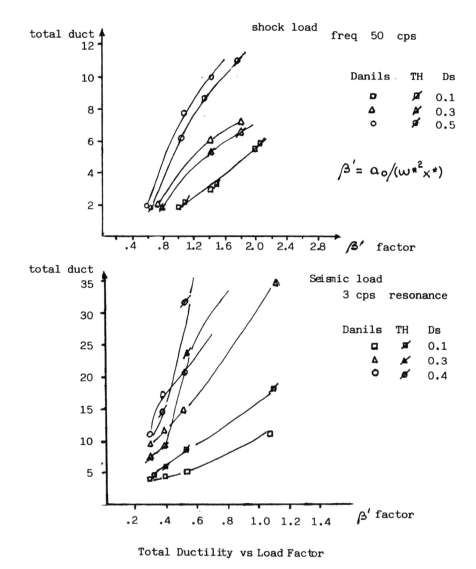

Total Ductility vs Load Factor

Fig.1

Evaluation of a German Standard Problem on the statics and dynamics of a large pipe

M.Firnhaber

Gesellschaft für Reaktorsicherheit, Köln, FR Germany

W.Ch.Müller

Gesellschaft für Reaktorsicherheit, Garching, FR Germany

1.   INTRODUCTION

The German Ministry of Interior (BMI) is sponsoring a series of German Standard Problems to test the capability of codes to predict safety and licencing related problems in the nuclear field adequately. In a German Standard Problem engineering organizations working in the nuclear field are asked to perform "blind" pretest calculations on a large scale test and to submit the results to the GRS before the experimental data are distributed.

As part of the HDR test program static and dynamic tests on a recently constructed pipeline (length ≈ 20m, diameter ≈ 400mm, wall thickness ≈ 20mm) have been performed at the HDR test facility. These tests have been chosen as reference tests of the new German Standard Problem on piping systems. The objective of this new Standard Problem is to investigate the static modal and dynamic behaviour of a large piping system loaded in the elastic range.

Eleven German engineering firms took part in the Standard Problem, among these were reactor manufacturers and experts working in the licencing procedure. The codes they used are standard piping analysis codes.

The Standard Problem was divided into two parts:

- Part 1   investigates the static and modal behaviour of the piping system. The participants were asked to predict the static tests in which the piping system was displaced both horizontally and vertically by a force of 70 kN. In addition they were asked to give the first five eigenfrequencies and eigenmodes of the model used for these calculations.

- Part 2   investigates the structural dynamic response of the pipeline to pressure waves caused by check valve closure. The participants were provided with measured pressures and had to predict the time functions of the same data as in the static test. It was left up to the participants how to determine the load functions and where to apply them.

2.   DESCRIPTION OF THE EXPERIMENTS

The three tests which are analyzed in this paper have been performend at the HDR test facility. It allows to run full scale tests under nuclear reactor conditions. Fig.1 shows the piping configuration.

The three tests selected for this paper are:

- HDR-tests T21.013 and T21.023 /1/

  This are static tests, in which the piping system filled with cold water has been loaded horizontally and vertically by 70 kN. After the static displacement the piping system was released and the free vibrations were measured. In these tests displacements and strains were measured. The free vibrations were used to determine eigenvalues and eigenmodes.

- HDR-test T21.1 /2/

  In this test the transient of the pipe system filled with subcooled water (p = 70 bar, T = 220°), caused by a pipe break and the following check valve closure, were studied. In this test pressures and other fluid quantities as well as structural displacements and strains were measured.

3.   MODAL ANALYSIS

The first five experimental eigenfrequencies derived from free oszillations of the system are 4.46, 6.20, 7.52, 12.7 and 16.8 Hz. In fig. 1 the location and direction of the maximal amplitudes is given. The first eigenfrequency represents the horizontal motion of the complete system and the second eigenform represents the corresponding vertical motion. Thus the first eigenform is the displacement pattern of the horizontal load case while the second eigenform is the displacement pattern of the vertical load and the dynamic transient. In the transient the substantial effect of the pressure transient comes from the difference force on the vertical pipe.

Fig. 2 compares measured and calculated eigenfrequencies by giving the ratios f(calculated)/ f(measured).
Fig. 2 shows that most of the calculated eigenvalues are within an error of 10% and about 50% of the calculated eigenvalues are within an error of 5%. But the error varies from each participant and eigenvalues.
Typically for each participant the curve of ratios is not a straight line but U-shaped. The ratios for the first eigenvalue are higher than for the second and third which is almost the same while the ratio is increasing again with the fourth and fifth eigenvalue.
The second eigenvalue is approximated best by all models while there is a tendency to overestimate the first. This can be interpreted that the horizontal stiffness of the system is lower than predicted by a typical beam modell for some reasons that have not been identified.

4.   STATIC ANALYSIS

In the static tests the piping system has been loaded in x-direction and z-direction by a force of 70 kN. In this paper we will consider the following data for comparison:

226

MN4201 = x-displacement (fig.1)
MN4203 = z-displacement (fig.1)
MK2121 = vertical bending stress (fig.1)
MK2222 = horizontal bending stress (fig.1),

The following table gives an overview over the results:

Participant	horizontal		vertical	
	MN 4201	MK 2222	MN 4203	MK 2121
EXP	-26,4	26,3	12,5	-50,9
A	-26,9	24,5	11,25	-51,7
B	-27,82	26,	11,78	-53,5
C	-35,12	25,2	13,69	-56,3
D	-25,42	24,7	10,99	-51,5
E	-35,6	29,1	16,4	-58,7
F	-28,4	0	13,91	0
G	-34,12	29,4	13,94	-58,9
H	-27,6	22,6	13,4	-42,0
I	-23,93	21,7	9,99	-44,1
J	-25,0	22,9	10,4	-47,1
K	-27,46	24,4	12,16	-48,6

This shows that the vertical direction is much stiffer resulting in higher bending stress.

In the horizontal case most participants are within 10% of the measured displacement while three participants overestimate grossly by about 30%. For the corresponding stress MK2222 most participants are within 10% of the measured result, while even those participants which overestimated the displacement are within 15% of the measured data.

In the vertical case the general agreement is somewhat poorer than before. Only half of the participants are within the 10% range, while three underestimate the experimental results between 10 and 20% and three overestimate between 10% and 30%. For the corresponding stress MK2121 the agreement is better, two overestimate the bending stress (maximum +16%) and two underestimate (maximum -12%) by more than 10%.

From the above table it can be derived that the discrepancies in displacements and stresses are related but the error in the stresses is much lower. In no case displacements or stresses were grossly underestimated and some participants achieve a quite accurate prediction.

5.   DYNAMIC ANALYSIS

The blind calculations of the dynamic test were performed after the test had been run but the structural results were kept secret. The only information the participants were provided were the pressure time histories at various locations along the pipe. The participants had to do the calculations in two steps

1) derive load functions from the measured data by interpolation

2) perform a dynamic piping analysis.

227

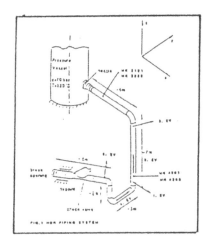

FIG. 1 HDR PIPING SYSTEM

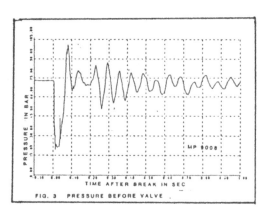

FIG. 3   PRESSURE BEFORE VALVE

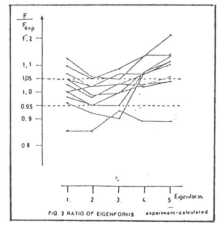

FIG 2 RATIO OF EIGENFORMS   experiment-calculated

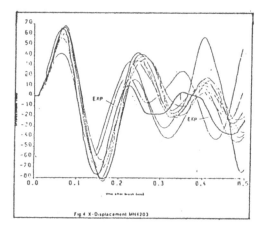

Fig 4 X-Displacement MN4203

Most participants calculated load functions for the midpoints of the elbow by linear interpolation between pressure points and determining the forces by the formula

$$F = p \cdot A \ (F \ \text{force}, \ p \ \text{pressure}, \ A \ \text{area})$$

The substantial force acting on the piping system in the transient is the pressure difference along the vertical section of the pipe. A typical pressure is given in fig. 3. It shows that the main frequencies of the main force are essentially higher than the eigenfrequencies of the structure.

The highest displacements were reached by MN4203 with a maximum of -77.3 mm. Unfortunately at about 200 msec a slight plastification of the piping system occurred. Therefore the calculated and measured displacements are not comparable after that time.

Fig. 4 show the measured and calculated displacements MN4203. In the first 200 msec the agreement is qualitatively good but after this time the agreement between the participant is decreasing rapidly. For the first maximum all participants with one exception are in a range of + 36% to - 14.5% of the measured value. There are two important factors which lead to the gross differences at times after 200 sec:

228

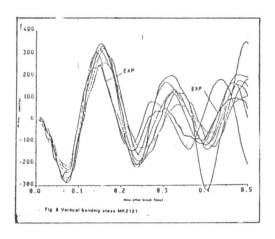

Fig 5 Vertical bending stress MK2121

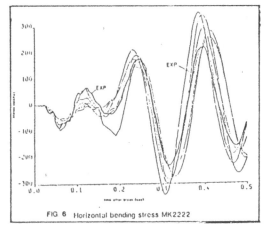

FIG 6   Horizontal bending stress MK2222

1) The errors in the eigenfrequencies accumulate and affect both
   the height of the maxima and minima and the times at which
   they occur.

2) The effects of the different techniques of calculation and application of loads accumulate.

Due to the local plastification at 200 msec the calculated curves overestimate the measured data. The same holds for the stresses, where
maximum values of stress are measured at MK2121 (fig. 5). For the
first minimum all participant do quite well. But at later times the
agreement is rather poor as could be expected from the displacements.
The main result from the transient analysis is that the first oscillation
is adequately predicted by most participants but afterwards the differences in the eigenforms and eigenfrequencies of the models accumulate and lead to rather poor agreement of the calculated transients.
If the main loading of the structure occurs at a later time as it is the
case for MK2222 (fig. 6) this can lead to a gross under- or overpre-
diction of the stresses. But even in this case the transient is predicted qualitatively, i.e. it can be seen from the calculation that the
main loading occurs at a later time and the time span when it occurs
is predicted adequately.

229

Conclusion

The HDR piping system under consideration is typical for a nuclear reactor. The techniques used to predict the structural behavior of the system are state-of-the-art. The comparison of measured and calculated data is a test for reliability of the calculational prediction in this environment.

The comparison demonstrates:
1) Eigenfrequencies and eigenforms are predicted adequately. Many participant's predictions are within 5% of the measured results.
2) Static displacements, too, are predicted adequately well, but the variation range is somewhat higher, with most participants within a 10% error bound.
3) The prediction of the bending stress is qualitatively comparable to the displacement but the differences between measured and calculated data is about half the amount that was found for the displacements.
4) The prediction of the transient is adequate only for the first 200 msec (first vibration). After this time the errors accumulate. As a consequence the agreement of the calculated curves degrades rapidly. After 200 msec the special features of the test results are predicted only in a qualitative sense while the quantitative agreement is rather poor.

The main sources of discrepancies are:
1) In the transient analysis several steps have to be carried out "by hand", and are not carried out automatically by the computer.
2) Several steps especially the evaluation of the pressures and the calculation of the load functions are not well established and left to the judgement of the analyst.

This work was performed with support of the German Ministry of Interior. And the test was sponsored by the German Ministry of Research and Technology.

References
/1/   R. Grimm
       Snapback-Versuche an einer Rohrleitung (ROR) Versuchsgruppe
       RORB-Versuchsprotokoll
       PHDR-Arbeitsbericht Nr. 2.197/85

/2/   M. Tenhumberg
       Blowdownversuch mit Ventilschießen Versuchsgruppe RORB-Versuchsprotokoll
       PHDR-Arbeitsbericht NR. 2.204/85

Environmentally assisted cracking II: Time domain assessment and stress corrosion

Time domain analysis for quantitative evaluation of EAC and its relevance to life evaluation procedure of RPV

H.Takahashi & T.Shoji

R.I.S.F.M., Tohoku University, Sendai, Japan

T.Kondo & N.Nakajima

Japan Atomic Energy Research Institute, Ibaraki

J.Kuniya

Hitachi Research Laboratory, Hitachi Ltd, Ibaraki, Japan

1 INTRODUCTION

Although recent significant advance in mechanistic understanding of EAC in pressurized high temperature water can give some prediction for EAC growth rate under a given condition, some uncertainties are still remaining in, for example, threshold behavior, size effects and a part through crack problem. It is also needed that an actual engineering circumstance should be examined first of all in terms of specific mechanical, metallurgical and environmental factors.

One of the recent understandings of crack tip mechanochemical reaction is shown illustratively in Fig.1(Shoji 1986), where the complicated interrelations among crack tip strain rate, crack tip environment and crack tip material for characterizing the degree of crack growth enhancement in the LWR water are demonstrated. In this illustration emphasis is also placed on the possible significant influence of the environment on the crack growth enhancement in a flaw contained in the free surface of a structure.

This paper discusses a methodology of life prediction procedure recently developed in Japan which is a modification of present ASME flaw evaluation technique taking into consideration a more realistic situation in flaw shape, sulfur content and loading conditions. A use of the predicted crack growth curve in a time domain analysis of life of vessels is discribed.

2 EFFECTS OF SPECIMEN SIZE AND CYCLIC FREQUENCY ON EAC GROWTH RATE CHARACTERISTICS AND ITS MECHANISTIC UNDERSTANDING

The materials employed here were two SA533B steel with different sulfur contents, which were prepared for the JCF round robin program. The basic data for those heats are summarized in Table 1.

The crack growth test specimens were machined in the "T-L" direction. The specimens were of ASTM compact tension type, the most common size being 1TCT, while in the study on the size effect 2TCT and 4TCT were also employed. The loading condition was that with the stress ratio, R=0.2, the loading frequency f=1 cpm (17 mHz) with the sinusoidal wave form. In some cases, frequencies were f=0.05 cpm(0.85 mHz) and f=12 cpm (0.2 Hz). The test environments were simulated BWR water (288°C, $\sim$ 200 ppb O_2) and PWR water (320°C, $<$ 5 ppb O_2) unless otherwise stated. In the tests conducted in ambient air environment, little difference

in the da/dN versus ΔK relation was recognized among specimens of
different thickness. The plots follow nearly the ASME code air line.
Similarly, the results obtained from the tests conducted in the typical
BWR type environment(288°C, ~200 ppb O_2) showed no significant effects
of the plate thickness on crack growth rate as shown in Fig.2 where a
scatter band obtained from 1TCT is also illustrated. The results
obtained by a 2TCT specimen of heat M showed slightly higher growth rate
than 1TCT at higher ΔK range but not so remarkable. Further work is
needed to draw more conclusive words, specially in connection with a
part through crack growth behavior where stress state and local water
chemistry effect may become important for crack growth enhancement.

In the study of examining the effect of loading frequency, tests were
made with BWR type environment with high dissolved oxygen (8 ppm). Tests
with lowering the frequency to 0.05 cpm (0.85 mHz) yielded some
increased da/dN values well exceeding the ASME Section XI water line as
shown in Fig.3. It is also noted that the effect was seen in both
materials. The other extreme in terms of the loading frequency, 12 cpm
(0.2 Hz) gave the values falling nearly on the ASME air line.

As shown in Fig.3, there is no appreciable difference in crack growth
rate between low and medium sulfur steels in the oxygenated water, which
is quite different from those in deoxygenated water. The data shown in
Fig.3 can be interpreted in terms of crack growth rate in water
environment denoted a_{ENV} and crack growth rate in air environment, a_{AIR},
and results are shown in Fig.4 with those of steels containing low,
medium and high sulfur at low (R=0.2) and high (R=0.7) stress ratio. The
data at low R of low and medium S steels are from JCF round robin tests
and those of high S and some of medium S are after W.H.Bamford (1977)
and ICCGR, respectively.

Based on the analysis on crack tip strain rate with crack growth by
FEM (Shoji 1985), these experimental crack growth data can be replotted
on a $da/dt]_{ENV}$ - $\dot{\varepsilon}$ diagram which is a more general expression of crack
growth rate of a mechanistic view point as shown in Fig.4. High sulfur
steel showed less sensitivity to the strain rate and vice versa. In the
figure, crack growth data at R=0.7 for the medium sulfur steel are
plotted together with the JCF round robin data for comparison.

The preliminary trend lines shown in Fig.4 as a function of S contents
in steels can be given by

$$a]_{ENV} = (4.61 \times 10^{-3} + 5.55 \times 10^{-5} \times S) \times \dot{\varepsilon} \begin{smallmatrix} 0.64 - 0.15S \\ \text{crack tip} \end{smallmatrix}$$

These terend curves obtained for sulfur contents of 0.004%, 0.014% and
0.019% are transposed again to da/dN - ΔK curves, respectively, assuming
the applied frequency and loading stress ratio. The results for
f=0.1cpm, 1cpm and 1Hz at R=0.2 and 0.7 are shown in Fig.5 for S=0.014%.
Based upon the growth data shown in Fig.4, crack growth rate can be
estimated for various combination of ΔK, R and frequency as shown in
Fig.5, and are used in the following life prediction procedure.

3 NUMERICAL EVALUATION OF LIFE TIME FOR RPV

As mentioned previouly, there can be a 3-dimensional effect on the crack
growth acceleration due to the influence of actual thickness and flaw
size of components, fluid flow characteristics and crack closure
behavior on the crack tip electrochemical reaction.

In this paper a model of corrosion fatgue crack growth shown in Fig.6
is assumed, where two typical crack profiles are considered, based on
possible different growth rates in the directions into the thickness and

parallel to the surface. A factor, $F=\Delta L/\Delta a$, is introduced to characterize the EAC growth behavior, where ΔL is the crack extension in length measured on the plate surface, and Δa the crack extension in depth direction. Thus the factor $F=\Delta L/\Delta a$ can be used as a parameter for characterizing the surface effect of EAC. Numerical data and evaluation lines in a form of da/dN vs. ΔK used in this study are given also in Fig.6, 7 and 8.

In this computation, major emphasis is placed on the crack growth prediction under high R with small stress fluctuation conditions in LWR belt-line components. Evaluation of the postulated flaw has been carried out using the method of ASME Sec.XI standard procedure. A corrosion fatigue crack growth analysis was firstly undertaken to determine how much the given crack would grow in a plant operating period, under such high R ratio stress cycles as shown in Fig.9, where R is fixed as 0.82, and the repetition of high R stress cycles 10^5 in a unit block. Operational period is counted by a number of the block. Next, critical flaw sizes were determined for emergency or faulted condition, by using the fracture toughness values(K_{IC}) which depends on a degree of irradiation for the RPV. Fast neutron fluence for a block is 10^{18} n/sq cm(time scale of one block is about a year).

At the time corresponding to the maximum stresses, membrane stress and bending stress are σ_m=196, σ_b=7.4 for the normal condition and σ_m=196, σ_b=440MPa for the emergency condition(Marston 1978).

The numerical caluculation is continued till the flaw size reaches to the depth of 1/4 plate thickness, A=T/4(T: plate thickness, 180mm in this study), or K_f=K_{IC}(K_f: stress intensity factor at the end-of-life flaw size). The flaw shape parameter Q and correction facors M_m and M_b are determined according to the ASME Sec.XI guide line. The change in crack depth with operation time is shown in Fig.10. Numerical results are summarized in Table 2 where combined effects of the surface flaw growth rate factor $\Delta L/\Delta A$ as defined previrously and the threshhold value for EAC growth are examined in relation to life time. Either increasing $\Delta L/\Delta A$ or decreasing the threshold value ΔK_{th} results in a significant reduction of the life time in comparison with the estimation by use of the ASME reference wet line. This evidence mentioned above suggests that an experimental determination of $\Delta K_{th(E)}$ as a function of R ratio is urgently needed. Based on a mechnistic understanding of EAC in the combination of LWR steels/evironments, a reasonable estimation procedure should be developed.

4 CONCLUSIONS

JCF round robin tests data are re-evaluated on a da/dt]$_{ENV}$ - $\dot{\varepsilon}$ (or da/dt]$_{air}$) diagram, and the significance of sulfur content in crack growth enhancement is emphasized quantitatively. The importance of low frequency/high ΔK combination and high frequency/high R and low ΔK combination are also emphasized in relation to the ASME wet lines.

Furthermore, based upon the predicted growth data by the time domain analysis the evaluation of subcritical crack growth of surface flaw in LWR environments was performed, taking into consideration the change in the shape of surface crack with crack growth. Finally, research works needed in future are pointed out for establishing a structural integrity assessment code of RPV and piping components containing a flaw in the LWR environment.

REFERENCES

Shoji, T. 1986. Sub-Crirical Crack Growth and Structural Integrity of Light Water Reactor. J. JSME. 89(808):37-43.

Bamford, W.H. 1977. Effect of Pressurized Water Reactor Environment on Fatigue Crack Propagation of Pressure Vessel Steels. The Influence of Environment on Fatigue, I Mech E, p.51-56. London.

Shoji, T. 1985. Quantitative Prediction of Environmentally Assisted Cracking Based on Crack Tip Strain Rate. Predictive Capabilities in Environmentally Assisted Cracking. R.Rungta (ed.) ASME PVP-Vol.99: 127-142

Marston, T.U. (eds.) 1978. Flaw Evaluation Procedure. EPRI NP-719-SR

Table 1 Chemical composition and tensile properties.

Material	C	Si	Mn	P	S	Ni	Cr	Mo	Cu
JCF Low Sulfur	0.17	0.24	1.37	0.003	0.004	0.80	0.07	0.46	0.02
JCF Medium Sulfur	0.21	0.29	1.45	0.007	0.014	0.65	0.03	0.51	0.03

Material	Temp. (℃)	0.2% Yield Strength MPa (kgf／mm²)	Tensile Strength MPa (kgf／mm²)	Elongation (%)	Reduction of Area (%)
JCF Low Sulfur	20	468 (47.7)	622 (63.4)	23.7	72.3
	300	416 (42.4)	586 (59.8)	20.8	70.0
JCF Medium Sulfur	20	475 (48.4)	629 (64.2)	21.9	60.8
	300	422 (43.0)	608 (62.0)	18.3	57.2

Table 2 Effect of F and ΔKt level on life time (Blocks).

New Parameter $F(=\Delta L/\Delta A)$	ΔKt (MPa√m)				ASME SEC.XI	JCF-M	JCF-L
	2.0	3.0	4.0	5.0			
0.5	0.45	0.74	3.85	13.68	48.42	334.8	334.8
1.0	0.41	0.67	3.25	11.57	40.99	290.0	290.2
2.0	0.35	0.59	2.66	9.34	33.14	238.6	239.0
3.0	0.32	0.54*	2.34	8.05	28.04*	212.2	212.6
4.0	0.23*	0.43*	2.11*	7.24	25.09*	195.2	195.4
5.0	0.19*	0.39*	1.95*	6.75	22.96*	182.7	182.8
A/L=0.1 CONST.	0.14*	0.31*	1.62*	5.48*	18.99*	140.6	153.5
K_{IC} (MPa√m)	93					124	202

* $K_f > K_{IC}$

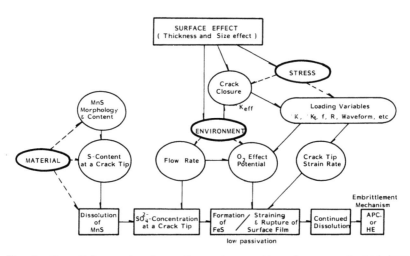

Fig.1 Possible mechanism of crack growth enhancement and key parameters for characterizing corrosion fatigue crack growth including surface effect.

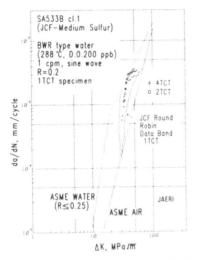

Fig.2 Effect of specimen thickness on crack growth rate.

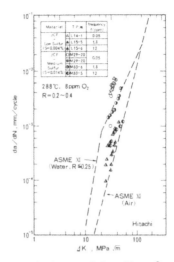

Fig.3 Effect of loading frequency on crack growth rate.

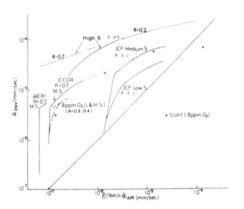

Fig.4 Crack growth rate diagram for S content and stress ratio.

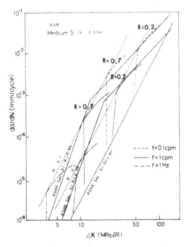

Fig.5 Predicted crack growth rate for medium sulfur steel.

SURFACE FLAW GROWTH RATE FACTOR

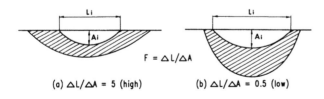

$F = \triangle L / \triangle A$

(a) $\triangle L / \triangle A$ = 5 (high)

(b) $\triangle L / \triangle A$ = 0.5 (low)

Numerical Example
in this study

A_i = 5,10 mm
L_i = 25,50,100,200 mm
A_i/L_i = 0.1 (fixed)
$\triangle L/\triangle A$ = 0.5 – 5.0
T = 180 mm

Fig.6
Geometry of
surface flaw.

237

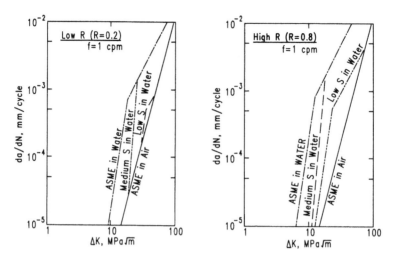

Fig.7   Assumed reference curves for low R ratio and high R ratio.

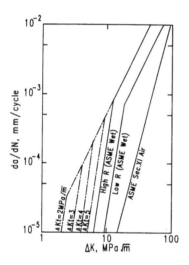

Fig.8   Modification of ASME sec.XI reference curve.

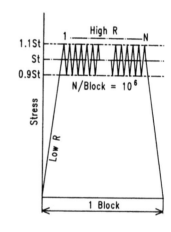

Fig.9   Schematical diagram for load spectrum of this calculation.

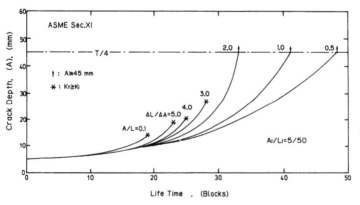

Fig.10
Change in crack depth with operation time.

Enhancement of fatigue crack growth rates in pressure boundary materials due to light-water-reactor environments

W.A.Van Der Sluys & R.H.Emanuelson
Babcock & Wilcox Research and Development Division, Alliance, Ohio, USA

The high level of reliability required of the primary-coolant pressure boundary in a nuclear reactor system leads to a continuing interest in the interaction among the coolant, pressure boundary materials, and service loadings. One area of concern involves the possible enhancement of the growth rate of fatigue cracks due to the coolant.

Advances have occurred recently toward a better understanding of the variables influencing the material/environment interactions and methods of addressing this interaction.

Sulfur now appears to be one of the principal agents responsible for the observed enhancement of the fatigue crack growth rates in light-water-reactor (LWR) environments. This paper presents the results of investigations on the effect of sulfur in the steel, in the bulk water environment, and at the crack tip.

A time-based format of data presentation is used in this paper along with the conventional crack growth rate based on cycle format. The time-based format is a useful method of data presentation. When presented in the conventional format, an apparent substantial amount of scatter in the data is eliminated and the data fall within a relatively narrow scatter band. This model permits extrapolation from the frequency and ΔK regions where experiments were conducted into previously unexplored regions.

The results of many experiments, mostly conducted under constant ΔK loading conditions, have been used in developing this paper. These experiments were conducted (Van Der Sluys 1986) over the last nine years under a contract with the Electric Power Research Institute (EPRI). Compact fracture specimens were tested in simulated, 288°C, boiling-water-reactor (BWR) or 288°C, pressurized-water-reactor (PWR) environments. Materials tested include four heats of SA 533 GrB material. The sulfur content of the SA 533 material varied from 0.004% to 0.025%. Applied ΔK levels ranged from 6.6 MPa$\sqrt{m}$ to 88 MPa$\sqrt{m}$, with load ratios (R) varying between 0.1 and 0.9. Because loading frequency is an important variable, a wide range (from 10 Hz to 0.000025 Hz) was studied. All waveforms were sinusoidal.

Figures 1 and 2 present the results from a series of experiments conducted on a steel with a high sulfur content (0.025%) in a simulated BWR environment. These results are typical for a material-environment combination in which environmentally assisted cracking (EAC) is present. These figures show three different ways in which the results of these experiments can be presented.

A significant amount of data scatter is apparent in the da/dn versus ΔK plots. However, closer examination of the individual data points reveals

239

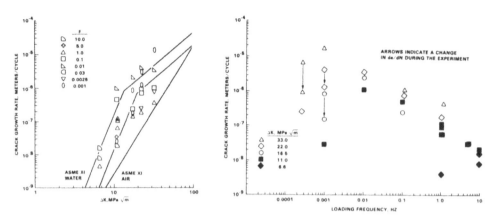

Figure 1. Fatigue crack growth rate results for experiments conducted on
SA 533-1 material with 0.025% sulfur in BWR environment.

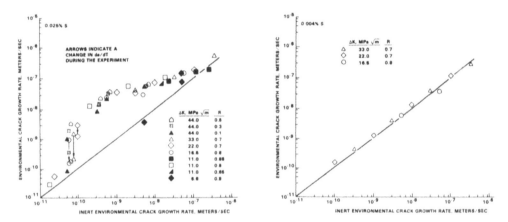

Figure 2. Fatigue crack growth rates plotted in time-based format for
SA 533-1 material with 0.025% and 0.004% sulfur.

that the data are ordered in accordance with the loading frequency. Also,
Figure 1 shows three reference flaw growth lines representing the flaw
growth lines recommended by Section XI of the ASME Boiler and Pressure
Vessel Code. The second graph in this figure shows this frequency dependence. A critical frequency exists for each ΔK level studied. This critical frequency is one where the highest crack growth rate is observed.
At loading frequencies higher than critical, slower crack growth rates
are observed; while at lower frequencies, the high EAC rates cannot be
sustained. These conditions are shown in these figures by arrows pointing
downward. In these experiments, the frequency was lowered to the current
test frequency. At frequencies below critical, the crack starts growing
at a high rate that cannot be sustained, and the crack growth slows.
    The same data are shown in Figure 2 in a time-based format. In this
format, the crack growth rate as a function of time is plotted versus the
predicted inert environment crack growth rate. All of the apparent data
scatter collapses into a single well behaved line. The only point outside
this line is one at an extremely low ΔK value, where no EAC was observed.

The data presented in the second graph in Figure 2 were obtained on an SA 533 material containing only 0.004% sulfur. This material exhibits no susceptibility to EAC. The time-based plot shows this lack of susceptibility.

All data presented in Figure 2 were obtained using a frequency-decreasing test approach. The crack was propagated under constant ΔK and constant loading frequency conditions to obtain a crack growth rate. After enough crack extension to obtain a good measure of the crack growth rate, the frequency is lowered and another crack growth rate determined. This procedure is repeated until the critical frequency is found. A few frequency-increasing experiments have been conducted immediately following a frequency-decreasing experiment. The results from one such experiment are shown in Figure 3. The crack growth rate was high at a loading frequency of 0.0005 Hz. When the frequency was reduced to 0.0001 Hz the crack growth rate slowed substantially. When the frequency was increased back to 0.0005 Hz, the crack growth rate increased. This experiment series of was performed twice. The results from both of these experiments are presented in the time-based plot in Figure 3.

All data from the four heats of SA 533 material in PWR and BWR environments are summarized in Figure 4. The time-based data are summarized in the first graph, while the second graph shows the critical frequency measured in the experiments versus the applied ΔK level.

The time-based fatigue crack growth rates obtained in the BWR environment are ordered directly by sulfur content, while the corresponding critical frequency is insensitive to sulfur. No EAC was observed in the 0.004% sulfur material but was significant in the 0.013% ,0.021%, and 0.025% sulfur material, along with a possible ordering by sulfur content.

The results in the PWR environment are not as orderly. The 0.013% sulfur material exhibits the most EAC, with the 0.025% and 0.021% sulfur materials showing decreasing enhancement, respectively. The critical frequency information in the PWR environment is also not as clear as in the BWR case.

As suggested by Shoji (1981), the time-based presentation is based on the concept that the crack tip strain rate is directly proportional to the inert crack growth rate. Therefore, the data presented in Figure 2 indicate that there is a critical range of crack tip strain rates over which EAC can occur. Furthermore, it appears that the crack growth rates when EAC is present are nearly constant when one considers the crack growth rate as a function of time rather than cycles. The range in crack tip strain rates in which EAC was observed in the BWR environment, although a function of the material sulfur content, did not vary substantially. However, in the PWR environment this range varied substantially. The strain rate at the high end of the range was fixed but the low strain rate end of the range varied among materials.

These results show that in the PWR environment, sulfur content alone is not sufficient to explain all the results. The microstructures of the test materials along with the sulfur prints for some of the materials have been studied (Van Der Sluys 1986). The sulfur prints and sulfide morphology observed in optical micrographs do not fully explain the differences observed in the EAC susceptibilities of these materials. More work is needed to explain why one material is susceptible while a second is not. The experimental test results and a material's microstructural information must be coupled with a mechanistic model to obtain a better understanding of this phenomena.

The sulfur concentration in the environment at the crack tip is important to the mechanism. This sulfur comes from the dissolution of mangan-

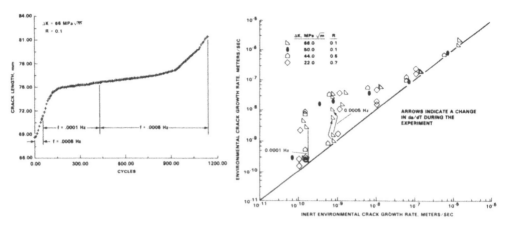

Figure 3.   Results from fatigue crack growth experiments in which loading frequency was increased and decreased.

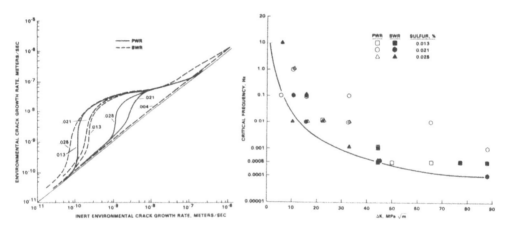

Figure 4.   Summary of fatigue crack growth rate results for PWR and BWR environments.

ese sulfide inclusions. A series of experiments was conducted to determine if sulfur introduced from a different source can also cause EAC.

In the initial experiment, the environment in a PWR autoclave was contaminated by introducing a number of different sulfur species. The contaminants included sodium sulfide, sodium sulfate, and sulfuric acid. The results from these experiments are presented in Figure 5. The test material was the same SA 533 material tested to produce the data shown in Figure 2, where no evidence of EAC was present. The sulfur content of this material (0.004%) is too low to cause EAC. The crack length versus cycles plot in Figure 5 shows no effect on the crack growth rate from the various contaminants in the environment. Either the sulfur concentration was not high enough, or the sulfur in the bulk environment did not enter the crack tip region.

Since there was some question whether the sulfur reached the crack tip, an experiment was designed to introduce the contaminated environment directly to that area. A small hole (2.4 mm in diameter) was drilled into the back of the specimen on the same plane as the crack. The hole was

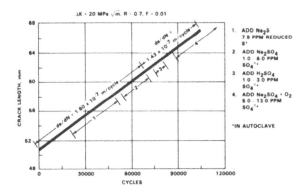

Figure 5. Effect of sulfur additions to
PWR environment on crack growth rate in
SA 533-1 with 0.004% sulfur at
$\Delta K = 20$ MPa$\sqrt{m}$, $R = 0.7$, and $f = 0.01$ Hz.

deep enough so that the crack would encounter the hole at a crack length
of 56 mm. An environmental-supply system was constructed to supply pressurized water of the desired composition to the hole in the specimen at
an extremely low flow rate.

The results from the first experiments using this system are presented
in Figure 6. These experiments were conducted on the same SA 533 material
used to produce the results shown in Figure 2. This material exhibits no
susceptibility to EAC. The results from nine separate crack growth rate
determinations are presented in this figure. In the first two experiments, the crack tip was not exposed to the contaminated water and the
measured crack growth rates were the same as reported in Figure 2.

In the third and fourth experiments, the crack tip was contaminated
with water containing 500 ppm H_2S. In these experiments, the crack
growth rates were high, of a magnitude usually observed when EAC is
present (see Figure 1).

In the fifth and sixth experiments, the feed system supplying water to
the crack tip was switched to supply noncontaminated PWR water to clean
the sulfides from the crack tip. The crack growth rate did not decrease
but remained at a rate similar to that measured during the contaminated
experiments. The crack tip and water feed system were then flushed with
H_2O_2 at 660°C for 60 hours.

Experiment seven was conducted after this cleaning and with PWR water
being circulated through the crack. The measured crack growth rate in
this experiment is only slightly faster than previous noncontaminated
determinations.

Experiments eight and nine were then conducted with 50 ppm of H_2S-con-
taminated water in the crack. The crack growth rate under those conditions was again high, of a similar magnitude as measured with the 500-ppm
H_2S contamination.

The loading frequency was a second variable that was studied in these
experiments. The effect of the loading frequency changes is difficult to
observe in the da/dn versus ΔK presentation format but easy to see in the
time-based presentation. The loading frequency did not have the strong
effect on the measured fatigue crack growth rates in the contaminated
crack tip experiments as it did in the high-sulfur-steel experiments.

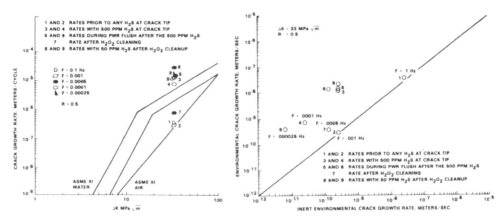

Figure 6. Results from fatigue crack growth experiments in which H_2S-contaminated PWR water was injected at crack tip.

Future experiments are planned to determine the critical level of H_2S needed in the crack to induce EAC and to determine if introducing H_2S to the crack tip lowers the observed ΔK thresholds for EAC.

The results from the series of experiments described here are useful in evaluating various mechanisms that have been proposed to explain EAC. There are three observations discussed that must be explained by a mechanistic model:

- Existence of critical frequency
- Reversibility of critical frequency
- Lack of critical frequency when sulfur is introduced into crack tip

The existence of the critical frequency and the lack of it when the sulfur is added to the crack tip can both be explained by a model based on a critical sulfur level being required at the crack tip for EAC to occur. But the reversibility of the phenomenon is not as easy to explain. It is difficult to postulate how the same critical sulfur level is achieved in a frequency-increasing condition as obtained in a frequency-decreasing experiment if the sulfur level is a function of the time-based crack growth rate. The predictive model will have to include not only the sulfur content of the material and the crack growth rate but also some relationship to describe the transport of sulfur out of the crack tip region.

REFERENCES

Van Der Sluys, W. A., "Corrosion Fatigue Characterization of Reactor Pressure Vessel Steels," RP1325-1 (Progress Report, May 1983 - June 1986), Electric Power Research Institute, Palo Alto, Calif., Oct. 1986.
Shoji, T., and Takahash, H., "Role of Loading Variables in Environment Enhanced Crack Growth for Water Cooled Nuclear Reactor Pressure Vessels," International Atomic Energy Agency Specialists Meeting on Subcritical Crack Growth, May 13-15, 1981, Freiburg, FRG, July 1981.

Assessment of environmentally assisted cracking in PWR pressure vessel steels

D.R.Tice

UKAEA Springfields, Salwick, Preston, UK

1 INTRODUCTION

The low alloy ferritic steel used for construction of the reactor pressure vessel and other major circuit components of light water reactors such as steam generator shells, has been subject to extensive study with regard to susceptibility to environmentally assisted cracking (EAC). Two forms of EAC are generally distinguished, viz stress corrosion cracking under steady loading, and corrosion fatigue when subjected to a cyclic load. Recent studies have shown that for the well controlled deoxygenated conditions relevant to PWRs, static stress corrosion cracking is most unlikely (Hurst et al 1985) so that the most probable form of environmental cracking in the reactor pressure vessel is the extension of pre-existing flaws by cyclic crack growth due to plant transient loading. Assessment of cyclic crack growth is currently made using the reference crack growth curves given in Section XI Appendix A of the ASME Boiler and Pressure Vessel Code (1980). Separate curves (Fig. 1) are given for 'dry' buried defects and for 'wet' surface breaking defects. In the latter case the presence of the austenitic clad layer on the walls of the reactor pressure vessel (but not for example in PWR steam generators) is ignored in the assessment. The code recognises an observed influence of mean stress on 'wet' crack growth by defining separate curves for high and low stress ratios, R.

There has been a large degree of inter-laboratory variability in reported crack growth data under simulated PWR conditions (eg Fig. 1) (Scott and Truswell 1983). It is now known that this is due to the synergistic effects of several variables which were not well controlled in early laboratory tests. Thus the sulphur content of the steel, water quality (especially dissolved oxygen and sulphur-containing contaminants) and water flow rate interact to control the local concentration of sulphur-rich ions within the crack crevice, and hence EAC susceptibility as shown in Fig. 2 (Tice et al 1986). It should be noted, however, that the bulk sulphur content of the steel is only a crude indication of susceptibility and local effects due to segregation of sulphide inclusions may be important.

Under conditions conducive to environmental cracking there is a large observed influence of the cyclic loading rate on crack growth (Fig. 3) (Swan 1985). Since the database for the ASME curves was obtained over a narrow range of loading rates, at the upper limit of plant transients, it is not obvious that the code is necessarily con-

servative under all likely loading conditions. Recent experimental
work has aimed at addressing these uncertainties, and has indicated a
need for an alternative assessment route (Scott 1986).

2   ALTERNATIVE ASSESSMENT METHODS

There has been significant progress in recent years in the development
of predictive methods which recognise the time dependent nature of
environmental cracking (Scott 1986, Ford 1986 and Gilman 1986). Using
these models, the environmental crack growth rate expressed with
respect to time ($\dot{a}_e$) is plotted as a function of either the crack tip
strain rate or inert crack velocity. The latter quantity is used here
since it is simpler to use for applications purposes.

Assuming Paris Law behaviour for inert crack growth, the inert
growth rate converted to a time base, $\dot{a}_i$, is given by:

$$(1) \qquad \dot{a}_i = 2\upsilon\frac{da}{dN} = 2\upsilon C \Delta K^m$$

where C and m are the usual Paris Law coefficients, υ is the loading
frequency and crack growth is assumed to occur only on the rising load
portion of the loading cycle.

A recent review (Eason et al 1985) has indicated that the ASME
dry line is a poor representation of air data and that a better description of mean growth is given by the use of a value of m of 2.6 and
$c = 1.2 \times 10^{-11}$ (da/dN in m/cycle, ΔK in MPa$\sqrt{m}$). (Note that for
assessment purposes a 95% confidence upper bound is recommended, with
$c = 3.4 \times 10^{-11}$). Using this representation of mean inert growth, the
corrosion fatigue data of Fig. 3 are replotted in Fig. 4. This presentation removes the frequency dependence of the data, and indicates
that over this range of loading frequencies there is a common threshold
crack velocity $\dot{a}_T$ above which time dependent EAC is observed and that
the rate of environmental cracking above the threshold can be described
by:

$$(2) \qquad \dot{a}_e = 5 \times 10^{-5} \dot{a}_i^{\;\frac{1}{3}} \quad (\dot{a} \text{ in ms}^{-1})$$

Below this threshold for time dependent cracking, there is still a
small degree of enchancement of crack growth compared to that in air.
This 'base rate wet growth' can be described by a Paris Law with
$c = 6.8 \times 10^{-11}$ and m = 2.6.

Equation (2) and the threshold value, $\dot{a}_T$ essentially define a
'window' for environmental cracking. The value of the threshold
velocity is controlled by the material and environmental variables
illustrated in Fig. 2. Thus a lower threshold is observed for high
sulphur steels or in the presence of sulphur contamination or dissolved oxygen in the water. There are at present only limited
experimental data at the low frequencies relevant to many plant
transients. The available data generally indicate that the threshold
velocities determined at higher frequencies are still applicable,
although transient high rates of cracking are sometimes observed below
the threshold (i.e. at lower ΔK) following changes to loading
conditions. There is also a lack of data covering the full range of
stress ratios relevant to plant.

The need to define the threshold velocity is crucial in terms of
assessment of crack growth under plant loading conditions using this

model, since only those transients whose frequency and ΔK combinations lie within the EAC 'window' will result in EAC. In contrast, the present ASME reference crack growth curves apply the same environmental factor irrespective of the transient loading frequency.

3 CALCULATION METHOD

Indicative calculations have been performed to assess the consequences of the new assessment model for crack growth in a PWR reactor pressure vessel during operational service. The aim of the calculations was to predict the likely extension of various postulated initial defects which may be present in the wall of a reactor pressure vessel when it enters service, or which may be discovered during in-service inspection. It should be noted that scrupulous manufacturing procedures combined with modern methods of non-destructive examination minimise the possibility of such defects being present in current generation vessels.

In order to compute the extent of propagation of assumed defects under operational and transient loading conditions, the following information is required:

i) specification of the transients expected to be experienced by the plant during its design life.

ii) stress analysis for various locations in the RPV under the specified transient loading. Knowing the appropriate stresses will allow the relevant stress intensity factor (K) for any given flaw to be calculated.

iii) Knowledge of the rates of crack growth as a function of the applied stress or stress intensity. This information is usually presented in the form of Paris type crack growth laws, but time dependent descriptions as discussed above were also used in the case of environmental cracking.

The calculations described here were performed for a Westinghouse 4 loop PWR vessel, assumed to be subjected to a spectrum of loading transients as specified by Westinghouse (Marshall 1982). 'Normal', 'upset' and 'test' transients were considered in the analysis. Stress analysis for several regions of the vessel was performed using appropriate finite element methods (Gardner and Hughes 1985) including three dimensional analysis where geometrical discontinuities such as nozzles necessitated it. It was assumed that overall crack growth over a defined period could be described as the summation of the individual crack growth increments due to each transient. The transients were divided into five blocks per year, each assumed to start with a heat-up/cool down transient. The specified number of transients in the Westinghouse design specification was assumed to occur uniformly over a forty year design life. The crack tip stress intensity range for each transient within the block was calculated for the initial assumed defect size 'a'. Crack growth for each transient was then calculated using one of several crack growth 'laws' as described below. The summation of all crack growth increments calculated for each transient occurring δN times during one block of transients (δ a) was added to the current crack size to give an updated crack size (a+δa) for the start of the next block of transients. This process was iterated for each block of transients in turn, up-dating the crack depth at each iteration until the design life of 40 years was reached.

Several alternative descriptions of crack growth have been used in these calculations. For dry cracks, calculations were performed using

247

either the ASME 'dry' reference line or the alternative 95% confidence upper bound to air data, with $c = 3.4 \times 10^{-11}$ m/cycle, $m = 2.6$. Calculations for wet cracks used either the variable R ASME 'wet' curves, or the alternative time dependent model described above. In the latter case overall crack growth was computed as the sum of 'base rate wet' crack growth ($da/dN = 6.8 \times 10^{-11} \Delta K^{2.6}$) and, if the relevant inert crack velocity for the transient under consideration was above the threshold velocity $\dot{a}_T$, time dependent cracking described by equation 2. However, time dependent cracking was not included for those test transients, such as proof tests, for which the vessel temperature was much lower than the operating temperature, since there is no evidence for time dependent EAC under these conditions. Because of uncertainty in knowledge of the most relevant threshold velocity under PWR conditions, the sensitivity to several assumed values of threshold velocity was assessed.

4 RESULTS OF CRACK GROWTH CALCULATIONS

The calculations were performed for a range of defect sizes, and for several values of flaw aspect ratio. As expected, the highest crack growth was predicted in those regions of the vessel where local stresses are enhanced by geometrical discontinuities, such as in the nozzle corner. For defect sizes which are readily detectable by non destructive examination the extent of crack extension under 'dry' conditons is of no concern. For the most highly stressed region, the 95% upper bound to air data predicts lower crack growth than the ASME 'dry' curve.

A wider range of predicted crack extensions was obtained for wet cracks, depending on the assumed description of crack growth. Illustrative results for a relatively large assumed initial defect size of 25 mm with a 6:1 width to depth aspect ratio are shown in Fig. 5. Crack extension for smaller defect sizes will be considerably less than shown here. If environmental conditions are such that a threshold above 5×10^{-10} ms^{-1} is applicable, crack growth predicted using the new model is much less than if the ASME reference curves are used for assessment. This should be the case for modern steels operated in well controlled PWR primary coolant chemistry where growth is described by the 'base rate wet' equation for most transients. However, if less ideal conditions (e.g. contaminated chemistry, high sulphur steels or stagnant flow) result in a lower threshold, then very much higher crack growth in the nozzle corner is predicted (Fig. 5a). This indicates that when time dependent EAC occurs, it proceeds at an unacceptably high rate in practice. The lower stresses in the main cylindrical regions of the vessel away from geometrical discontinuities result in unacceptable rates of cracking only if threshold velocities are as low as 10^{-11} ms^{-1} (Fig. 5b). No indication of threshold velocities as low as this has been reported for either PWR or BWR operating conditions.

The specific results reported here are directly relevant only to a specific design of PWR reactor pressure vessel subjected to the design transients specified by Westinghouse. Moreover, the rates of loading during these transients are of crucial importance in the assessment. Test transients, such as proof tests and in-service leak tests have been assumed not to result in time dependent EAC, since they take place at temperatures outside the current database, where the mechanism of cracking may be different. There is a need to ensure that

this assumption is valid, i.e. that substantial time dependent EAC
cannot occur at low temperatures.

5 CONCLUSIONS

1) Since environmentally assisted cracking (EAC) is a time dependent
process, assessment should be based on time rather than cycle dependent
parameters. Thus an $\dot{a}_e$ vs $\dot{a}_i$ (or strain rate) basis for assessment
should be used in preference to da/dN vs ΔK.

2) The threshold strain rate or velocity for the onset of EAC is
controlled by material and environmental factors (e.g. steel sulphur
content and water chemistry), and possibly by mechanical loading
factors such as R ratio and load interaction effects. Above the
threshold, crack growth rates are usually unacceptably rapid.

3) Sample calculations show that predicted crack growth rates using
a time based model can be below or above those calculated using ASME
XI depending on the value of the EAC threshold velocity but that for
normal PWR operating conditions rates are likely to be below those
predicted by the ASME code.

REFERENCES

Eason, E., Andrew, S.P., Nelson, E.E., Youssefi, K.Y. and
  Warmbrodt, S.B. 1986. IAEA Specialists meeting on Sub Critical
  Crack Growth, Sendai, Japan May 1985. NUREG CP-0067, Vol. 2, p.309.
Ford, F.P. 1986. Corrosion 86, Paper 327, Houston:NACE.
Gardner, C & Hughes, M. 1985. Private communication, UKAEA Harwell
Gilman, J.D. 1986. IAEA Specialists meeting on Sub Critical Crack
  Growth, Sendai, Japan May 1985. NUREG CP-0067, Vol. 2, p.365.
Hurst, P., Banks, P., Pemberton, G. & Raffel, A.S. 1985. Proc. 2nd
  Int. Symposium on Environmental Degradation of Materials in Nuclear
  Power Systems - Water Reactors. Monterey CA:NACE.
Marshall, W. 1982. An Assessment of the Integrity of PWR Pressure
  Vessels. London:UKAEA.
Scott, P.M. 1986. IAEA Specialists meeting on Sub Critical Crack
  Growth, Sendai, Japan May 1985. NUREG CP-0067, Vol. 2, p.293.
Scott, P.M. & Truswell, A.E. 1983. J. Pressure Vessel Tech, 105, 245.
Swan, D.I. 1985. Data cited by D R Tice, Corros. Sci. 25, 705.
Tice, D.R., Atkinson, J.D. & Scott, P.M. 1986. IAEA Specialists
  meeting on Sub Critical Crack Growth, Sendai, Japan May 1985.
  NUREG CP-0067, Vol. 1, p.251.

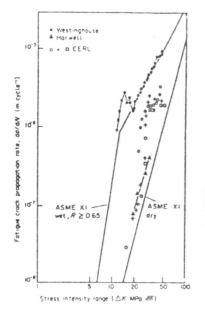

Fig 1 Corrosion fatigue crack growth data
for an A533B Steel in 288°C water.

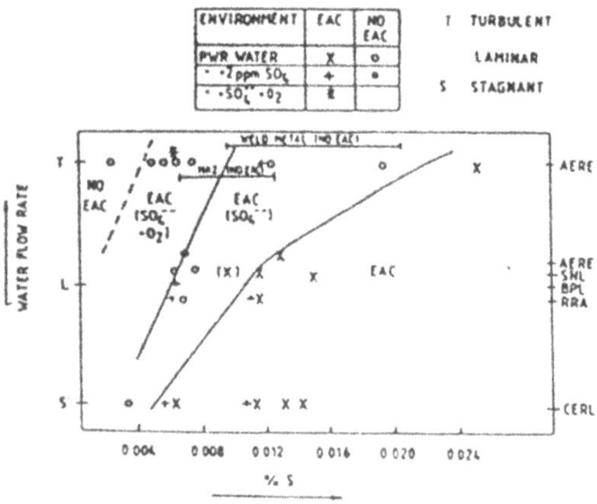

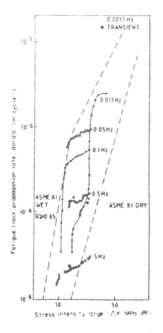

Fig 3 Effect of loading frequency on
corrosion fatigue crack growth of
RPV steels in 288°C PWR water

Fig 2 Effects of steel sulphur content,
flow rate and water chemistry on
EAC susceptibility of RPV steels

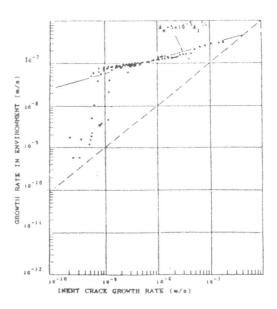

Fig 4 Data of Fig. 3 expressed as a
function of inert crack velocity.

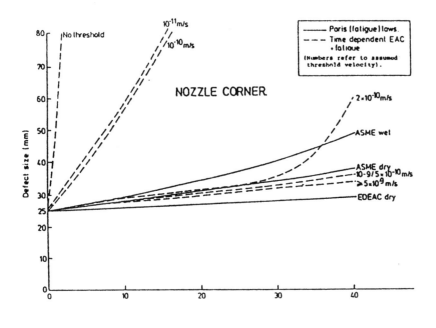

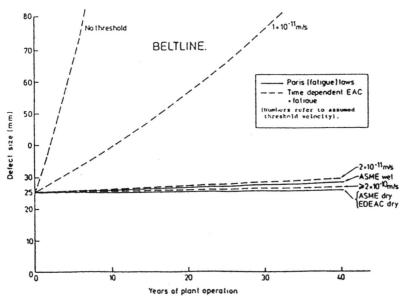

Fig 5 Calculated crack growth in a PWR
pressure vessel :
(a) nozzle corner
(b) beltline

Verification of a model for predicting corrosion-fatigue crack-growth rates in austenitic stainless steels in light water reactor environments

J.D.Gilman
Electric Power Research Institute, Palo Alto, Calif., USA

R.Rungta, P.Hinds & H.Mindlin
Battelle, Columbus, Ohio, USA

1 INTRODUCTION

Stress corrosion cracking of austenitic stainless steels has been studied extensively in recent years, as a result of the widespread incidence of intergranular stress corrosion cracking (IGSCC) at welds in BWR piping systems. Fatigue crack growth in these same materials and environments has received less research emphasis because cyclic loading is not considered a fundamental cause or contributor to the IGSCC problem.

Nevertheless, there are some incentives for examining fatigue crack growth rates in stainless steels for reactor environments. Some IGSCC repair techniques leave existing cracks in place, with added reinforcement or reduced stress, and these flaws must be fully evaluated against several criteria including fatigue crack growth (Section XI Task Group 1986). When piping is replaced with more crack-resistant materials, a crack growth analysis is useful to show flaw tolerance of the improved material. Finally, reliable flaw assessment methods are needed generally, for both BWR and PWR applications, to perform flaw evaluations in support of the in-service inspection program.

The purpose of this paper is to present available corrosion fatigue data in a format suggested by other investigators, to answer several questions:

(1) Does this format for presentation aid data interpretation and correlation?
(2) Does the corrosion fatigue behavior exhibit strong time-dependent or frequency-dependent effects which should be considered in flaw assessments?
(3) Is a conventional Paris-law approach applicable?
(4) Are there enough data to answer these questions?

2 BACKGROUND

In the case of austenitic stainless steels, the predominant view is that environmentally assisted cracking occurs by the anodic dissolution process (Ford et al. 1986). But whether the mechanism of cracking is anodic dissolution or hydrogen embrittlement, the rate of cracking is believed to depend upon the rate at which fresh surface is exposed at the crack tip. Based on a dislocation emission rate at the crack tip, Shoji et al. (1983,1981) suggested that the time based crack growth rate in an inert environment could be chosen as a parameter quantifying fresh metal surface creation rate at the crack tip. A comparison of time based crack growth rate in the inert environment and in an aggressive environment, for identical loading conditions, will then provide a measure of the acceleration in crack growth rates due to the environment. Shoji et al. (1983) further suggested that the time based crack growth rate in air (a relatively inert environment) could also be used as a descriptor of the crack tip strain rate which governs surface exposure through oxide film rupture. Using this approach, Ford et al. (1986) have proposed that environmentally assisted crack growth rates are a single function of the crack tip strain rate for static load, cyclic load, and constant extension rate tests. In cyclic loading, the crack tip strain rate is proportional to the time based crack growth rate in air. Using Ford's approach, there should be a relationship of the form $\dot{a}_e = \dot{a}_a + A\dot{a}_a{}^n$ between the crack growth rates in air, $\dot{a}_a$, and in the water environment, $\dot{a}_e$. This study seeks such a correlation for available crack growth data for stainless steels.

3 EFFECTS OF ENVIRONMENT

Figure 1 shows expected relationships between crack growth rate and crack tip strain rate, based on Ford et al. (1986), for four different reactor environments. Figure 2 defines those environments, again consistent with Ford et al. (1986), as a function of the electrochemical potential (ECP) and the room-temperature conductivity. The ECP depends on oxidizing species in the water (primarily dissolved oxygen) and the conductivity is an indication of ionic species which may accelerate crack growth. Sulfate and chloride are known to be detrimental.

Currently, BWR operators seek to control water purity to maintain conditions in the upper left box in Figure 2. Typical operating conditions are close to the "clean" line. Conditions on the "nominal" line are typical of past practice without these controls. Some plant owners are adopting hydrogen water chemistry (HWC) to reduce oxygen levels and maintain conditions in the lower left-hand box. Figure 1 shows the expected benefits of these water chemistry improvements for weld sensitized Type 304 stainless steel.

The PWR primary system operates normally at a very low ECP; typically -600mVshe, for which very little influence of environment on crack growth is predicted by the model just described.

4 DATA SOURCES

A large number of fatigue crack growth data have been compiled in EDEAC (EPRI Database for Environmentally-Assisted Cracking), a computerized database operated by Battelle (Mindlin et al. 1986). Data in EDEAC were solicited directly from numerous laboratories, or were obtained in some cases through the MPC-PVRC Task Group on Crack Propagation Technology. A few data not in EDEAC are also included in this study.

These laboratory data generally do not include measured ECP and water conductivity. For this reason, the only variable characterizing environment in the present study is the dissolved oxygen content. It is quite possible that these data are influenced by variations in conductivity and ECP. Unfortunately, there is not presently a significant body of fatigue crack growth data for stainless steels in which the water environment has been controlled and characterized with respect to ECP and with respect to very low-level impurities.

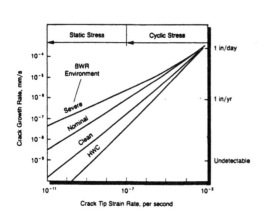

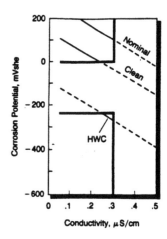

Figure 1. Crack growth in weld sensitized stainless steel depends on the crack tip strain rate and water chemistry, according to the model developed by Ford et al. (1986).

Figure 2. BWR water chemistry is characterized by the corrosion potential (usually controlled by dissolved oxygen) and the conductivity (which may indicate harmful ionic species).

254

5 DATA ANALYSIS METHOD

The crack growth rate for stainless steel in air, which is the key parameter for this data correlation, has been presented by James and Jones (1985) in the form

$$\text{(1)} \qquad \frac{da}{dN} = C_o \, S_r \Delta K^n$$

where C_o is a temperature-dependent constant and S_r depends on the Stress ratio R. For $288^\circ C$,

$$\text{(2)} \qquad \begin{aligned} C_o &= 3.428 \times 10^{-9} \\ n &= 3.30 \end{aligned}$$

$$\begin{aligned} S_r &= 1.0 + 1.80 \, R & 0 \leq R \leq 0.79 \\ S_r &= -43.35 + 57.97 \, R & 0.79 \leq R \leq 1.0 \end{aligned}$$

The time-dependent crack growth rate is defined as

$$\text{(3)} \qquad \dot{a} = \frac{da}{dN} \cdot \frac{1}{T_r}$$

where T_r is the load rise time. This is based on the assumption that the environmental influence occurs predominantly during rising load conditions.

In the data plot to follow, the ordinate $\dot{a}_e$ is the observed value of da/dN, divided by T_r. The abscissa $\dot{a}_a$ is the calculated value of da/dN for air divided by T_r.

6 CRACK GROWTH RATES IN OXYGENATED WATER

Figure 3 presents crack growth data for sensitized stainless steel in air-saturated water containing 8 ppm of oxygen (Kawakubo et al. 1980, 1986). Although there is considerable scatter, the upper bound of the data can be fit by a curve of the form $\dot{a}_e = \dot{a}_a + A\dot{a}_a^n$. In this case, A = .0025 and n = 0.446. Under slow-cycling conditions, for $\dot{a}_a = 2 \times 10^{-8}$ mm/s, the observed fatigue crack growth rate in water exceeds the predicted rate in air by a factor of about 50. Predictions based on extrapolation of the curve lead to higher factors at slower cyclic rates.

7 CRACK GROWTH RATES IN LOW-OXYGEN WATER

Figure 4 shows crack growth data for sensitized materials in water containing 0.2 ppm of dissolved oxygen (Hale et al. 1978), the normal level for the BWR. The curve which bounded the 8 ppm data (Figure 3) also follows the trend of the 0.2 ppm data. It is possible that some sensitivity to oxygen level is masked by variations in conductivity (ionic impurities) and ECP, both of which affect crack growth independently of oxygen content.

Also shown in Figure 4 are four data points obtained in water con-taining only 0.015 ppm of oxygen, with conductivity controlled to less than 0.2 μ8/cm (Gordon et al. 1985, Jewett et al. 1986). The materials are, respectively, solution annealed 304, weld sensitized 304, furnace sensitized 304, and Type 316NG. Differences in fatigue crack growth rates among these four materials are not significant. All four exhibited transgranular cracking in this low-oxygen environment. Crack growth rates exceed the air rate by factors of 18 to 25. Unpublished data from another source (Schmidt et al. 1986) indicate crack growth rates for Type 316NG material in 0.2 ppm water that are close to the broken line in Figure 4.

8 CRACK GROWTH RATES IN SOLUTION ANNEALED MATERIAL

Figure 5 shows data (Hale et al. 1978) from EDEAC for solution-annealed materials in 0.2 ppm oxygenated water. The trend of these data is almost parallel to the "air" line, which could indicate an environmental effect that is not time dependent. However, this is not supported by Figure 4, which presents data on the broken line (reproduced in Figure 5) for solution annealed material in 0.015 ppm oxygenated water. These particular data are insufficient to establish whether solution annealed material performs better than sensitized material (the solid line in Figure 5) under slow cycling conditions in 0.2 ppm oxygenated water.

255

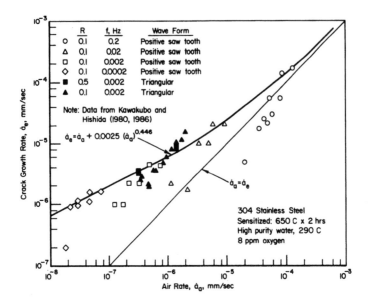

Figure 3. Time based presentation of corrosion fatigue data for a sensitized
stainless steel.

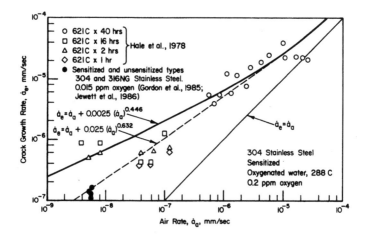

Figure 4. Time based comparison of crack growth data for a stainless
steel with varying degree of sensitization.

9 CRACK GROWTH RATES IN DEOXYGENATED (PWR) WATER

Available data (Cullen et al. 1984, Bamford 1977, Landerman et al. 1986) for stainless
steels in the PWR primary water environment are shown in Figure 6. These data are for
cast materials, which may result in crack growth behavior different than previously
discussed wrought stainless steels. None of the data are for the slow-cycling
conditions where significant time-dependent effects are observed in oxygenated
environments. The transient crack growth measurements are much higher than the solid
line representing trends for wrought materials in oxygenated water environments.

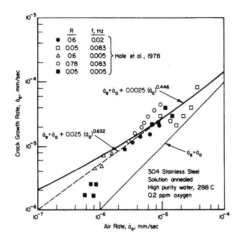

Figure 5. Time based presentation of cyclic crack growth data for a solution annealed stainless steel in high purity water containing 0.2 ppm oxygen.

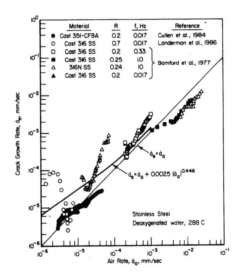

Figure 6. Time based presentation of cyclic crack growth data for stainless steels in deoxygenated (PWR) water at 288C.

10 DISCUSSION AND CONCLUSIONS

Solutions to the IGSCC problem in BWR piping focus on material changes to avoid grain boundary sensitization, and environment controls to reduce the oxidizing potential. IGSCC can be eliminated by these means, according to other studies (Jewett et al. 1986), but fatigue loading in high temperature water still produces transgranular crack growth at rates considerably higher than in air. In oxygen-containing water, the data shown clearly that the corrosion fatigue crack growth rate is time-dependent. The data analysis method or model presented here is a useful way to examine the time dependence, but it would be an overstatement to say that these data verify the model. Available data exhibit a great deal of scatter, which may result from uncertain conditions of water chemistry or microstructure.

An ASME Task Group (1986) has recommended corrosion fatigue crack growth rates in austenitic piping materials that are higher than the air rate by constant factors. A multiplier of 10 is recommended for the BWR environment, and 2 for the PWR environment. Figure 4 shows that crack growth rates in the normal BWR environment will exceed the air rate (the 45° line) by factors greater than 10 when the rate of cycling is low, such that the air rate is less than about 10^-7mm/sec. This is true even for unsensitized material in high purity, low oxygen (.015 ppm) water.

Many PWR data exceed the air rate by factors greater than 2, according to Figure 6. As noted previously, these data are for cast materials which may have different characteristics than wrought material. The erratic behavior could result from residual stress in the cast test specimens.

As noted previously, the data in Figure 6 are insufficient to show whether crack growth in PWR water is time dependent. However, Prater et al. (1985) measured crack growth rates for sensitized Type 304 stainless steel in deoxygenated water at very slow cyclic rates and found no indication of time dependence or frequency dependence. The electrochemical potential in those tests was below -400 mV. There is the possibility, still to be confirmed, that a constant multiplier on the air rate can be justified for corrosion fatigue in the PWR primary environment as well as in the BWR Hydrogen Water Chemistry environment (Figure 2).

REFERENCES

Bamford, W. H. 1977. Fatigue crack growth of stainless steel reactor coolant piping in a pressurized water reactor environment. Westinghouse Report WCAP-8953.
Cullen, W. H., R. E. Taylor, K. Torronen and M. Kemppainen 1984. The temperature dependence of fatigue crack growth rates of a 351-CF8A cast stainless steel in LWR environment. USNRC Report NUREG/CR-3546.
Ford, F. P., D. F. Taylor, P. L. Andresen and R. G. Ballinger 1986. Environmentally-controlled cracking of stainless and low-alloy steels in light-water reactor environments. Electric Power Research Institute Report NP-5064M.
Gordon, B. M. et al. 1985. Hydrogen water chemistry for BWRs. Electric Power Research Institute Report NP-3959M.
Hale, D. A., J. L. Yuen and T. L. Gerber 1978. Fatigue crack growth in piping and RPV steels in simulated BWR water environment. USNRC Report NUREG/CR-0390.
James, L. A. and D. P. Jones 1985. Fatigue crack growth correlation for austenitic stainless steels in air. In R. Rungta (ed.), Predictive Capabilities in Environmentally Assisted Cracking, PVP Vol. 99, P. 363-414. American Society for Mechanical Engineers: New York.
Jewett, C. W. and A. E. Pickett 1986. The benefit of hydrogen addition to the boiling water reactor environment on stress corrosion crack initiation and growth in Type 304 stainless steel. Journal of Engineering Materials and Technology 108: 10-19.
Kawakubo, T., M. Hishida, K. Amano and M. Katsuta 1980. Crack growth behavior of Type 304 stainless steel in oxygenated 290C pure water under low frequency cyclic loading. Corrosion 36: 638-647.
Kawakubo, T. and M. Hishida 1986. Toshiba Corporation, EDEAC files 96, 176 and 190.
Landerman, E. I. and W. H. Bamford 1986. EDEAC file 121.
Mindlin, H., R. Rungta, K. Koehl and R. Gubiotti 1986. EPRI database for environmentally-assisted cracking (EDEAC). Electric Power Research Institute Report NP-4485.
Prater, T. A., W. R. Catlin and L. F. Coffin 1985. Influence of dissolved hydrogen and oxygen on crack growth in LWR materials. Electric Power Research Institute Report NP-4183.
Schmidt, C. G., R. D. Caliguiri and L. E. Eiselstein 1986. Stress corrosion susceptibility of Type 316 nuclear grade stainless steel and XM-19 alloy in simulated BWR water. To be published by Electric Research Institute for BWR Owners Group for IGSCC Research.
Section XI Task Group for piping flaw evaluation 1986. ASME code, Evaluation of flaws in austenitic steel piping. Journal of Pressure Vessel Technology 108:352-366.
Shoji, T., H. Takahashi, M. Suzuki and T. Kondo 1981. A new parameter for characterizing corrosion fatigue crack growth. Journal of Engineering Materials & Technology 103:298-304.
Shoji, T., H. Takahashi, H. Nakajima and T. Kondo 1983. Role of loading variables in environment-enhanced crack growth for water-cooled nuclear reactor pressure vessel steels. In Proceedings of the IAEA specialists Meeting on Subcritical Crack Growth, USNRC Report NUREG-CP-0044, P. 143-172.

Modelling of environmentally assisted cracking in the stainless steel and low alloy steel / water systems at 288°C

F.P.Ford & P.L.Andresen

General Electric Corporate Research & Development Center, Schenectady, N.Y., USA

1 INTRODUCTION

Failure of structural components by stress corrosion or corrosion fatigue presents a possible life limiting factor in the operation of Light Water Reactors. The objective of this investigation (Ford et.al. 1987) was to gain a quantitative knowledge of the mechanism of environmentally assisted cracking of ductile, structural steels in high temperature water. The specific uses of this knowledge were: (1) to provide a theoretical under-pinning to the empirically derived remedies for intergranular stress corrosion cracking of Type 304 stainless steel piping in Boiling Water Reactors, and (2) to evaluate the fundamental validity of the ASME XI (1980) life evaluation code for corrosion fatigue of low alloy pressure vessel steels in Light Water Reactor environments.

The approach taken was to propose a working hypothesis for the mechanism of environmentally assisted crack propagation and then to independently evaluate the parameters of fundamental importance in this mechanism. On the basis of these data, theoretical values of the subcritical crack propagation rate could be predicted for different steady state and transient conditions of environment (e.g., degree of aeration, impurity content), material (e.g., degree of grain boundary sensitization, sulphur content), or stress, (e.g., static load, cyclic load). These predictions could then be compared with observed data in order to evaluate the quantitative validity of the original working hypothesis before applying it to the practical analyses referred to above.

2 WORKING HYPOTHESIS AND ITS QUANTIFICATION

The slip dissolution/film rupture mechanism of environmentally assisted cracking was chosen as the working hypothesis for crack propagation. The reason for this choice was that previous investigation (Ford 1982) at temperatures <115C had indicated that it was valid quantitatively for low alloy and stainless steels in various aqueous environments; moreover, extrapolation of these low temperature data to 288C yielded predictions of crack growth rates in water which were of the right order of magnitude compared with the observed rates. The model relates crack advance to the oxidation reactions occurring at the crack tip where a thermodynamically stable oxide (or protective film) is ruptured by an increase in the strain in the underlying matrix (Fig. 1). The periodicity of this rupture event is related to the strain rate in the metal matrix and this, in turn, is controlled by either creep processes under constant load or applied strain rates under monotonically increasing or cyclic load conditions. Thus, the model is potentially applicable to not only stress corrosion (under constant stress) but also to corrosion fatigue under a variety of stress amplitude, mean stress, frequency, etc. combinations. In the ultimate analysis, the average crack propagation rate, V_T, may be related to the crack tip strain rate, $\dot{e}_{ct}$, via the power law relationship:

$$(1) \qquad V_T = A \, \dot{e}_{ct}^{\,n}$$

where A and n are parameters dependent on the material and environment compositions at the crack tip. There are limits to the validity of the relationship in Eqn. 1 which are observed at high and low crack tip strain rates (Fig. 2). At high crack tip strain rates (~0.01/s) a bare surface is maintained continuously at the crack tip, and the environmentally assisted crack propagation rate becomes independent of the crack tip strain rate, since it cannot exceed the Faradaic equivalent of the bare surface dissolution rate. Various phenomena may contribute to a lower validity limit for Eqn. 1. The most general, however, is the fact that sharp cracks cannot be maintained when the average crack tip propagation rate, V_T, approaches the oxidation rate on the crack sides, V_s; under this particular condition, the crack propagation rate will slow down with exposure time and crack arrest will occur due to blunting.

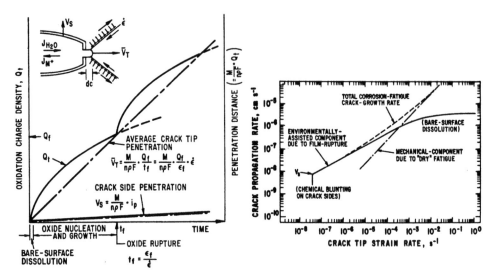

Figure 1. Schematic oxidation charge density/time relationships for a strained crack tip and unstrained crack sides.

Figure 2. Strain rate dependence of the crack growth rate due to film rupture model.

Under either constant or monotonically increasing load conditions, the stress corrosion crack propagation rate is defined by Eqn. 1 (within the limits discussed above). Under cyclic loading conditions, however, the crack may also be moving forward by irreversible cyclic plastic deformation; this mechanical component of crack advance is additive to the oxidation related component, as shown by the dotted line in Fig. 2 for corrosion fatigue conditions.

In order to quantify such a model, the following processes must be defined: **1. The steady state and transient compositions** of the environment at the crack tip as a function of that in the bulk (external) solution; **2. The oxidation rates** for the material/environment system expected at a strained crack tip; and **3. The oxide fracture strain and the crack tip strain rate**, defined in terms of engineering parameters such as ΔK, R, frequency, K, etc.

These tasks have been completed and reported fully elsewhere (Ford, et.al. 1987, Andresen & Ford 1985, Andresen 1985, Ford & Emigh 1985, Taylor and Caramihas 1982, Taylor & Caramihas-Foust 1985). It is the purpose of this paper to illustrate the validity and practical use of prediction algorithms of the form in Eqn. 1. It is important to underline, however, that these algorithms are based on the fundamental observations in the tasks itemized above, and are in no way empirical in nature.

3 COMPARISON OF OBSERVED VS PREDICTED CRACK PROPAGATION RATES

Observed crack propagation rates for various combinations of stainless steel and low alloy steel microstructure, dissolved oxygen content, solution purity and stressing mode have been compared with the values predicted by the slip dissolution/film rupture model. The theoretical values have been derived from Eqn. 1, where the crack tip strain rate has been calculated for various stressing modes, and the values of A and n have been derived from oxidation rates on bared surfaces in the relevant material/environment systems calculated to exist at the crack tip. An example of the correlation between observed and predicted crack velocity / strain rate relationships is shown in Fig. 3 for furnace sensitized 304 stainless steel which has been stressed in aerated water at 288C under static ($\mathbb{O}$,etc.), monotonically increasing (α,β,etc.) and cyclic loading ($\bullet,\times$,etc.) conditions. Similar agreements between observation and theory were obtained when the environmental conditions were changed; for instance, the synergistic effects of corrosion potential (dissolved oxygen content) and solution conductivity (and hence, anionic activity) on the cracking susceptibility of sensitized stainless steel under constant load in water at 288C are clearly predicted (Fig. 4). The level of agreement between observation and theory for the stainless steel/water system is illustrated in Fig. 5 via the distribution of the ratio (calculated/observed) crack propagation rates over a wide range of material, environment, and loading conditions. The mean value of this ratio is 1.17, while the variance in the distribution can be directly (Ford et.al. 1987) correlated with the uncertainty in the system definition.

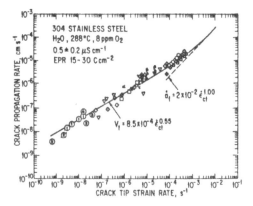

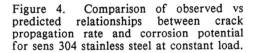

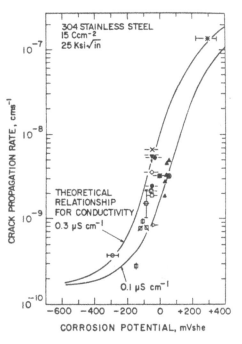

Figure 3. Observed and predicted crack propagation rate vs crack tip strain rate relationship for 304 stainless steel.

Figure 4. Comparison of observed vs predicted relationships between crack propagation rate and corrosion potential for sens 304 stainless steel at constant load.

In the case of pressure vessel steels in water at 288 C, it was predicted that environmentally assisted cracking might be expected under certain system conditions, especially where a "high" dissolved sulphur content is maintained at the crack tip (Fig. 6). The achievement of such an environmental condition is predictable from considerations of liquid mass transport in the crack (Andresen 1987, Ford et.al. 1987). The accuracy of the predictions in Fig. 6 are shown in Fig. 5 via the distribution of the (calculated/observed) crack propagation rate ratio for A533B, A508 and SA333 low alloy steels in water at 288C under a variety of environmental, material and stressing conditions. It was also predicted that, in these particular alloy steel/water systems, a limiting condition for the maintenance of sharp crack propagation would be reached and this would be evidenced by "threshold" stress intensity amplitudes and "critical" loading frequencies. In this investigation, (Ford et.al. 1987) an argument is put forward that this limiting condition is associated with the onset of chemical blunting of the crack at certain combinations of frequency, stress intensity and mean stress. Although such a "blunting" argument gives reasonable predictions of the onset of environmentally assisted cracking, it is recognized that other limiting criteria might be operating.

261

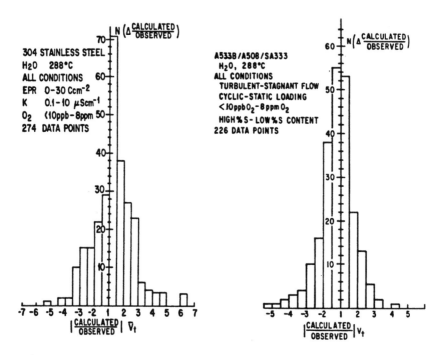

Figure 5. Frequency distributions of the calculated/observed propagation rate ratio for 304 stainless steel and A533B, A508, or SA333 low alloy steel in 288C water.

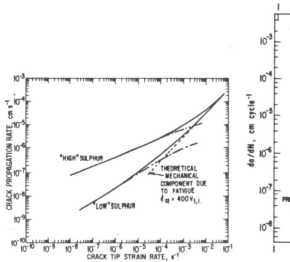

Figure 6. Theoretical crack growth rate vs crack tip strain rate for the low alloy steel/water system at 288C with material and environmental conditions where "high" or "low" activities of dissolved sulphur species exist at the crack tip.

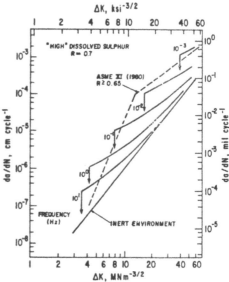

Figure 7. Theoretical da/dN vs ΔK relationships for A533B/A508 pressure vessel steel cyclically loaded at various frequencies under R = 0.7 conditions.

4 CONSEQUENCE OF A QUANTITATIVE MECHANISTIC UNDERSTANDING OF CRACK PROPAGATION PROCESSES

The fact that the environmentally assisted crack propagation rates for stainless and pressure vessel steels in water at 288C can be predicted with reasonable accuracy from first principles has considerable practical impact. First, it gives a methodology whereby the fundamental validity of empirically derived remedies and life time evaluation codes can be examined, and, second, it presents the beginning of a process for optimizing plant operating conditions vis a vis resistance to environmentally assisted cracking.

In many cases, remedies are formulated on the basis of "accelerated" tests, whereby early failure is achieved by, for instance, constant extension rate testing, intermittent fatigue loading, or the introduction of aggressive anions. The soundness of such acceleration methods have been assessed by the (validated) slip dissolution/film rupture model (Ford et.al. 1987).

With respect to the validity of the current ASME XI (1980) lifetime evaluation codes for low alloy steel pressure vessels and piping undergoing corrosion fatigue, it is apparent that the maximum theoretical (da/dN) vs Δ K values approximate those in that evaluation code (Fig. 7). It is also apparent, however, that the use of the cyclic based code can be conservative. This is especially the case if the pressure vessel is operating under the majority of the stressing, material, and environmental conditions shown in Fig. 7. Under these corrosion fatigue conditions, it makes more engineering sense to utilize a time based code (Gilman 1985).

With respect to the empirically derived remedies for sensitized stainless steel piping in Boiling Water Reactors, it was verified (Ford et.al. 1987) that the actions of reducing the degree of tensile stress, decreasing the corrosion potential (by, for instance introducing hydrogen into the feed water), increasing the water purity or decreasing the degree of grain boundary sensitization should decrease the crack propagation rate by the amounts observed in qualification tests performed by the reactor manufacturers and confirmed by other laboratories. Moreover, the theory predicts the observed variation in cracking susceptibility of piping with, eg., water conductivity and suggests the combinations of corrosion potential and solution conductivity that must be achieved to ensure satisfactory stress corrosion resistance (Fig. 8).

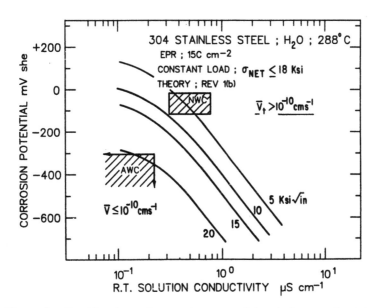

Figure 8. Combinations of corrosion potential, solution conductivity and stress intensity values which will give crack growth rates on sensitized 304 stainless steel which are either > or $<10^{-10}$ cm s^{-1}.

5 CONCLUSIONS

1.   The slip dissolution/film rupture model for environmentally assisted crack propagation is capable of predicting the observed crack propagation rates for the stainless steel and low alloy steel/water systems at 288C for a wide range of material, environment, and loading conditions. The absolute accuracy of the prediction depends critically on the accuracy of the system definition. Thus, there is a strong practical argument for combining the model prediction capabilities with system monitors which accurately define the actual operating conditions.
2.   The material variables that can be accounted for are the degree of grain boundary sensitization in Type 304 stainless steel and the sulphur content in the low alloy A533B or A508 pressure vessel steel. The environmental variables that can be accounted for are the oxygen content, the solution flow rate, and the anion content as monitored by the solution conductivity. The effects of these variables are fundamentally related to their effect on the crack tip environment and oxidation rates on a bared surface, and how they alter the values of A and n in Eqn. (1).
3.   The mechanism correctly predicts the effect on the crack propagation rate of a wide range of stressing parameters, spanning constant load, monotonically increasing load, and cyclic load for various load amplitudes, mean load and frequency conditions. These stress effects are accounted for through changes in the crack tip strain rate.
4.   The validated mechanism gives a theoretical underpinning to the empirically derived remedies for intergranular stress corrosion cracking of sensitized stainless steel in oxygenated water, and confirms that, for extended time periods, the ASME XI (1980) lifetime evaluation code for corrosion fatigue of low alloy pressure vessel steels is probably conservative.

6 ACKNOWLEDGEMENTS

Acknowledgement is extended to General Electric for permission to publish and to the Electric Power Research Institute (J.D. Gilman & R.L. Jones) who funded much of this work and who have provided advice and encouragement.

7 REFERENCES

Ford, F.P., Taylor, D.F., Andresen, P.L., & Ballinger,R.G. 1987, "Corrosion Assisted Cracking of Stainless and Low Alloy Steels in Light Water Environments", Final Report NP-5064M, EPRI Contract RP2006-6, Feb 1987.

Ford, F.P. 1982, "Mechanisms of Environmental Cracking in Systems Peculiar to the Power Generation Industry", Final Report NP-2589, EPRI RP1332-1, Sept 1982.

Andresen, P.L. & Ford, F.P. 1985. "Modeling & Life Prediction of Stress Corrosion Cracking in Sensitized Steel in High Temperature Water", Proc, ASME Symp on "Predictive Capabilities in Env Assisted Cracking", Miami, Nov 1985, pp 17-38.

Andresen, P.L., 1985. "Transition & Delay Time Behavior of High Temperature Crack Propagation Rates Resulting From Water Chemistry Changes", Proc, ANS/NACE/AIME Conf on "Environmental Degradation of Materials in Nuclear Power Systems - Water Reactors", Monterey, Sept 1985.

Ford, F.P. & Emigh, P.W. 1985. Corr Sci 2, 8/9, 673-692, 1985.

Taylor, D.F. & Caramihas, C. 1982. J Electrochem Soc 129, 2458, 1982.

Taylor, D.F. & Caramihas-Foust, C. 1985. J Electrochem Soc, 132, 1811, 1985.

Andresen, P.L. 1987, "Modeling of Water and Material Chemistry Effects On Crack Tip Chemistry and the Resulting Crack Growth Kinetics", Proc ANS/NACE/AIME 3rd Int Conf, Degradation of Matls in Nuclear Power Ind, Traverse City, Aug 1987, NACE.

Gilman, J.D. 1985. "Application of a Model for Predicting Corrosion Fatigue Crack Growth in Reactor Pressure Vessels in LWR Environments", Proc, ASME Symp on "Predictive Capabilities in Env Assisted Cracking", Miami, Nov 1985, pp 1-16.

Slow strain rate testing of RPV steels in high temperature water

J.Congleton
University of Newcastle upon Tyne, UK
T.Shoji
Tohoko University, Sendai, Japan

1 INTRODUCTION

Slow strain rate tests can be used to assess the susceptibility of materials to stress corrosion cracking. The tests are rapid and relatively inexpensive but it is difficult to meaningfully quantify the resulting crack growth rate data and relate it to probable crack growth rates for components. Additionally, it is not yet clear whether such data can be used to predict corrosion fatigue crack growth rates for low cyclic frequencies at which conventional corrosion fatigue tests would take excessively long times to perform.

Reactor pressure vessel (RPV) steels will crack in high temperature pure water if the oxygen content is high and in PWR primary side water chemistry if the potential is raised to above about -200mV s.h.e. Congleton, Shoji and Parkins (1985), Congleton and Hurst (1986), Congleton and Parkins (1987). The high resistivity of pure water makes potential control difficult in that environment but cracking can be prevented in high oxygen content pure water or initiated in low oxygen pure water by appropriate potential control. The minimum potential for cracking is affected by outcropping plate-like sulphide inclusions, Hurst et al (1985) which can reduce the necessary potential by as much as 200mV. As the open circuit potential of an RPV steel in water varies with temperature and oxygen content, Indig et al (1982), Macdonald et al (1983), there is an obvious interest in establishing the minimum potential for cracking because such information indicates desirable ranges of water chemistry to prevent crack initiation.

Slow strain rate tests are often performed at a strain rate of the order of $10^{-6}s^{-1}$ and typically last a few days. The fractured specimen is examined for stress corrosion cracks on the fracture surface and for the existence of secondary cracks in the gauge length. For screening materials and/or environments, a cracking-no cracking classification is adequate but in many cases it would be advantageous if meaningful crack growth rates were obtained. However, crack growth rates are difficult to calculate accurately for two reasons. First, there is rarely any clear indication of crack initiation on a load-time curve, so the time during which the cracks are growing is uncertain. Second, the specimens usually neck prior to failure, and the high strains and strain rates in the neck may affect the stress corrosion crack growth rate. Certainly, longer cracks usually occur on the fracture surface and in the neck than elsewhere on the specimen, Congleton and Parkins (1987) and in some cases small cracks are observed on the fracture surface but no secondary cracking occurs in the gauge length. Thus, as there are uncertainties in both the maximum crack length and time during which the crack was growing, a range of average crack growth rates can be calculated for a single specimen. The lowest values, which are clearly an underestimate, result if the maximum crack length measured on a microsection is divided by the total test time. The highest values result if the maximum crack depth measured on the fracture surface is divided by (time to fail-time to initiation). The latter is often unknown but can be approximated by (time to fail-time to maximum load) on the assumption that cracks only initiate after necking. The calculated crack velocities can vary by an order of magnitude depending upon the method used. This is inconsequential for cracking-no cracking criteria but of significant importance if the data is to be used for modelling or life assessment calculations.

2 EXPERIMENTAL

SSRT data has been obtained on A533B and A508 in oxygenated high temperature pure water and in lithiated and borated water with low and with high oxygen contents. Tests were also performed under potential control in $LiOH/H_3BO_3$ doped water. The tests were usually taken to complete failure but some tests were interrupted after various amounts of plastic strain had been imposed.

3 RESULTS

The SSRT data obtained at Newcastle is presented in Tables 1 and 2. The crack growth rates quoted in Table 1 were calculated from the maximum secondary crack depth divided by (time to fail-average time for

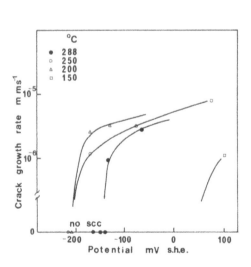 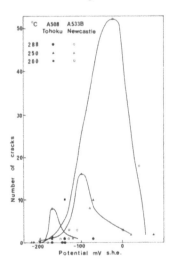

Figure 1. Average crack growth rate versus open circuit potential for slow strain rate tests on A508 in oxygenated pure water at various temperatures.

Figure 2. Number of secondary cracks in gauge length versus open circuit potential for slow strain rate tests on A508 and A533B in oxygenated pure water.

TABLE 1.   Interrupted slow strain rate test data; PWR primary
side water chemistry with various oxygen additions.

	Strain %	Range of oxygen content ppb	Potential mV she	Crack Velocity mms^{-1}
A533B	2	330–123	−280	3.6 x 10^{-6}
	4	372–266	−300	1.2 x 10^{-5}
	8	335–108	−400	3.5 x 10^{-5}
	6	223–53	−230	nil
	8	266–64	−230	7.9 x 10^{-6}
	10	181–29	−270	2.0 x 10^{-5}
	4	126–39	−240	4.2 x 10^{-6}
	6	158–46	−240	2.9 x 10^{-5}
	12	130–64	−240	2.6 x 10^{-5}
	6.5	64–15	−270	nil
	6.0	47–7	−480	1.1 x 10^{-6}
	9.5	57–11	−250	9.5 x 10^{-6}
A508	3.5	115–26	−310	nil
	7	100–55	−230	1.7 x 10^{-5}
	9	105–45	−170	2.3 x 10^{-5}
	10.6	52–16	−390	5.1 x 10^{-6}
	10.8	45–4	−420	nil

TABLE 2.   Slow strain Rate data for A533B and A508 in PWR
primary side water chemistry at 250° and 288°C for tests at
various applied potentials.

Material	Applied Potential mV (she)	Crack growth rate a_{max} on fracture surface $t_f - t_{max}$ load	Temperature °C
A533B	−65	5.2 x 10^{-5}mm/s	250
	−85	1.2 x 10^{-5}mm/s	"
	−120	7.3 x 10^{-5}mm/s	"
	−180	2.8 x 10^{-6}mm/s	"
A533B	−65	1.9 x 10^{-5}mm/s	288
	−120	nil	288
A508	−115	1.0 x 10^{-4}mm/s	288
	−170	7.2 x 10^{-5}mm/s	288
	−215	1.1 x 10^{-4}mm/s	288
	−265	nil	288

initiation) as more fully described previously (Congleton and Hurst, 1986). The data indicates a minimum oxygen content of about 50ppb to induce cracking even for strains of the order of 10% and a minimum strain to initiate cracks of about 2% even in > 200ppb oxygen content water. Data for tests in PWR primary side water chemistry at various applied potentials are presented in Table 2. The minimum potential for cracking is -215mV, but it appears from this data as well as from work on a series of iron base alloys (Congleton and Parkins 1987) and from work performed at Tohoku University that cracking susceptibility (in terms of the maximum crack growth rate and the number of cracks formed) is greater at potentials just above the minimum cracking potential than at slightly higher potentials.

Data obtained at Tohoku University for A508 in oxygenated pure water are shown in Figure 1. That data gives similar minimum cracking potentials to the Newcastle data obtained in PWR primary side water at various applied potentials; the lower crack velocities result from total test time being used to calculate the average crack velocity. Also, as shown in Figure 2, the number of cracks initiated is greatest just above the minimum cracking potential.

4 DISCUSSION

The minimum cracking potential for A533B and A508 steels is of the order of -215mV she, but varies with temperature. This is apparent in both the Tohoku University data, Figure 1 and the Newcastle University data, Table 2. Oxygen control in the low flow Newcastle rigs was difficult and there were large differences between the inlet and outlet oxygen contents. Thus, the data in Table 1 are somewhat imprecise but suggests that as well as there being a minimum strain for initiating stress corrosion cracks that increases with decreasing oxygen content of the water, there is also the possibility that cracks can initiate at lower potentials in oxygen containing PWR water than in potentiostatically controlled tests. The fastest crack growth rates were of the order of 10^{-4}mms^{-1}.

Corrosion fatigue data for RPV steels in high temperature water show plateau crack growth rates of about 3 m/cycle for 0.01Hz sinusoidal loading. This corresponds to a time based crack growth rate of 6 x 10^{-5}mms^{-1}, i.e. reasonably close to the stress corrosion crack growth rates. However, for higher frequency tests the time based plateau growth rates are somewhat greater than those that would be predicted from the stress corrosion crack growth rate. This may result from only a fraction of the crack front exhibiting EAC features, as found by Atkinson et al (1986) and by Torrenon (1986), with enhanced stress levels in the ligaments giving fast, non EAC influenced crack growth in those regions, Congleton (1986). There are clear indications that the fan shaped stress corrosion cracks on slow strain rate specimens originate at manganese sulphide inclusions. The data in Table 1 indicates a minimum plastic strain for initiating cracks of about 2% and it is possible that cracks initiate when inclusions become debonded from the matrix and dissolve in the crevice so formed, giving a local acidic environment. Clearly, with strains of 2% or more being necessary, the inclusions causing crack initiation in slow strain rate tests could be just subsurface, and there is ample fractographic evidence showing examples of this type of initiation. There is also fractographic evidence linking fan shaped EAC features formed during corrosion fatigue crack propagation with manganese sulphide inclusions so it seems possible that the EAC occurring during slow strain rate tests is phenomenologically similar to EAC that occurs during corrosion fatigue crack propagation, despite the fact that enhancement of fatigue crack growth can apparently occur in low oxygen content water when the potential of the specimen should be about −700mV s.h.e., i.e. much lower than the minimum cracking potential of about −200mV found in slow strain rate tests. However, outcropping MnS inclusions lowered the minimum cracking potential for A508 to about −400mV, Hurst et al (1985) and sulphate additions to the water induce a similar effect, Shoji (1987). Thus, the EAC may depend upon acidification in the enclave and the dissolution of sulphides, although conclusive experiments separating the possible roles of hydrogen assisted cracking and anodic dissolution during corrosion fatigue have still to be performed.

As well as there being a discrepancy between the probable potentials at which EAC has been reported for corrosion fatigue and the measured minimum potential for cracking in slow strain rate tests, there is also a difference in the strain rate at the crack tip, both in magnitude and in the fact that cyclic strains are generated during fatigue loading. It is difficult to calculate the effective crack tip strain rate for cyclic loading, but Scott's (1984) $\dot{\varepsilon}_{average}$ seems a reasonable estimate. For R = 0.7 loading, Scott's equation reduces to $\dot{\varepsilon}_{average}$ = 0.1ν, where ν is the cyclic frequency. Thus, for a loading frequency of 0.01Hz, $\dot{\varepsilon}_{average}$ = $10^{-3}s^{-1}$. Slow strain rate tests are often performed at $10^{-6}s^{-1}$ for tests that last one or two days. The specimens often show secondary cracking, so the effective crack tip strain rate depends upon the number of such cracks, Congleton et al (1985), Congleton et al (1987) according to

$$\dot{\varepsilon}_{crack\ tip} = 3d\,\dot{\varepsilon}_{applied} + \frac{\dot{a}}{5}\ln\,(33.6d)$$

where $\dot{a}$ is the stress corrosion crack growth rate in mms^{-1}, d the mean crack spacing in mm and $\dot{\varepsilon}_{applied}$ is the applied strain rate. Even if only a single crack occurs in the necked region of the gauge length, the maximum crack tip strain rate is unlikely to increase to $10^{-4}s^{-1}$ for an applied strain rate of $10^{-6}s^{-1}$, the more likely value being about $10^{-5}s^{-1}$. It is impractical to increase the applied strain rates in slow strain rate testing to any appreciable extent because the tests would then be of very short duration and EAC would be limited on that account. There is therefore an incompatibility between SSRT and corrosion fatigue tests related to matching the crack tip strain rates. It may therefore be sensible to consider using cyclic loading in conventional slow strain rate rigs to match crack tip strain rates for SSRT and corrosion fatigue. The difference in the test procedures would allow cyclic SSRT to be performed under potentiostatic control, which is technologically difficult for conventional corrosion fatigue tests on compact tension specimens.

5 CONCLUSIONS

Slow strain rate tests on RPV steels in high temperature water suggest that environment assisted cracking only occurs above a minimum potential and the minimum potential for cracking is influenced somewhat by the water temperature, the presence of outcropping sulphide inclusions and by sulphate added to the water. A minimum strain of about 2% is required for cracking, even in relatively high oxygen content water, suggesting that the initiation of stress corrosion cracks is associated with the debonding of sulphide inclusions from the matrix. The maximum crack growth rates are probably about $10^{-4}mms^{-1}$ but care must be taken in how such values are calculated from slow strain rate data if they are to be used in quantitative modelling or predictive procedures.

REFERENCES

Atkinson, J.D., Bulloch, J.H. and Forrest, J.E. 1986. Proc 2nd IAEA Specialists' Meeting on Subcritical Crack Growth. Sendai, Japan, 2: 269-290, NUREG/CP-0067, MEA-2090

Congleton, J. 1986. Proc 2nd IAEA Specialists' Meeting in Subcritical Crack Growth. Sendai, Japan, 2: 119-129, NUREG/CP-0067, MEA-2090.

Congleton, J. and Hurst, P. 1986. Proc of the Second IAEA Specialists' Meeting on Subcritical Crack Growth. Sendai, Japan 1985, 1: 439-464, NUREG/CP-0067, MEA-2090.

Congleton, J. and Parkins, R.N. 1987. Paper No 105, Corrosion 87. NACE, San Francisco, USA.

Congleton, J., Parkins, R.N. anad Hemsworth, B. Paper No 106, Corrosion 87, NACE, San Francisco, USA.

Congleton, J., Shoji, T and Parkins, R.N. 1985. Corrosion Science, 25: 633-650.

Hurst, P., Appleton, D.A., Banks, P. and Raffel, A.S. 1985. Corrosion Science 25: 651-671.

Indig, M.E., Weber, J.E. and Weinstein, D. 1982. Reviews on Coatings and Corrosion (R.N Parkins (ed)) 5: 173. Freund Publ House Ltd, Israel.

Macdonald, D.D., Smialowska, S. and Pednekar, P. 1983. Final Report, EPRI-NP-2853.

Torrenen, K., Kemppainen, M. and Hanninen, H. 1984. EPRI, Report No 3483.

Scott, P.M., Truswell, A.E and Druce, S.G. 1984. Corrosion 40: 350-357.

Shoji, T. 1987. To be presented at 9th SMIRT Conference, Lausanne.

ACKNOWLEDGEMENTS

The work was jointly supported by the UKAEA and the Procurement Executive, Ministry of Defence, U.K.

Environmentally assisted cracking of Inconel X750

P.Skeldon & P.Hurst

UKAEA, Risley Nuclear Power Development Laboratories, Warrington, UK

ABSTRACT

The resistance of different heat treatments of Inconel X750 to
environmentally assisted cracking in simulated PWR primary water at
340°C has been assessed by slow strain rate, U-bend and bent beam
tests. At the corrosion potential (ca -670mV (Ag/AgCl)), in low
oxygen conditions ($\leq$2ppb), a single-stage ageing (704°C/20h) gives
much improved resistance compared with two-stage ageing (885°C/24h +
704°C/20h). However, material given the former ageing treatment can
be susceptible to cracking at highly anodic potentials
(>-200mV(Ag/AgCl)) if the alloy is significantly sensitized.

This work has been wholly funded by the CEGB under the Thermal
Reactor Agreement.

1   INTRODUCTION

Inconel X750 is a precipitation-hardened, nickel-based alloy used
within the primary circuits of Pressurised Water Reactors (PWRs) and
Boiling Water Reactors (BWRs) for components requiring high strength
and relaxation resistance. Operating experience, however, has shown
the alloy to be susceptible to intergranular environmentally assisted
cracking (EAC) in both reactor systems (McIlree 1983). As a result
laboratory programmes have been set up to investigate the cause of
failures and seek palliatives. An important outcome has been the
development of a high temperature (ca 1100°C) solution anneal and
single stage ageing (704°C/20h) heat treatment which confers greater
resistance to EAC than the lower temperature solution anneal/two
stage ageing treatment used previously (Yonezawa 1983).
   In the UK three modifications likely to be applied to Sizewell 'B'
Inconel X750 guide tube support pins are a high temperature (1093°C)
solution anneal prior to a single-stage (704°C) ageing treatment;
lowering of the torque level to reduce maximum shank stress; redesign
to reduce the stress concentration factor. The effectiveness of the
modified heat treatment and, to a lesser extent, the influence of
stress level are being assessed in U-bend, bent beam and dynamic
strain (monotonic and cyclic) tests. This paper describes initial
results from U-bend, bent beam and slow strain rate (SSR) tests on
two heats of Inconel X750, one manufactured to the preferred heat

treatment, and the other laboratory heat-treated at low (980°C) and high (1093°C) solution annealing temperatures, followed by either one-stage (704°C/20h) or two-stage (885°C/24h + 704°C/20h) ageing treatments.

2 EXPERIMENTAL

2.1 Materials and Heat Treatment

Two commercial heats, A and B, of Inconel X750, (compositions in Table 1), are being evaluated. Heat A was supplied as 28.6 mm dia. bar in the preferred heat treated condition (solution annealed 1107°C/1.3h and aged at 704°C/21h) and tested as-received. Heat B, obtained as 28.6mm dia. bar, was cold-rolled or swaged, using inter-stage annealing at 1050°C, to produce 2mm thick strip or 8mm dia. bar, then solution annealed at either 980°C or 1093°C; specimens were machined then aged either in one (704°C/20h) or two (885°C/24h + 704°C/20h) stages. Materials from both heats were gas-quenched following solution annealing, and air-cooled after ageing.

2.2 EAC Tests

SSR tests were carried out on specimens with a 15mm gauge length of 3mm diameter using a 100kN Schenck servo-electric machine and a recirculating loop facility (Connolly 1984). Specimens passed through a small volume stainless steel test section. Tests were conducted at nominal strain rates of $1.5 \times 10^{-6} s^{-1}$ and $2.0 \times 10^{-7} s^{-1}$ in simulated PWR primary water (<2ppb O_2) at 340°C. An external Ag/AgCl reference electrode (0.1M KCl) enabled measurement and control of potential; iR compensation was applied in potentiostatically controlled tests. The flow rate was 14 1/h corresponding to a linear velocity in the test section of 0.12 m s^{-1} and a Reynolds number of ca 5000. Table 2 lists typical environmental conditions. Failed specimens were examined by scanning electron microscopy

U-bends (formed to an outer diameter of 12.5mm from 37.6 x 10.0 x 1.2mm strip, with the arms held parallel by an X750 nut and bolt) and bent beams (prepared from 67.0 x 10.0 x 0.8mm strip using four-point bending, the separation of inner and outer pairs of loading points being 27 and 54mm, respectively) were exposed to simulated PWR primary water at 340°C in replenished stainless steel autoclaves (~1 l capacity) at a flow rate of 2 1/h. Specimens were isolated from the autoclave walls. Conditions were similar to those of Table 2 except for the pressure being ca 16.9 MPa. Additionally, some specimens were exposed to oxygenated water (150-200 ppb O_2, 300μS m^{-1}) at the same temperature and pressure.

3 RESULTS

3.1 Slow Strain Rate

__Strain Rate $1.5 \times 10^{-6} s^{-1}$__: The free corrosion potential in 340°C PWR primary water containing <2ppb O_2 was ca. -670mV after 1h exposure (potentials are quoted with reference to the external Ag/AgCl electrode). Experiments were performed at this potential and at more anodic potentials up to 0mV. Specimens were also tested in either N_2 or He, at 340°C and atmospheric pressure, and susceptibility to EAC determined by comparison of failure times in the water and inert environments, and from fractographic evidence. The relative failure times are shown graphically in Figure 1.

Heat B, solution annealed at 980°C and single-stage aged, gave failure times similar to those for the control specimen at all potentials (fig 1c), the fracture surface being of the dimpled, ductile type. When a 1093°C solution anneal was combined with the single-stage ageing treatment the failure times were slightly lower than that of the control (fig 1e), but the fracture surfaces were again dimpled. Observations of the gauge lengths revealed heavy distortion and a profusion of slip steps causing difficulty in identifying small cracks due to EAC for either solution anneal temperature. In contrast the heat A alloy, solution annealed at 1107°C and single-stage aged, clearly showed susceptibility at potentials above -200mV (fig 1a), with intergranular EAC on the fracture surface and secondary cracking on the gauge length. At more cathodic potentials failure times were generally slightly lower in PWR water than in the inert environment but EAC was not observed on the fracture surface. However, more detailed inspection for possible shallow secondary cracking on the gauge length is required. A significant difference between heat A and all heat treatments of heat B was the high proportion of intergranular failure on fracture surfaces of the former alloy tested in an inert environment; in contrast heat B exhibited a mostly dimpled surface.

Heat B with the 980°C solution anneal and two-stage ageing was susceptible to EAC at potentials in the region of -500mV (fig 1b). At this potential intergranular EAC was found on the fracture surface with secondary cracking on the gauge length. At -400 and -600mV secondary cracking was also observed with small regions of intergranular failure on the fracture surface. There was no clear evidence of EAC at the free corrosion potential or at potentials >-300mV. Material of the same heat, solution annealed at 1093°C prior to the two-stage ageing treatment, did not reveal the maximum in susceptibility at -500mV (fig 1d). Failure times were generally slightly lower than the control's, the reductions being greater towards the free corrosion potential. Only the specimens tested at this potential and at -600mV clearly showed a small region of intergranular failure, possibly due to EAC, on the fracture surface. Shallow, secondary cracking may be present on some samples but requires further study.

Strain Rate 2.0 x 10^{-1} s^{-1}: Both heat A and heat B, were tested at the free corrosion potential; the results are given in Table 3. The failure time of heat A alloy in PWR water was slightly lower than the control's. However, no EAC was observed on the fracture surface although there appeared to be shallow cracking on the gauge length. Heat B, similarly heat treated, had the longest failure time in the PWR environment. The same alloy with a 980°C solution anneal and single-stage ageing showed no EAC on the fracture surface although again some shallow cracking may be present on the gauge length. However, both the 980°C and 1093°C solution annealed heat B alloy with two-stage ageing revealed intergranular EAC on the fracture surface and secondary cracking on the gauge length. The extent of cracking was greater on the high temperature solution annealed material, which had the shortest time to failure.

3.2 U-Bends and Bent Beams

Two autoclaves contained specimens exposed to simulated PWR primary water at the free corrosion potential: eight U-bends of heat A in the first, and duplicated U-bends and bent beam specimens of both heat A and heat B (in the four heat treatments) in the second. Bent beams were loaded to 0.6, 0.8 and 1.0 times the room temperature 0.2% proof stress (PS). No failures have occurred for heat A U-bends after 3888h exposure in the first autoclave. Inspection of specimens in the second autoclave after 360h exposure revealed that duplicated U-bends of both the 980°C and 1093°C solution annealed heat B with the two-stage ageing had failed. After 1392h the duplicated bent beam specimens of the same heat loaded to the 0.2%PS with the 980°C solution anneal and two-stage ageing had also failed, together with a single specimen with similar ageing but solution annealed at 1093°C.
   A single autoclave was operated with oxygenated water and contained duplicated U-bends and bent beam specimens (loaded to the room temperature 0.2% PS) of both heats, including the four heat treatments of heat B. Initially problems were encountered in controlling the oxygen level and during the first weeks the oxygen concentration was low ($\leq$30ppb). After 696h the rig stabilized to operation at 150-200 ppb oxygen and exposure continued to 1032h. In between these exposure times the duplicated heat B U-bends, annealed at 980°C and 1093°C, given two-stage ageing had failed.

4   DISCUSSION

In general the results show the importance of heat treatment on the EAC behaviour of Inconel X750. For low oxygen PWR primary water the single rather than two-stage ageing treatment increases the resistance to stress corrosion cracking; material receiving the latter treatment cracked at, or near, the corrosion potential. In SSR tests at $1.5 \times 10^{-6} s^{-1}$ two-stage aged material with the low temperature solution anneal was most susceptible at ca -500mV; however, at $2.0 \times 10^{-7} s^{-1}$ cracking occurred for both low and high temperature solution anneals at the corrosion potential. Similarly in the U-bend and bent beam (0.2% PS) tests two-stage aged material, with both solution anneals, failed after relatively short exposures.
   In contrast there was no clear evidence of EAC in low oxygen conditions, close to or at corrosion potential, for X750 given the single-stage ageing treatment. In some SSR tests slight reductions (~7%) in time to failure were observed in the water environment, compared with control tests; no evidence of EAC was found on the fracture surfaces, but detailed examination of the gauge length has still to be carried out. In addition no single aged specimens have failed in U-bend and bent beam tests.
   At anodic potentials, heat A was readily cracked; this susceptibility correlates with significant sensitization of this heat (as indicated by modified Huey testing), for which there appears to

be a material variability probably related to processing history. Heat A has not shown susceptibility to EAC in oxygenated water tests with constant strain specimens but a calibration of oxygen content versus potential is required to compare results of SSR and constant strain tests.

The effect of solution annealing temperature was unclear. At the free corrosion potential and a strain rate of $2.0 \times 10^{-7} s^{-1}$, the two stage aged alloy was more susceptible when annealed at 1093°C; in contrast, at $1.5 \times 10^{-6} s^{-1}$ and -500mV greater susceptibility occurred for annealing at 980°C. In single-stage aged heat B X750 the failure time for 1093°C annealing was some 7% greater than for 980°C annealing but control data are lacking to substantiate its significance. Although no effect of annealing temperature has yet been revealed in the present work, benefits from a high temperature solution anneal are anticipated if formation of a semi-continuous distribution of carbides at alloy grain boundaries is key to the production of a resistant microstructure; more carbon will be taken into solution for subsequent precipitation at grain boundaries during ageing (Kekkonen 1983). This will be studied in future work.

5   CONCLUSIONS

1.   In simulated PWR primary water Inconel X750 subjected to a single-stage ageing treatment (704°C/20h) has greater resistance to EAC than material given a two-stage ageing treatment (885°C/24h + 704°C/20h).

2. EAC has been observed for single-stage aged material under more oxidising conditions ($\geq$-200mV Ag/AgCl). This correlates with sensitisation which appears to be dependent on the cast of material and/or prior fabrication history.

3. Little evidence has been obtained so far on the effects of solution annealing temperature.

REFERENCES

McIlree, A.R. 1983. International Symposium on Environmental Degradation of Materials in Nuclear Power Systems - Water Reactors. South Carolina, USA.
Yonezawa, T., Onimura, K., Sakamoto, N., Sasaguri, N., Nakata, H., and Susukida, H. 1983. International Symposium on Environmental Degradation of Materials in Nuclear Power Systems - Water Reactors. South Carolina, USA.
Connolly, H., Hurst, P., Knowles, P.J.F. and Appleton, D.A. 1984. UKAEA Report ND-R-884(R).
Kekkonen, T. and Hanninen, H. 1983. "Corrosion and Stress Corrosion of Steel Pressure Boundary Components and Steam Turbines" (ed J Forsten), IAEA, Finland.

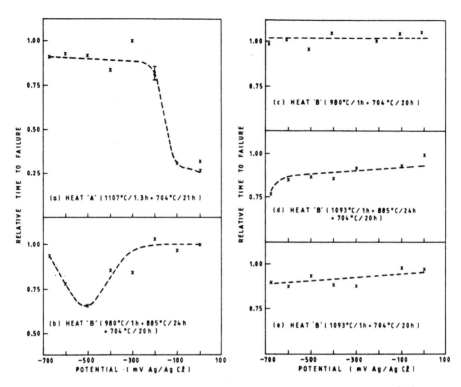

Figure 1. SSR tests on Inconel X750 in 340°C PWR primary water ($1.5 \times 10^6 s^{-1}$)

TABLE 1 - Chemical Composition of Inconel X750 Alloys (w/o)

	V	C	Mn	Si	P	S	Cr	Ni	Cu	Co
Heat A	.02	.04	.03	.01	.007	.001	15.43	71.79	.02	.03
Heat B	-	.036	.25	.25	-	.003	15.07	71.37	.33	.49

	Al	Ti	Nb+Ta	Zr	Fe	Sn	As	Bi	Sb
Heat A	.71	2.50	.92	.04	8.21		<10ppm		10ppm
Heat B	.76	2.54	.93	-	7.10				

TABLE 2 - Test Conditions

Pressure MPa	18.7
Temperature (°C)	340
Conductivity ($\mu S\ m^{-1}$)	2000
Oxygen (ppb)	<2
H_3BO_3 (ppm)	6000
LiOH (ppm)	2.0
Cl^- (ppb)	<50
F (ppb)	<50
SO_4^{2-} (ppb)	<30

TABLE 3 - Results of SSR Tests at $2.0 \times 10^{-7} s^{-1}$

Heat and Condition	Environment	Failure Time
A (1107°C/1.3h + 704°C/21h)	He	489.3(h)
"	PWR	436.5
B (1093°C/1h + 704°C/20h)	PWR	487.4
B (1093°C/1h + 885°C/24h + 704°C/20h)	PWR	180.8
B (980°C/1h + 704°C/20h)	PWR	449.0
B (980°C/1h + 885°C/24h + 704°C/20h)	N_2	457.5
"	PWR	333.4

Components I: Stress analysis

Piping design and analysis: Comparison between the Belgian applications of French and American rules

P.H.Daoust, L.H.Geraets & J.P.Lafaille

Department of Mechanical Engineering, TRACTEBEL, Brussels, Belgium

1 SUMMARY

In the process of a feasibility study of a new nuclear power plant in Belgium, the French and American rules for piping design have been compared.

The Belgian method rests on the American nuclear set of rules and uses the ASME code. French rules were initially based on the American rules (1978). Subsequent individual development led to a differentiation of the rules. Presently the mechanical part of the French rules is mainly contained in the RCC-P ("Règles de Conception et de Construction relatives aux Procédés") and the RCC-M ("Règles de Conception et de Construction des matériels Mécaniques").

This paper compares the piping design rules from a general point of view; examples of applications allow to identify benefits or drawbacks of the use of ASME or RCCM codes.

2 INTRODUCTION

Belgium is involved in the peaceful use of nuclear energy for a very long time.

Different nuclear power plants of the PWR type have been constructed since 1960. The BR3 at Mol has acted as a pilot plant for the Belgian nuclear power industry. After participation in design for a 300 MWe PWR power plant (Chooz A - 1967), three other units were commissioned in 1975 (Doel 1, 2 and Tihange 1).

For the latest nuclear units of the PWR-900 type (NSSS - Westinghouse or Framatome), full power was reached in 1983 for Doel 3 and Tihange 2 or in 1985 for Doel 4 et Tihange 3. For all those plants, the applicable rules were strongly based on the US rules.

France has an important experience in the design and construction of PWR power plants (56 units constructed or in construction). French rules have been derived from American ones since 1978.

For the next nuclear power plant in Belgium (N8 - PWR - 1400 MWe), a comparison is made between the Belgian applications of French and American rules for the piping design. The first point of this paper considers the general rules that have been used in Belgium or in France for the latest power plants. The Belgian rules are mainly based on the ASME code while French engineers have their own safety rules

(RCCP see ref. 3) and calculation code (RCCM see ref. 2). A theoriti-
cal comparison gives the main differences between the ASME or RCCM
codes 1983 edition. Different examples show the impact of the use of
each code. The latest developments in the U.S. have led to the edition
of new seismic rules.(codes cases N411 and N397). Gains obtained on
different piping systems by the use of those new rules are given
hereafter. Modifications that could appear in a near future for
American codes and rules are also evaluated by different examples.

3 DEFINITION

The h coefficient is defined as "a design margin" obtained on the different equations of the codes : η = (1 - stress ratio) wherein the
stress ratio is the calculated stress divided by the allowable stress
(see fig.1).

4 COMPARISON OF GENERAL RULES

The table 1 compares the main American-Belgian rules and the corresponding French rules applicable to piping analysis. In the past, those
general rules have led to design specifications and stress reports
with many similarities.

However, the evolution of codes and guides in the U.S. (see ref. 4
to 10) will generate some differences in the piping design. As emphasized in the following paragraphs, those differences appear mainly in
the consideration of the seismic effects.

5 ASME 83 VERSUS RCCM 83

For piping design, volumes B, C, D - Tome 1 from RCCM are very similar
to the subsections NB, NC, ND of the Asme code Ed. 77. The subsequent
evolution of the ASME code has not appeared in the RCCM code. The main
differences or evolutions in the ASME code are :
1) the suppression of the $|\Delta T1|$ contribution in the equation 10 of
   Class 1 (Summer 79)
2) the introduction of B1, B2 coefficients in Cl 2/3 (Ed. 81)
3) the introduction of higher Sh allowable stresses (Ed. 81)
4) the introduction of allowables based on Sy (Ed. 81)

Modifications 1 and 3 yield a decreased conservatism in the calculations. Modifications 2 and 4 yield calculations closer to the reality
with either increased or decreased conservatism. Different systems
were chosen and calculated with both codes. Table 3 summarises some
results obtained in the examples.

In Class 2/3, the design margins (η) obtained on the verification of
primary stresses are generally higher in the ASME code than in the
RCCM. Higher allowables (+ 50 %) and introduction of B1 coefficient
(0 < B1 < 0,5) yield calculations closer to the reality. In some cases, the introduction of coefficient B2 (B2 > 0,75 i) and the lower
allowables (stainless steel at high temperature) give lower design margins in equations 9 of ASME 83. This last remark is mainly valid for
thin elbows and tees : it indicates that for those cases RCCM 83 but
also ASME 80 might be less conservative than ASME 83.

In class 1, results differ in the fatigue analysis. The presence of the term $E\alpha|\Delta Tl|/1-\nu$ in Eq 10 of RCCM penalizes heavily the usage factors. RCCM 83 is more conservative than ASME 83.

6 IMPACT OF NEW CODE CASES

Table 2 compares the damping values for seismic design in both sets of rules. New code cases refer to the use of the CC N411 (variable damping - see ref. 5) and CC N397 (Peak Shifting - see ref. 6). In the examples, the benefit obtained with the use of CC N397 is marginal : a maximum value of 5 % approximately.

The application of code case N411 is extremely beneficial. For acceleration spectra taken from both sites (Tihange and Doel) peaks are reduced by a factor of 33 %. As a result, seismic stresses are also reduced by similar amounts depending upon the flexibility of the system (eigen frequencies and mode shapes), the type of seismic event (OBE or SSE) and the piping diameter.

Table 4 gives comparisons of primary stresses (Asme 83, Eq. 9). The gain obtained depends on the relative importance of the seismic event with respect to pressure, weight and other mechanical loads. The first system (CPE) is a small diameter (2") connected to the primary loop with 20 frequencies below 20 Hz. Seismic stresses equivalent to the sum of pressure and weight stresses are reduced by 45 %. Three of the four snubbers could be removed from that line. The second system (CIS) is a large diameter outside containment with 22 modes below 20 Hz. Seismic stresses represent one -fourth of the total primary stresses. The use of CC N411 allows a reduction of 25 % on the seismic stresses, which lowers in turn the primary stresses by 10 %.

7 COMPARISON BETWEEN ASME 86+ AND ASME 83

ASME 86+ refers to the modifications that could appear in the near future in the addenda of the ASME code Ed. 86. Observations made after seismic events have shown that the risk of the seismic inertia effects has been overestimated. Tests with seismic inputs have shown that the failure mode was fatigue ratchet with leakage instead of collapse, and that the stress behaviour is far closer to secondary than primary (see ref. 7).

As a result, ASME 86+ proposes to treat the differential anchor displacements as the most dangerous seismic effect. This could be achieved by the following measures : in Class 1 (see ref. 9), the withdrawal of the seismic inertia effect from Eq. 9 OBE and the consideration of the differential seismic anchor displacements for SSE in a new equation :

$$C2 \frac{Mam}{z} \leq 3 \ Sm$$

in Class 2/3 (see ref. 10) the withdrawal of the seismic inertia effect from Eq. 9B, the withdrawal of the seismic anchor displacements OBE from Eq 10 or 11 and the consideration of the seismic effects (inertia and displacements OBE and SSE) in a new equation :

$$i \ \frac{Md}{z} \leq \frac{1688}{(N/U)^{0,2}}$$

For the seismic event, the number of cycles considered in Belgium is
50 for OBE and 10 for SSE. Table 5 gives a selection of results obtained with two examples. The design margin (η) of an equation of ASME 83
must be compared with the design margin of the same equation in ASME
86+ but also with the value of h in the new fatigue equation. In the
examples, the design margin increases with the use of ASME 86+. The
system is a main steam line from the steam generator to the containment penetration. As seen in table 5, the new fatigue equation is not
dimensioning.

Another modification that could appear in ASME 86+ is the following : the allowable stress limits on code eq. 9 would no longer be
applicable provided that Su/Sy is at least 1.5 (Sy = specified minimum
yield strength and Su = specified minimum ultimate tensile strength).
The gain could be important for ferrictic steels (SA106 Gr B) used in
Cl 2/3.

In conclusion, the use of ASME 86+ could significantly reduce the
importance and the number of seismic restraints required to meet primary stress equations.

8 CONCLUSIONS

The comparison shows that the traditional Belgian set of rules (as
defined in table 1) is essentially similar to the French set of rules.
For practical applications however the differences could lead to substantially more penalizing designs using the French rules rather than
the Belgian rules. This arises from :
- the conservatism of the 83 edition of the RCCM close to the 77
  edition of the ASME.
- the current evolution of the ASME code in order to lower
  conservatism
- the trends of American rules in order to reduce the impact of the
  seismic inertia effects.
As a consequence, it has been decided to keep using the US based
rules adapted to the Belgian environment for the design of the
piping of the prospective new nuclear plant. The reason for this
has been that too much conservatism in the design rules leads to
unnecessarily binding the piping system, which induces stresses.
Consequently, overconservatism in design would reduces the global
safety of the system.

9 ACKNOWLEDGMENT

The authors wish to thank all the members of the piping analysis
branch of TRACTEBEL involved in this comparison. The contribution
of MM. Bastiaanse and Claeys is peculiarly acknowledged.

10 REFERENCES

1. ASME BOILER AND PRESSURE VESSEL CODE - Section III - Division 1
   Editions 77, 80, 83, 86
2. Règles de conception et de construction des matériels mécaniques des îlots nucléaires PWR (RCC-M) (Ed 83 + Modif. janvier
   1985) éditées par l'Association française pour les règles de

conception et de construction des matériels des chaudières
électronucléaires (AFCEN)

3. Règles de conception et de construction des centrales nucléaires REP. RECUEIL DES REGLES RELATIVES AUX PROCEDES (RCCP - rév.
   2 - Sept. 85)

4. Report of the U.S. Nuclear Regulatory Commission Piping Review
   Committee NUREG 1061 - Volume 2 (April 1985) Evaluation of
   Seismic Designs - A review of Seismic Design - Requirements for
   Nuclear Power Plant Piping

5. Code case N411 - Alternative damping values for seismic
   analysis of Class 1, 2 and 3   Piping Sections Section III,
   Division 1   (Sept. 84 - ASME BPVC)

6. Code case N397 - Alternative rules to the spectral broadening
   Procedures of N-1226.3 for Classes 1, 2 and 3 Piping, Section
   III, Division 1 (Febr. 84 - ASME BPVC)

7. Minutes of Committee Meeting ASME - WG on Piping Design and
   Dynamic Analysis (Feb. 86, May 86, Sept. 86)

8. S.E. Moore and E.C. Rodabaugh : Background for changes in the
   1981 Edition of the ASME Code for Controlling Primary Loads in
   Piping Systems (Journal of Pressure Vessel Technology - Nov.
   1982, Vol 104, p 351 to 361)

9. Proposed Code case NXXX - Alternative Rules for the analysis of
   piping products under seismic loading, Section III, Division 1,
   Class 1 (ASME - BPVC).

10. Proposed Code Case NXXX - Alternative Rules for the analysis of
    piping products under seismic loading, Section III, Division 1,
    Class 2 and 3 (ASME - BPVC).

FIG. 1 <u>Definition of</u>

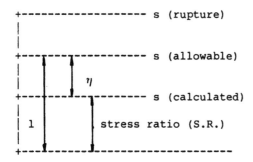

283

Table 1. Comparison of general rules

US/Belgian rules	French rules
10 CFR	Arrêtes ministeriels (A.R.) et règles fondamentales de surete (RFS)
ANSI standards R.G. S.R.P. NUREG Reports	RCCP
Transposition sur le plan belge de l'ASME Section III Subsections NB, NC, ND Appendix I Appendix N Code Cases CCA (Conditions Complémentaires d'application de l'ASME)	RCCM Tome 1 Volumes B, C, D Annexe ZI
ASTM standards	Normes AFNOR
ANSI, MSS-SP standards	ANSI, MSS-SP standards

Table 2. Damping for seismic design.

Type of piping	Type of seismic event	France (RCCP or USAGE)	USA (Belgium) (RG1.61 or USAGE)	USA CC N411
Primary loop	OBE <--> SNA SSE <--> SMS	FRA 4 4	W 2 4	VARIABLE 5 to 2 5 to 2
Ø > 12"	OBE SSE	2 3	2 3	5 to 2 5 to 2
Ø ≤ 12"	OBE SSE	2 2	1 2	5 to 2 5 to 2

Table 3. Comparison ASME 83 - RCCM 83 - CL 2/3
   Design margin (η) on equations 8,9 (ASME) or 6,10 (RCCM)

System	diameter (inches)	Design pres. (bars)	Design temp (°C)	Points	ASME 83 Eq. n°	η (ASME 83)	η (RCCM 83)	RCCM 83 Eq. n°
CPE	2" sch 40S	173	343	Ten most stressed points	8	0.403	0.265	6
					9B	0.363	0.254	10B
					9D	0.387	0.490	10D
CIS	10, 12, 14" sch 40 S	11	150	Ten most stressed points	8	0.838	0.835	6
					9B	0.823	0.838	10B
					9D	0.867	0.902	10D

Table 4. Impact of CCN411
   Design margin (η) on equations 9

System	Diameter	Design Pres. (bars)	Design temp (°C)	Points	Levels	η ASME 83 + CCN411	η ASME 83 + RG1.61
CPE	2" sch 40S	173	343	Ten most stressed points	Eq 9B	0.432	0.363
					Eq 9D	0.522	0.387
CIS	10,12,14" sch 40S	11	150	Ten most stressed points	Eq 9B	0.835	0.823
					Eq 9D	0.879	0.867

Table 5. Comparison ASME 86+ - ASME 83 - CL 2/3
Design margin (η) on equations NC/ND 3600

Equations	Rigid model : N8 MS1		More flexible model : N8 MS1 BIS	
	η ASME 83	η ASME 86+	η ASME 83	η ASME 86+
Level 0 : Eq. 8 : Pressure + Weight + Other Mechanical Loads $\leq$ 1.5 Sh	0.636	0.636	0.636	0.636
Level B : Alt. equation : OBE inertia and displacements $\leq$ 487 MPa	N.A.(*)	0.945	N.A.	0.910
Eq. 9 : Pressure + Weight + Other Mechanical Loads $\leq$ 1.8 Sh (+ OBE inertia)	0.464	min (0.555; 0.945)	0.4	min (0.484; 0.910)
Eq. 10 : Thermal Expansion + Anchor displacements $\leq$ SA	0.034	min (0.166; 0.945)	0.061	min (0.159; 0.910)
Eq. 11 : Pressure + Weight + Mech. Loads + Th. expansion + Anchor displacements $\leq$ SA + Sh	0.198	min (0.285; 0.945)	0.217	min (0.278; 0.910)
Level D : Alt. équation : SSE inertia and displacements $\leq$ 585 MPa	N.A.	0.915	N.A.	0.826
Eq. 9 : Pressure + Weight + Mech. Loads + SSE inertia	0.767	min (0.767; 0.915)	0.618	min (0.826; 0.618)

(*) N.A. : NOT APPLICABLE

Simplified structural design of piping systems

K.Moczall & M.Labes
Kraftwerk Union AG, Erlangen, Offenbach, FR Germany

1 INTRODUCTION

The stability, integrity, and operability of piping systems must be ensured for operational loading conditions and safety-related postulated loading conditions. This may be achieved with the help of design-by-analysis methods. However, such an approach is impractical for most piping, such as small-bore piping of up to DN 50 and cold systems (see Fig. 1). Accordingly, KWU developed a simplified and reliable method for designing such piping systems which basically avoids the use of computer analyses (see Fig.2) while making sure that the span lengths are short enough in order to resist seismic loads and flexible enough for operational loads.

2 APPROACH

This approach is incorporated in a Pipe Routing Guideline (PRG) the main aspects of which are

- a reliable, balanced approach used in designing and dimensioning piping and support components, based essentially on conventional technology which avoids complex parts such as snubbers,
- a systematic, detailed planning on the basis of a piping outline or scale model, including the generation of isometric drawings: the planning follows guidelines concerning piping layout and support configuration, including the specification of loads on supports, nozzles, etc. and simple rules for achieving sufficient structural flexibility,
- an in-plant quality inspection of piping systems after installation and before pressure testing with the purpose of checking whether the PRG is satisfied and if the proper degrees-of-freedom have been restrained at the supports; during the commissioning phase piping systems are inspected with respect to movements and vibration phenomena, the latter being evaluated by simple means.

A summary of the procedure followed in applying this simplified design approach is shown in Fig. 3.

3 VERIFICATION

The verification of the simplified approach to piping design has been based on a number of independent methods. Supporting analytical work included computer calculations of several new systems designed according to such rules and the statistical evaluation of numerical data from earlier power plants.

Considering inherent safety and design margins, the results confirmed that the simplified approach leads to support loads and stresses that are acceptable.

Supporting experimental work by KWU included a number of small- and large-bore piping subjected to intense dynamic movements exceeding predicted design loads by large factors. Some of these piping systems were supported over larger span widths than required by the simplified approach. Testing of a system supported according to the PRG leads to rather low stresses in the piping wall and to support loads comparable to PRG-loads.

As another kind of experimental evidence, inspection of power plant piping following actual strong earthquakes has revealed that no damage occurred in well-designed systems. Calculations of some systems which had survived a strong earthquake revealed that the stresses predicted had been exceeded considerably the allowable ones according to rules, thus showing the conservatism of current methods of dynamic piping analysis.

A detailed description of the full verification scheme concerning the simplified design of piping systems is given in Fig. 4.

4 CONCLUSION

In conclusion, the safety level of piping systems falling in the range of application of and designed according to the routing guideline for operational and postulated loading conditions is comparable to that of systems designed by analysis. The routing guideline has been approved by the authorized expert, tested extensively in actual practice, and found to allow a relatively straightforward piping design, including the influence of extreme loading conditions such as earthquakes.

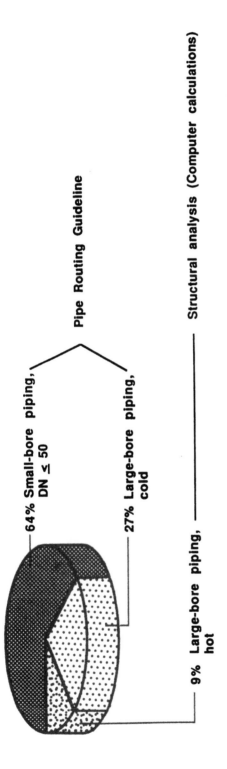

64% Small-bore piping,
DN ≤ 50

27% Large-bore piping,
cold

9% Large-bore piping,
hot

Pipe Routing Guideline

Structural analysis (Computer calculations)

Fig. 1   Scope of piping designed according to the simplified method and with the help of computer calculations relative to the total length of piping in a nuclear power plant

289

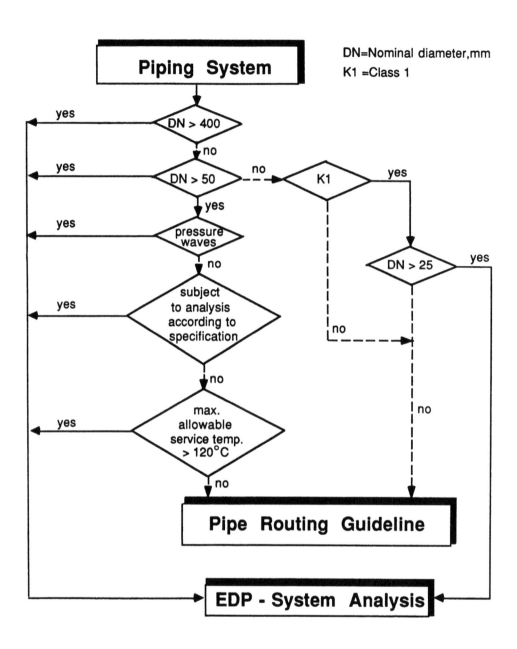

Figure 2: Decision Criteria

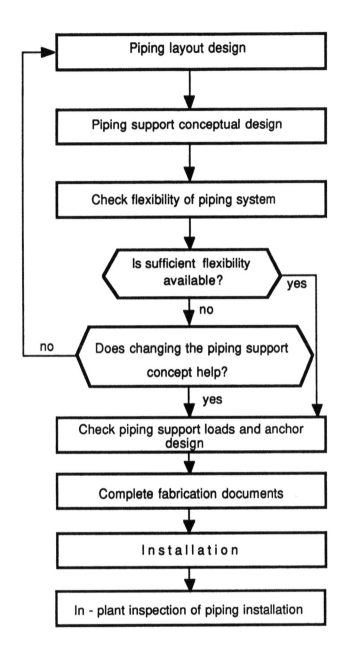

Fig. 3 Summary of procedure followed during simplified design

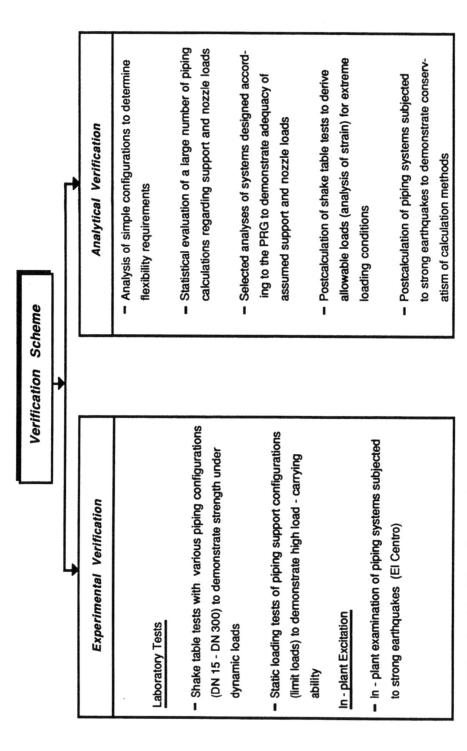

Verification Scheme

Experimental Verification

Laboratory Tests

- Shake table tests with various piping configurations (DN 15 - DN 300) to demonstrate strength under dynamic loads

- Static loading tests of piping support configurations (limit loads) to demonstrate high load - carrying ability

In - plant Excitation

- In - plant examination of piping systems subjected to strong earthquakes (El Centro)

Analytical Verification

- Analysis of simple configurations to determine flexibility requirements

- Statistical evaluation of a large number of piping calculations regarding support and nozzle loads

- Selected analyses of systems designed according to the PRG to demonstrate adequacy of assumed support and nozzle loads

- Postcalculation of shake table tests to derive allowable loads (analysis of strain) for extreme loading conditions

- Postcalculation of piping systems subjected to strong earthquakes to demonstrate conservatism of calculation methods

Fig. 4   Verification scheme for the simplified design of piping systems

Classification of stresses in locally thinned tubes

W.W.Teper

Ontario Hydro, Toronto, Canada

ABSTRACT

Classification criteria for stresses in locally thinned regions of tubes are examined with respect to the intent of Section III of the ASME Code for Class 1 components[1].
Depending on the size of the thinned region, the membrane stresses caused by pressure loading may be classified as general primary, primary local or secondary[2].
The criteria for classification of membrane stresses were investigated for some simple geometry and load cases, where plastic instability is the governing mode of failure due to primary loads.

1 INTRODUCTION

The local reduction of the wall thickness of piping components to a size below the minimum required by the basic formula for general primary stresses (NB-3641.1 equation 1) may result from either fabrication processes (e.g., excessive grinding) or from operating conditions (e.g., fretting). The definition of local primary membrane stresses is rather arbitrary. According to Article NB3213.10, the membrane stresses caused by pressure load may be considered local if the distance over which they exceed $1.1\ S_m$ does not extend more than $1 \times \sqrt{Rt}$ in the meridional direction. The allowable limit for this stress intensity is $1.5\ S_m$. No limitation is given for the extent of the region over which the membrane stresses may exceed $1.0\ S_m$. The question arises as to what size of thinned area may be considered local and its stresses considered as category P_L, when the requirement for $P_m < 1.1\ S_m$ within $\sqrt{Rt}$ is satisfied, but $P_m > 1.0\ S_m$ over a large distance.

Local membrane stresses occur where the structure is required to bear locally a greater share of the external mechanical loads. This is most often due to the removal of some material from the pressure boundary, and occurs in such locations as manways and nozzles.

One of the questions that this paper attempts to answer is how to classify stresses in regions with locally thinned tubes (Figure 1) subjected to internal pressure and axial tension. For the purpose of this study the yield strength of the tubes' material was $S_y = 36500$ psi with low true strain hardening $H = 146000$ psi. The

tubes' nominal dimensions were: mean radius R_m = 2.145", wall
thickness t_n = 0.150".

The stress classification must depend on the length of the thinner
section L. For large values of L, the stresses must be considered to
be general primary; for small values of L, the membrane stress may be
considered primary local. The critical length L was determined on
the premise that the plastic collapse load of the locally thinned
tubes was equal to the plastic collapse load of a uniform tube with
minimum required thickness.

2 ANALYSIS PROCEDURE

All work was carried out using the nonlinear finite element code
ABAQUS, with the axisymmetric shell element SAX1 and the axisymmetric
solid element CAX4.

2.1 Verification Tests

As a first step, the following three tests were carried out to verify
that the axisymmetric element SAX1 is suitable for the intended
analyses:
   1. A single SAX1 element in tension was tested. An axisymmetric
element with the cross-sectional area equal to 1 in^2 (R = 1",
t = .15915), was axially loaded. The plastic collapse load was
67666 lb. The corresponding theoretical collapse load is 68945 lb
(see Appendix B).
   2. Two models of a cylinder with wall thickness t = 0.130" and
mean radius R_m = 2.145" were set up, one using the axisymmetric
element SAX1 and the other using the solid elements CAX4. Both
models were subjected to simultaneous axial and pressure loading, so
that initially $\sigma_\theta = \sigma_\ell$. Constant ratio of pressure to the
axial load was maintained until collapse occurred at p = 3365 psi for
the solid model and at p = 3351 psi for the shell model.
   3. An axisymmetric cylinder model of R_m = 2" and t = 0.150" was
subjected to internal pressure only. The computed collapse pressure
was 3883 psi versus the 3905 psi theoretical pressure (see
Appendix A).
   On the basis of these tests, it was concluded that the
axisymmetric shell element SAX1 gave satisfactory results and it was
used in the analysis that followed.

2.2 Analysis of Locally Thinned Tubes

The design pressure of the tubes in question was 1475 psi, mean
radius R_m = 2.145", wall thickness t_n = 0.150". The collapse
pressures of locally thinned tubes were computed for various lengths
of the thinned regions. Thus, through successive iterations, the
critical lengths of thinned segments were found such that the
collapse loads of the locally thinned tubes were equal to the
collapse loads of smooth tubes, whose thickness is equal to the
minimum required, i.e.,

$$t_r = pR_o/(S_m + 0.5p) = \frac{pR_m}{S_m} = \frac{1475 \times 2.145}{24333} = 0.130".$$

294

where R_o = outside radius of the tube.

The loads were applied so that constant ratios of pressure to axial load were maintained through each loading sequence. This resulted in the reduction of $\sigma_\ell/\sigma_\theta$ as the loading progressed and the tube radii grew. Since the axial tensile stress reduces the radial growth with increasing pressure, the reduced ratios of $\sigma_\ell/\sigma_\theta$ resulted in underestimated values of critical lengths of thinned segments (L_{crit}) for cases where $\sigma_\ell/\sigma_\theta$ tends to be constant, such as for tubes subjected to internal pressure alone.

3   RESULTS AND CONCLUSIONS

For two tubes, one with t_t = 0.100" and another with t_t = 0.120", the critical lengths were found to be equal to 2" and 5.8" respectively for the case with initial $\sigma_\ell = \sigma_\theta$. The critical lengths of the thinned regions corresponding to other ratios of $\sigma_\ell/\sigma_\theta$ were also determined for the first tube as follows:

	initial	1.	0.5	0.207	0.1035	0
$\sigma_\ell/\sigma_\theta$						
	final	0.624	0.3	0.111	0.05	0
P_{crit} (psi)		3365.	3219.	3139.	3127.	3115.
L_{crit} (in)		2.	2.	1.07	0.6	0.22

where P_{crit} corresponds to the collapse loads of smooth, 0.130" thick tubes.

The stresses corresponding to the critical length of the thinned region, computed on an elastic basis for the design pressure p = 1475 psi are shown in Figures 2 (a) and 2 (b).

We may expect the plastic collapse to occur when the elastic stresses in the thinned segment reach the value of pR_t/t_t since at this stress level the thinned portion may carry the full pressure load. Figure 2 (a) shows that collapse occurred when the elastic hoop stresses reached or even exceeded the theoretical value of pR_t/t_t over a considerable distance along the tubes' axes. On the basis of Figure 2 (b), it is concluded that for ductile materials and for the geometry under consideration, the stresses in the thinned region may often be considered local even when the stress intensity does not fall below the value of 1.1 Sm beyond the distance of $\sqrt{Rt}$ as required by the Code.

REFERENCES

ASME BPV-III-1-NB.
F. Ferry, P. Bance, P.H. Toupin, "Analysis of Primary Membrane Stress in a Pipe Affected by Local Under-Thickness Discontinuity". F18/1, 8th SMIRT Conference, 1985.
W. Johnson, P.B. Mellor "Engineering Plasticity", Van Nostrand, Section 10.6.

Plastic Collapse of Open Ended Cylinder   Under Internal Pressure

Notations:

R = mean radius
t = wall thickness
σ' = deviatoric stress components
$d\varepsilon_\theta$, $d\varepsilon_t$, $d\varepsilon_\ell$ = increments of plastic strain

From Levy - Mises equation:

$$\frac{d\varepsilon_\theta}{\sigma'_\theta} = \frac{d\varepsilon_t}{\sigma'_t} = \frac{d\varepsilon_\ell}{\sigma'_\ell}$$

where $\sigma'_\theta = \dfrac{2pR}{3\ t}$, $\sigma'_\ell = \sigma'_t = \dfrac{-1}{3}\dfrac{pR}{t}$

Therefore,

$$\frac{d\varepsilon_\theta}{2} = \frac{-d\varepsilon_t}{1} = \frac{-d\varepsilon_\ell}{1} \qquad (1)$$

The effective strain: $\varepsilon_f^2 = \dfrac{2}{9}\left[(\varepsilon_\theta - \varepsilon_t)^2 + (\varepsilon_\theta - \varepsilon_\ell)^2 + (\varepsilon_\ell - \varepsilon_t)^2\right]$

thus $\varepsilon_f = \varepsilon_\theta$

As the thickness and radius change with the increasing pressure, the following holds:

$$P = \frac{\sigma_\theta t}{R}$$

At instability: $dP = \dfrac{d\sigma_\theta t}{R} + \dfrac{\sigma_\theta dt}{R} - \dfrac{\sigma_\theta t dR}{R^2} = 0$

Thus, $\dfrac{d\sigma_\theta}{\sigma_\theta} + \dfrac{dt}{t} - \dfrac{dR}{R} = 0$ or $d\varepsilon_\theta - d\varepsilon_t = \dfrac{d\sigma_\theta}{\sigma_\theta}$ $\qquad (2)$

by combining (1) and (2) we obtain:

$$d\varepsilon_\theta = \frac{2}{3}\frac{d\sigma_\theta}{\sigma_\theta} \quad \text{or} \quad \frac{d\sigma_\theta}{d\varepsilon_\theta} = \frac{3}{2}\sigma_\theta \qquad (3)$$

For rigid, linear work hardening material:

$$\sigma = Y + H\epsilon; \quad \epsilon = (\sigma-Y)/H; \quad \frac{d\sigma}{d\epsilon} = H$$

where Y = yield stress, H = strain hardening factor

Thus from equation (3), $H = \frac{3}{2} (Y + H\epsilon_c); \quad \epsilon_c = (\frac{2}{3} H - Y)/H$

where ϵ_c = critical strain at instability

For Y = 36500 psi and H = 146000 psi $\quad \epsilon_c = .417, \quad \sigma_c = 97333$psi

From equation 1: $\frac{dR}{2R} = \frac{-dt}{t}$, upon integration we obtain

$$\epsilon_c = \ln [\frac{R_c}{R_o}] = 2 \ln [\frac{t_o}{t_c}] , \quad \text{or} \quad R_c t_c^2 = R_o t_o^2$$

where: R_o = initial mean radius = 2.0"; t_o = initial thickness = .150"
$\quad\quad R_c$ = mean radius at collapse; t_c = thickness at collapse
Therefore $R_c = R_o \exp (\epsilon_c) = 2 \exp (.417) = 3.0348$ in

$$t_c = t_o \sqrt{\frac{R_o}{R_c}} = .150 \sqrt{\frac{2}{3.0348}} = .1218"$$

The collapse pressure is $P_c = \frac{\sigma_c t_c}{R_c} = 3905$ psi

APPENDIX B

Plastic Collapse of Axially Loaded Cylinder

$$P = 2\pi \ tR \ \sigma_\ell$$

at collapse $dP = 2\pi (t\sigma_\ell dR + R\sigma_\ell dt + Rtd\sigma_\ell) = 0$,

hence $\frac{dR}{R} + \frac{dt}{t} + \frac{d\sigma_\ell}{\sigma_\ell} = 0$ or $d\epsilon_\theta + d\epsilon_t = \frac{-d\sigma_\ell}{\sigma_\ell}$ \hfill (1)

Levy - Mises equation:

$$\frac{d\epsilon_\theta}{\sigma_r'} = \frac{d\epsilon_t}{\sigma_t'} = \frac{d\epsilon_\ell}{\sigma_\ell'}, \quad \text{here} \quad \sigma_\ell' = \frac{2}{3} \frac{P}{2\pi Rt} ; \quad \sigma_t' = \sigma_\theta' = \frac{-1}{3} \frac{P}{2\pi Rt}$$

Thus, $\frac{1}{2} d\epsilon_\ell = -d\epsilon_\theta = -d\epsilon_t$ \hfill (2)

From (1) and (2) $d\varepsilon_\ell = \dfrac{d\sigma_\ell}{\sigma_\ell}$

For $\sigma = 36500 \, (1 + 4\varepsilon_\ell)$; $\varepsilon_{\ell c} = .75$; $\varepsilon_\theta = \varepsilon_t = -.375$

The radius and thickness at collapse are:

$R_C = 1. \exp(-.375) = .687"$; $t_C = .15915 \exp(-.375) = .1094"$

$P_C = 2\pi \, R_C \, t_C \, \sigma_C = 2\pi \times .687 \times .1094 \times 36500 \, (1 + 4 \times .75)$

$\quad = 68945. \, lb$

(a) Locally Thinned Tube

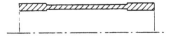

(b) Analysis Geometry

(c) Shell Analysis Model

FIGURE 1

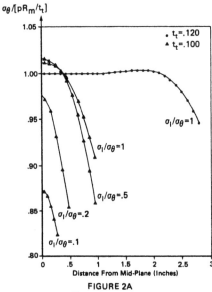

FIGURE 2A
Normalized Elastic Stresses Corresponding to
Critical Length of Thinned Region

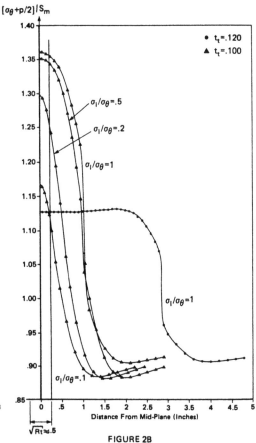

FIGURE 2B
Elastic Stresses for Design Pressure p=1475psi

298

Analysis of a pipe section due to static loading – A coupled numerical/experimental method

H.Diem & K.Kerkhof
Staatliche Materialprüfungsanstalt (MPA), University of Stuttgart, FR Germany
M.Shafy
Atomic Energy Establishment, Cairo, Egypt

1 INTRODUCTION

Stress analyses of safety relevant piping systems are generally performed with member programs based on the elementary theory of bending and providing mostly internal forces and displacements, resp. rotations. The results of such kind of stress analyses directly depend on the accuracy of the different boundary conditions

- stiffness of anchor points
- hangers
- snubbers
- gaps, friction effects
- geometrical differences, etc.

That means that the results of calculations are valid only as long as the boundary conditions will not be changed. In determining the stresses of highly loaded components, e.g. pipe bends, are evaluated by stress intensity factors /1/ which in some cases do not always yield sufficient results.

In this contribution a possibility is shown how to get a more detailed analysis of the deformation and stress behaviour of pipe-bends mounted in piping systems without any influence of unknown boundary conditions on the local analysis.

Partly with respect to this at the feedwater piping system of the decommissioned Heißdampfreaktor (HDR) near Frankfurt two static loading tests have been carried out. These tests were performed within the scope of Phase II of the HDR-Safety Program /2/ which is sponsored by the Ministry of Research and Technology of the Federal Republic of Germany (FRG). The global behaviour of the piping system could be determined by measured data of displacements and strains at several cross-sections. Additionally a detailed stress resp. strain evaluation could be performed by evaluating 76 strain gauges attached in one bend area. Parallel to the tests the piping behaviour was analyzed by finite element calculations.

2 TEST FACILITY

The static loading tests were performed at the feedwater
line of the HDR-Plant. The basic design concept of this
piping was similar to that of a typical feedwater line in a
boiling water reactor in the FRG having two fix points and
no other supports (Fig. 1). The piping system had a
length of about 23 m including four pipe bends and was
mainly manufactured by the low alloyed ferritic material
15 NiCuMoNb 5. The tests were performed at room
temperature without internal pressure. The piping was
loaded at two positions in horizontal (x) and vertical (z)
direction (Fig. 1) by a single force of about 70 kN.

3 TEST RESULTS AND ANALYTICAL DESCRIPTIONS

In a first step a reasonable member program (ASKA) /3/ for
both load cases was used in the theoretical process to
determine the global load-displacement resp. load-strain
behaviour of the whole piping system by a simple beam type
model using flexibility factors for the bends as defined in
the ASME-code /1/.
   In the second step the piping reactions were determined
with the advanced piping calculation code ABAQUS /4/.
There are essential differences between the usual member
programs and ABAQUS in the possible combined use of beam
and so-called pipe, resp. elbow elements for bent piping
sections. In case of piping geometry data for which the
validity of the shell theory is guaranteed
(thin-walled-bends) local stresses, resp. strains in the
elbow region can be calculated. Using this type of
modelling no flexibility factors are necessary to consider
the effect of ovalization.
   From the measured strain data in 4 cross-sections the
axial force, the two bending moments and the torsional
moment could be determined (Fig. 2). These data served
for comparison with theoretical results.
   The global calculations show a good agreement with the
measured data (Fig. 3 and 4). The global model in the
first member analysis proved - in a limited amount - to be
stiffer than the real structure, against which the
ABAQUS-results were in a good agreement with the measured
data.
   To perform a detailed analysis of a highly stressed
component in the third step elbow No.1 of the feedwater
piping system was regarded as a separate subsystem. This
subsystem had been modelled by a 3D finite element model
using thick shell elements from the ASKA program system /3/
(Fig. 5). The loading of this subsystem was first taken
from the results of the above mentioned beam type model for
the test snap x (ASKA-calculation 1).
   In the fourth step the loading of the considered subsystem
was taken from the measurement data of two cross-sections
close to the bend but in a distance out of the ovalization
zone /5/. Bending and torsional moments as well as axial
forces could be determined directly from the measuring

values of strain gauges. The missing shear forces are determined from the conditions of equilibrium (Fig. 6 and 7). This approach is here called "coupled experimental/numerical method" (ASKA-calculation 2).

In comparing measured data to the numerical results (Fig. 8) it is evident that the 3D-finite element model using loading data from a simple member program analysis provided a good demonstration of the local stresses in the elbow region both on the elbow outer as well as on the elbow inner surface. The calculation has especially proved that the maximum stressed area lies on the elbow inner wall at the bend flank.

The combined method of measurement and calculation also provided a good correlation between locally determined measurement values in the elbow region and the numeric results. However, it also showed the problems of this analysis method. Especially the absolute size of the measured strain values used for the calculation of the internal forces as well as the accuracy of the measured data are of great importance. These particular parameters will be discussed in detail and clearly elaborated in a final test report, demonstrated by example of test snap z.

As mentioned above using ABAQUS-program with special elbow elements local strains in the elbow region can be calculated. The comparison of these results with measured data is shown in Fig. 9. There is an excellent agreement between experimental and analytical results.

At last a detailed analysis of bend deformation behaviour at the middle cross-section was performed using the equations of the ASME-code subsection NB 3685 1.2 /1/. The input data for this analysis were taken from the ASKA member program calculation. As shown in Figure 10 the ASME-analysis provides a conservative description of pipe bend stressing in this case, but is not useful to consider local effects.

4 CONCLUSIONS

The presented "coupled approach" has the reasonable advantage of performing a detailed analysis of highly stressed components using measured strain data (e.g. during start up of a plant in operation). For analysing the local stress/strain behaviour of a bend in an extented piping system this procedure - in case of unknown or changed boundary conditions - can provide more accurate data than it is possible by the use of common member programs.

REFERENCES

1   ASME Boiler and Pressure Vessel Code, Section III, Nuclear Power Plant Components, Subsection NB 3685, Class 1 Components

2   HDR-Safety Program Phase II
    PHDR-Report No. 05.19/84,
    Kernforschungszentrum Karlsruhe, January 1984

3   ASKA-Part I: Linear Static Analysis,
    ASKA User Manual 202,
    Institut für Statik und Dynamik,
    Universität Stuttgart, 1979

4   ABAQUS: Theory and User Manual,
    Hibbit, Karlsson & Sorensen Inc., 1982

5   Diem, H., Müller, K.U.: Determination of Cross-sectional
    Loading of Piping Systems Using Strain Gages –
    Possibilities and Limitaions,
    9. GESA-Symposium, Berlin, May 1985, VDI-Report No. 522,
    pp. 139-152

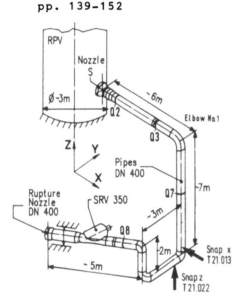

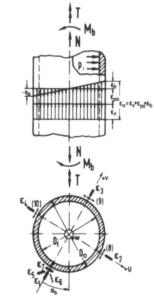

Fig. 1. Isometry of the investigated
       feedwater line

Fig. 2. Arragement of
       strain gages

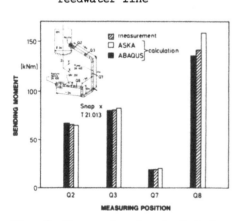

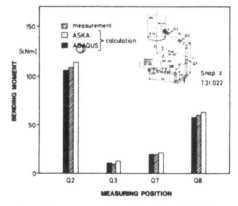

Fig. 3. Measured and calculated
       bending moments (T21.013)

Fig. 4. Measured and calculated
       bending moments (T21.022)

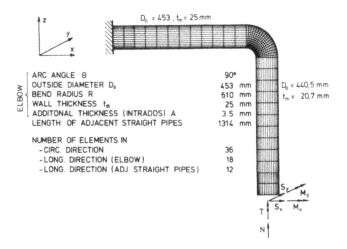

Fig. 5. Finite element modell using ASKA
shell elements

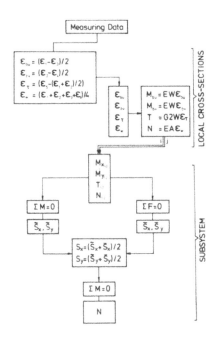

Fig. 6. Flow chart for the
evaluation of member
forces by means of
measured strains

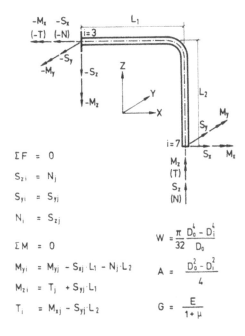

Fig. 7. Equilibrium conditions
of the considered
subsystem

303

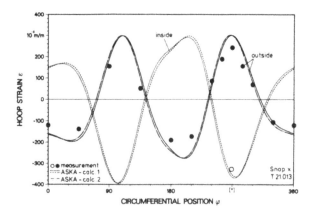

Fig. 8. Comparison of measured and calculated
hoop strains in elbow midsection (ASKA)

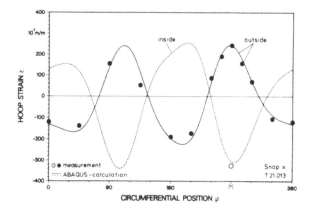

Fig. 9. Comparison of measured and calculated
hoop strains in elbow midsection (ABAQUS)

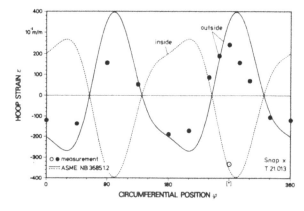

Fig. 10. Comparison of measured and calculated
hoop strains in elbow midsection (ASME)

Design criteria for piping components against plastic collapse
Application to pipe bend experiments

F. Touboul, M. Ben Djedidia & D. Acker
DEMT, CEA-CEN Saclay, Gif-sur-Yvette, France

INTRODUCTION

Recent years have witnessed developments in piping design criteria, initiated by the work of Rodabaugh and Moore (1), which led to significant differences between the criteria proposed by the latest design codes (2, 3, 4, 5). The requirements aimed to limit primary bending stresses in piping are expressed in the form :

$$(1) \qquad B_1 \frac{PD}{2e} + B_2 \frac{M}{Z} < S_A$$

The codes differ in the formulation of the indices B_1 and B_2 and in the value of S_A. For the ASME, indices B_1 and B_2 are by definition $B_1 = \sigma . \frac{2e}{PD}$, where σ is the elastic stress due to the pressure in the elbow, and $B_2 = \sigma Z/M$, where σ is the corresponding stress at the limit load of the elbow in bending. For applications, we use the values required by (2,3,4) and given tables 1, 2 and 3. Furthermore, the design of new reactors in which the piping components exhibit geometric characteristics quite different from those of pressurised water reactors, requires confirmation of the validity of the rules for these types of component. With these objectives in mind, we analyzed the experimental results of tests conducted at the CEA/DEMT in the past twelve years, in the sense of the codes and standards.

This paper presents an appraisement of level D design rules for elbows under in plane bending primary moments.

NOTATIONS

P = Internal pressure
D = Pipe Diameter
Z = Inertia modulus
Sy = Conventional yield strength
Sa = allowable stress
Su = conventional ultime strength
    at temperature
Rm = conventional ultime strength
    at room temperature

M = Bending moment
e = pipe wall thickness
α = Bend angle

305

EXPERIMENTS

The tests whose interpretation is presented here were performed between
1974 and 1986 at the CEA/DEMT in eight campaigns employing different
experimental rigs. For each of these campaigns, we have the geometric
measurements taken on the test components and the characteristics of
the material obtained on an identical component. This enabled us to
determine the characteristic parameter $\lambda = eR/r_m^2$ based on the
actual geometry of the components, as well as the allowable stress Sm,
from the properties measured on the material, thus avoiding an ultra
conservatism due to the tabulated properties. Combined these tests
enabled us to determine :
. The influence of the material (26 elbows of austenitic steel and
  8 elbows of ferritic steel).
. The influence of the characteristic geometric parameter λ. For our
  different series of tests, this parameter ranged between 0.076 and
  0.68, values that can be compared to those characterising nuclear
  reactor : λ = 0.16 for fast reactors and λ = 0.3 for pressurised
  water reactors.
. The influence of the bend angle between π and $\pi/6$ radians.
. The influence of pressure, characterised by the nominal stress
  σ_p = PD/2e varying between 0 and 0.72 Sy.
. Using equation (1), we accordingly calculated the critical moment
  allowable by the codes :

$$M_{cr} = Z/B_2 \ (\alpha \ Sm - B_1 \ PD/2e)$$

For each test campaign, we also have all the recordings made as well
as a precise description of the experimental rigs. For each test, this
enabled us to determine the limit moment M_L corresponding to a total
deflection of the elbow equal to twice the elastic deflection, and
the collapse moment M_R corresponding to the maximum moment on the
moment-deflection curve (figure 2).
    The safety margin offered by the rule is thus defined by the ratio
m = M_R/M_{cr}. If m is greater than 1, the collapse moment predicted
by the rule is lower than the actual moment, and the rule is safe. If,
by contrast, m is less than 1, the rule is unsafe.

RESULTS

RIGHT-ANGLE BENDS WITHOUT PRESSURE

Figure 1 shows the margin offered by the rules of the ASME section III
of the RCC-M and of RCC-MR, as a function of the characteristic geometric parameter of the material. It may be observed that, while the three
codes are prudent for ferritic steels, none of them is safe for austenitic steels. For the ASME, the average margin ranges between 0.71 and
0.88 for austenitic steels, and between 1.03 and 1.49 for ferritic steels.
This margin appears to be independent of parameter λ , which therefore
seems to be taken into account correctly in the expression of B_2.

ANGLE EFFECT

Figure 4 shows the variation in the margin as a function of bend angle.
This margin can be seen to decrease as the angle increases. While for
ferritic steel elbows it is still in the neighbourhood of 1 for a 180°
angle, for austenitic steel elbows it reaches this value for an angle
of about 60° and 0.68 for a 180° angle. This effect is not taken
into account by the present rules.

PRESSURE EFFECT

Figure 3 shows the influence of pressure, characterised by the parameter $\sigma_p/Sy = PD/2eSy$ on the margin offered by ASME section III.
It may be observed that the influence of this parameter increases
as the elbow becomes thinner.

The present formulation of index B_1 in NB 3680 (2) is hence quite
inadequate to take account of the pressure effect.

DISCUSSION

ANGLE EFFECT

In our tests, to obtain a constant margin as a function of the angle,
it suffices to multiply index B_2 by a coefficient of $(\pi/\alpha)^{0.4}$.
For right-angle bends, this amounts to assuming that $B_2 = 1.21\lambda^{-2/3}$,
by applying this correction to the index determined from the limit
analysis of Spence and Finley (6), which was carried out without
taking account of the straight parts, and is valid for an "infinite"
elbow. It is admit here that this index $B_2 = 1.6\lambda^{-0.6}$, is available
for 180° bends.

PRESSURE EFFECT

To describe the pressure effect, we assume that B_2 (P) can be
expressed in the form chosen by Dodge and Moore (7) for the flexibility indices :

$$B_2 \ (P) = B_2 \ (0)/(1+f(\lambda) \ g(P))$$

By also setting $B_1 = 0.5$, it is easy to infer from our tests that

$$g \ (P) = \sigma_p/Sy \quad \text{and} \quad F(\lambda) = 0.7\lambda^{-1}$$

ALLOWABLE LIMITS

The foregoing sections lead us to assume

$$(2) \quad B_2 = 1,6 \ \lambda^{-2/3} \ (\frac{\alpha}{\pi})^{0,4} \ (1 + 0,7 \sigma_p/\lambda Sy))^{-1}$$

This formulation does not improve the margins determined from equation
(1), and therefore raises the problem of allowable limits. By examining
the criteria proposed by the ASME (8) for structures other than piping
subjected to level D loadings, it can be pointed out that the maximum
allowable loading is 0.9 times the ultimate loading determined by a
limit analysis (such as the one carried out by Spence and Finley) for
which the flow stress adopted is Min(2.3 Sm, 0.7 Rm). For piping components, this rule leads to assuming that S_A = Min (2.07 Sm ; 0.63 Rm).

To avoid a distortion of the allowable limits at elevated temperature
in relation to the actual properties of the material, it is preferable
to write this equation in the form :

$$(3) \quad S_A = Min \ (1,4 \ Sy \ ; \ 0,63 \ Su)$$

If we decide to make an experimental analysis, the authorised
limit is M_L and the margin in relation to collapse, for the tests
without pressure presented here, has a mean value of 1.1 (figure 5).
We can compare this value to the one obtained on the overall tests

by applying equation (1) with $B_1 = 0.5$, with B_2 defined by equation (2) and the limits given by equation (3) (figure 6). The mean value of the margin obtained is thus 1.19, with a minimum value of 1.03 and a maximum value of 1.30 for austenitic steels, and 1.8 with a minimum value of 1.53 and a maximum value of 2.05 for ferritic steels.

This means that the margins obtained by these two methods are consistent. Our proposal is more accurate for pressurised elbows but more conservative for ferritic steel unpressurized elbows.

CONCLUSIONS

By interpreting the tests conducted on elbows at the DEMT in the past 12 years, we found that the design rules for level D loadings offered by the codes were not prudent and did not allow consideration of the pressure and bend angle effect. Based on the limit analysis carried out by Spence and Finley and on our experimental results we propose expression for a more suitable B_2 index and the use of ASME limits applicable to all components.

REFERENCES

1 Rodabaugh E.C., Moore S.E., 1968. Evaluation of the plastic characteristics of piping products in relation to ASME code criteria, NUREG report CR 0261.
2 ASME boiler and pressure vessel code, 1983, div.1, section III, sub-section NB article NB 3600.
3 RCCM, Règles de conception et de construction des matériels mécaniques des îlots nucléaires PWR, 1985, part 1, vol.B, chap.B 3600, pub. by AFCEN.
4 RCC-MR, Règles de conception et de construction des matériels mécaniques des îlots nucléaires RNR, 1985, part 1, vol.C, chap.C 3600, pub. by AFCEN.
5 ASME boiler and pressure vessel code, 1983, code cases nuclear components, case N 319, pub. by ASME.
6 Spence J.S., Findlay G.F., 1973, Limit loads for pipe bends under in plane bending, Proc.of the 2nd Int.Conf.of Pressure Vessel Technology, paper 1-28, pub. by ASME.
7 Dodge W.G., Moore S.E., 1972, Stress indices and flexibility factor for moment loadings on elbows and curved pipe, W.R.C.bulletin 79, pub. by Welding Research Council.
8 ASME boiler and pressure vessel code, 1983, div.1 section III, appendix F, pub. by ASME.

Table 1

Allowable Stress S_A

Code	RCCM	RCC-MR	ASME
Level D Limits	$S_A = 3\,Sm$	$S_A = min$ ($3\,Sm, 0.9\,Rm$)	$S_A = min$ ($3\,Sm, 2\,Sy$)

Table 2
Definition of Sm

Code	RCCM-ASME	RCC-MR
Aciers ferritiques	$Sm = min\ (\frac{2}{3}\ Re,\ \frac{1}{3}\ Rm,$ $\frac{2}{3}\ Sy,\ \frac{1}{3}\ Su)$	$Sm = min\ (\frac{2}{3}\ Re,\ \frac{1}{3}\ Rm$ $\frac{2}{3}\ Sy,\ \frac{1}{3}\ Su)$
Aciers austénitiques	$Sm = mini\ (\frac{2}{3}\ Re,\ \frac{1}{3}\ Rm,$ $0,9\ Sy,\ \frac{1}{3}\ Su)$	$Sm = mini\ (\frac{2}{3}\ Re,\ \frac{1}{3}\ Rm,$ $0,9\ Sy,\frac{1}{2.7}\ Su)$

Table 3
Definition of indices B

Code	RCCM	ASME-RCC-MR
B_1	0.5	$-\ 0.1 + 0.4\ \lambda\ 0.5$
B_2	$1.3\ /\ \lambda^{2/3}$	$1.3\ /\ \lambda^{2/3}$

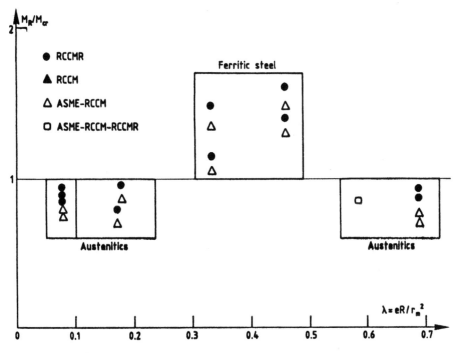

Figure 1 - λ effect on the Level D safety margins for 90° elbows
without pressure.

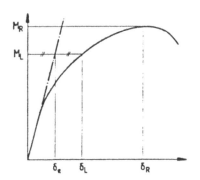

Figure 2. Limit and collapse moments definitions.

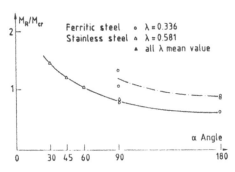

Figure 4. Effect of angle on ASME Level D Safety margins.

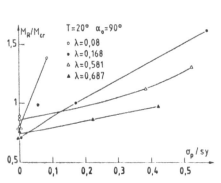

Figure 3. Effect of Pressure on ASME Level D Safety margins.

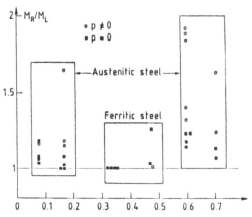

Figure 5. ASME Level D Safety margins by experimental analysis.

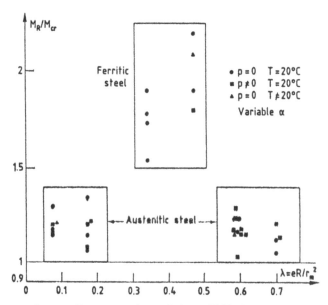

Figure 6. Level D Safety margins with modified B_2 and Limits for in-plane bending elbowsof angles between 30° and 180°.

310

The influence of geometric mismatch on steam generator U-bend stresses

A.P.Villasor, Jr.

The VILLA-4 Group, Pittsburgh, Pa., USA

1 INTRODUCTION

The realities of material properties, manufacturing processes and design practices cause geometric mismatches to occur between the steam
generator U-tube legs and their holes in the support plates. These off-
sets of centerlines give rise to additional stresses in the U-tube due
to the pinching or spreading of its legs. This paper presents the computational effort using a finite element program done to characterize
the stress changes attributable to the net geometric mismatch in a full
length row 1 U-tube 22.22 mm [0.875 in] O.D. and 1.27 mm [0.050 in]
uniform wall thickness under full operating temperature and pressure.
A previous paper by the author [1] covered the classical approach to
the U-bend analysis.

2 TUBE GEOMETRY AND MODEL

The row 1 U-tube height is 9117 mm [358.9375 in] from the top of the
tubesheet [TS] to the apex of the U-bend. The mean bend radius [R] is
55.56 mm [2.1875 in] and the tangent length above the mid-plane of the
top tube support plate [TSP] is 92.08 mm [3.625 in]. Seven TSP's at
approximately 1273 mm [50.125 in] center-to-center provide lateral support to the tube. Based on this design geometry, the U-tube model was
constructed as shown in figure 1. The generation of the model was in
accordance with the WECAN general purpose finite element program [2].
  The tube was generally modeled using STIF7 (3-D straight pipe) and
STIF29 (3-D curved pipe) elements. Because of its remoteness to the U-
bend region, the TS was simply modeled using two STIF8 (3-D spar) elements positioned at its top level. The spars were fixed to the tube and
were only allowed to expand longitudinally in the plane of the U-tube
under temperature loading. Their lengths were set to the mean bend radius of the U-bend; their joint at the centerline of the model was constrained to ground.
  At each TSP elevation, two STIF8 elements were used to simulate the
TSP and its thermal expansion in the horizontal direction. Again, each
spar length was made equal to the mean bend radius of the U-bend. The
TSP spars and the tube legs were connected by four STIF40 (1-D gap)
elements: two gap elements on the hot leg [HL] and two gap elements on
the cold leg [CL]. For each leg, one gap element represented the intrados TSP hole clearance while the other represented the extrados TSP

hole clearance. All gap elements were one-dimensional in the global X-direction. Each gap size was kept to the design clearance of 0.20 mm [8 mils].

With the coordinate origin located at the top of the TS midway from the leg centerlines, the 2-dimensional [X-Y] behavior of the U-tube was made to coincide with the plane of the U-tube for the 100% load condition and initial mismatch cases. To assure this situation in the analysis, the UZ, RX and RY degrees of freedom of the pipe nodes were set to zero. Moreover, since the spar displacements were limited to the global X, no deflections were allowed in the Y and Z directions. In the model, there were 176 pipe, 16 spar and 28 gap elements with 199 nodes.

3 LOADS ON THE MODEL

The 100% temperature and pressure load condition consisted of the temperature varying along the tube length from the HL side of the TS at 322°C [611°F] to the CL side of the TS at 284°C [543°F] and the pressure differential across the tube wall. The temperature profile along the tube was taken as the average between the outside diameter temperature variation and that on the inside diameter. This profile was applied with the assumption that the temperature distribution was linear through the tube wall. Both the TS and TSP spars were set to the steam temperature of 270°C [518°F]. The reference stress-free temperature was 21°C [70°F]. The steam generator tube is actually subject to an external (steam) pressure of 6.26 MPa [908 psia] and an internal (primary) pressure of 15.51 MPa [2250 psia]. In order to simplify the analysis, a pressure differential of 9.25 MPa [1342 psia] was imposed as an internal pressure on the tube elements in the model.

In addition to the temperature and pressure loads, the structural behavior of the tube is affected by the initial centerline mismatch between the tube and the TSP holes. This mismatch changes with temperature and results in a net mismatch which is the algebraic sum of the initial mismatch at the reference temperature, the thermal expansion of the TSP and TS, the design tube hole clearance and the global X-direction thermal expansion of the U-tube based on the average temperature change in the entire tube. The net mismatch produces either a pinching or spreading of the U-tube and therefore influences the type and magnitude variation of the stress in the U-bend region. A pinching of the U-tube due to an inward mismatch imposes tensile bending stresses on the extrados. On the other hand, a spreading of the U-tube due to an outward mismatch results in compressive bending stresses on the extrados.

Several cases of initial radial centerline mismatch of the top TSP at the reference temperature were considered with the 100% load condition so that the moment distribution along the U-bend could be determined. The shift in the location of the maximum moment was also found. All load cases included the same 100% load condition; however, the initial radial mismatch ranged in values from zero to 6.62 mm [300 mils] inward to 3.56 mm [140 mils] outward.

4 RESULTS

The moment distributions along the U-bend are plotted in figure 2 for the various initial radial mismatches of the centerlines of the top TSP holes. The sign convention used for the moment was based on the nodal

MZ moment for the ith node of the element and the generation direction of the pipe elements. When the bending stress was tensile on the extrados of the pipe, the moment was taken as positive.

For the zero mismatch case, the moment distribution is found to be anti-symmetrical about the apex. The maximum absolute values are equal and occur at the tangent points of both legs. This is due to the fact that the design TSP hole clearance gap is almost equal to the thermal expansion of half the U-bend and neither pinching nor spreading is initiated. The positive moments are toward the HL while the negative moments are toward the CL.

As the mismatch is increased inwardly (pinch), all moment values are increasing in the positive direction with the peak values shifting from the HL transition toward the apex. For the outwardly-increasing values of mismatch, all moments decrease with the peak negative moment drifting in location from the CL toward the apex. It is observed that for all cases of mismatch, the moment curves pass through the tangent points of the U-bend. The shifting of the maximum moment location in the U-bend is shown in figure 3.

The bending stress for each load case was calculated from the MZ moments by using the curved beam theory. The standard straight beam formula was adjusted by stress factors [3] in order to take into account the nonlinear bending stress distribution across the cross-section of the curved tube in the U-bend. Because of the curvature and the increasing flexibility in the U-bend, the bending stress increases in magnitude towards the intrados and decreases towards the extrados. Also, the neutral axis shifts toward the intrados. In the calculations, the stress factors used were: 1.127 for the intrados and 0.900 for the extrados. It is seen in table I that the initial mismatch dramatically affects the value of the maximum stress intensity and the von Mises stress in the U-bend when the mismatch is greater than 1.02 mm [40 mils]. It is also noted the intrados stresses are higher than those at the extrados. but this observation is expected to switch when actual local thickness due to tube bending are considered for correcting the stress values. In any case, the initial mismatch can be a strong contributing factor to tube cracking in steam generators.

REFERENCES

[1] Villasor, A. P. Jr. 1985. Stresses in bend transitions of steam generator U-tubes. Brussels. 8th SMiRT.
[2] WECAN - A general purpose finite element program. 1980. Westinghouse Electric Corporation. Pittsburgh, Pennsylvania (USA).
[3] Roark, R. J. & Young, W. C. 1975. Formulas for Stress and Strain. 5th Edition. New York. McGraw-Hill.

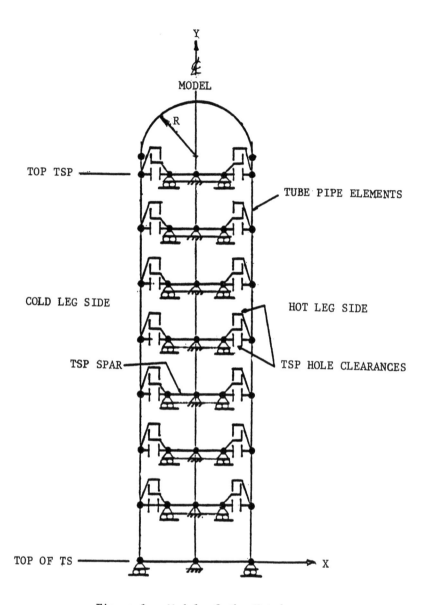

Figure 1 – Model of the U-tube

314

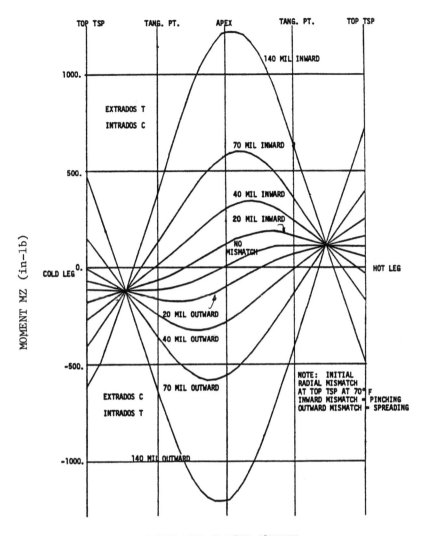

Figure 2 – Variation of moment (MZ) in U-bend as a
function of the initial radial mismatch

315

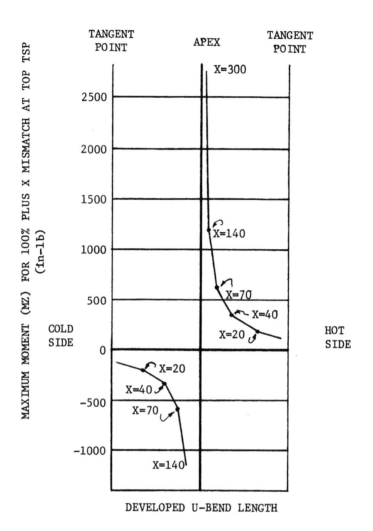

Figure 3 - Magnitude & location of maximum
moment (MZ) in U-bend due to 100%
load plus mismatch

Table I - Effect of mismatch on stresses
in the U-bend

INITIAL MISMATCH (mils)	INTRADOS		EXTRADOS	
	SI	VM	SI	VM
+140	61157	56677	46101	41881
+ 70	31777	28061	24873	21637
+ 40	19823	17174	16199	14216
+ 20	12921	11863	11120	10779
0	11187	10815	10401	9661
- 20	14394	12871	11121	10779
- 40	21532	18651	16357	14339
- 70	33573	29768	25088	21832
-140	62983	58480	46334	42108

Notes: + mismatch means outward or spread
         - mismatch means inward or pinch
         SI = maximum stress intensity (psi)
         VM = von Mises stress (psi)

Flexibility of trunnion piping elbows

G.D.Lewis
Duke Power Company, Charlotte, N.C., USA
Y.J.Chao
University of South Carolina, Columbia, USA

1. INTRODUCTION

Pipe support designs in nuclear power plants are often dictated by the obvious constraints such as piping loads, space restrictions, and available materials for fabrication. In certain instances, the piping stress analysis may require the placement of a support or anchor on a pipe fitting such as a tee or elbow. In the petrochemical industry, it is not uncommon to have supports attached to an elbow by means of a trunnion, particularly in the close proximity of equipment nozzles. This need to attach trunnions to elbows gave rise to the previous work of Williams and Lewis (1) in which basic stress indices were defined. The purpose of this work is to extend the previous analyses to define the in-plane and out-of-plane flexibility factors that are necessary to allow for greater use of the trunnion elbow attachment in the nuclear industry.

The ASME Boiler and Pressure Vessel Code (2), Section III, provides analysts and designers with a recommended analysis method for nuclear safety related piping. The objective of this piping analysis is to safely and economically guard against failure of the pipe material or pipe supports, leakage of flanged joints, and overstrain of connected equipment nozzles. To simplify the often enormous task of analyzing an entire piping system, the Code approved analysis method utilizes a rough finite element model of straight and curved one dimensional beam elements to represent the piping system. The actual complex behavior of the various local piping components within the beam model is represented by the use of flexibility factors and stress indices.

Flexibility factors and stress indices for piping component such as straight pipe, elbows, buttwelding tees, branch connections, and butt-welding reducers are contained in the Code, but many of the less common piping components, like the trunnion elbow, do not have flexibility factors or stress indices defined. The purpose of this paper is to identify the in-plane and out-of-plane flexibility factors in accordance with Code procedures for welded trunnions attached to the tangent centerlines of long radius elbows. Figure 1 illustrates the geometry and loading axes for a typical trunnion elbow support.

This work utilized the finite element method as applicable to plates and shells for calculating the relative rotations of the trunnion elbow-ends for in-plane and out-of-plane elbow moment loadings. These rotations are used to derive the corresponding in-plane and out-of-plane flexibility factors.

319

In order to cover as wide a range as practicable, a parametric investigation of various combinations of elbow and trunnion geometries was
performed based on the commercially available trunnion/stanchion assemblies given in the ITT Hanger Standards (3). In addition, the work was
limited to long radius elbows because these are by far the most commonly
utilized in the power industry. Furthermore, the relative component
sizes analyzed were restricted to trunnion attachments of approximately
one-half to one times the elbow diameter (D/2<d<D) and trunnion thicknesses of approximately one to two times the elbow thickness (T<t<2T).

2. FINITE ELEMENT ANALYSIS

The ANSYS (4) computer program was utilized in this study. The
quadrilateral shell element with six degrees of freedom at each node
was employed to model both the elbow and trunnion attachment. The
quadrilateral elements were utilized in the trunnion and elbow areas
away from the intersection, while triangular elements were used to re-
fine the mesh in the vicinity of the intersection. The aspect ratio
was limited to approximately 3:1 wherever possible and was nearer to
1:1 in the critical stress regions surrounding the trunnion-to-elbow
intersection.

A typical whole model element mesh is shown in Figure 2. The whole
model was used for all out-of-plane moment loading cases. A finer mesh
half model was utilized for the symmetrical in-plane moment loadings
and for verification of the adequacy of the full model mesh.

To determine the change in flexibility brought about by the inclusion
of the trunnion attachment, it is necessary to find the rotation of each
elbow end due to a moment load, M, and in the direction of the moment,
M. But first, one must ensure that the elbow ends are not adversely
affected by the boundary conditions or the applied loads. Therefore,
tangent lengths equal to one pipe diameter were included at each end
of the elbow with the boundary conditions and moments being applied to
the tangent ends. The upper tangent end was held fixed against rotation
and translation while moment loads (M_{x2}, M_{y2}, M_{z2} on Figure 1), were
individually applied at the lower tangent end. Similarly, the lower
tangent end could be held fixed while moments (M_{x1}, M_{y1}, M_{z1}) were
applied to the upper tangent end giving only slightly different local
stress results and essentially identical flexibility factors.

For simplicity of calculation, nominal bending moments $M=\sigma Z=1000Z$
(Z=section modulus) were applied by means of a statically equivalent,
linearly varying, axial nodal loading pattern as described by Tso and
Weed (5). Torsional moments were generated by means of a statically
equivalent tangential nodal loading pattern. In the area of the tangent
removed from the applied axial load and fixed boundaries, nominal values
of bending and torsional stresses as calculated by the familiar flexure
and torsional formulas (MD/2I and MD/2J, respectively) were observed
for verification purposes. Details of the imposed moment loads and
other dimensional parameters used in the flexibility calculations can
be found in Table 1 and ref 6.

All finite element analyses were performed using the basic assumptions
of homogeneous, isotropic thin plates and shells within the range of
linear elastic behavior. Other assumptions involve possible differences
between the finite element model and the actual trunnion-elbow assembly.
The model simulates a full penetration weld without a contouring fillet
weld since shell elements were used. However, the actual weld would
likely have a contouring fillet that would greatly reduce the stress
concentration but probably would have little effect on the flexibility.

320

Also, an actual elbow is not likely to have exactly uniform thickness or roundness of all cross sections due to the elbow fabrication process. Piping analyses experts believe these effects to be reasonably small.

3. FLEXIBILITY FACTOR DEVELOPMENT

The rotations Θ_{za} at section A-A and Θ_{zb} at section B-B are obtained from the average of all nodal rotations across each section. The rotations across each elbow end sections along with the calculated nominal rotation, $\Theta_{nom}=M(EBR)(\pi/2)/EI$, can be used in the flexibility formula:

$$\Theta_{za}-\Theta_{zb}/\Theta_{nom} = (\Theta_{ab})_z/\Theta_{nom} = k_z \qquad (1)$$

to arrive at the in-plane flexibility factors.

For the half model, Θ_x, Θ_y, d_z are zero for all nodes along the symmetry plane. The other non-zero terms d_x and d_y can also be used to derive displacement flexibility factors. However, for most piping systems, the rotational flexibility factor is of primary importance. The displacement flexibility factors would be significant only where the elbow constitute a major portion of the entire piping system (7). Furthermore, transverse or out-of-plane flexibility generally plays a small part in determining the overall system flexibility (8). Therefore, the in-plane rotational flexibility factor is considered to be of greatest importance to the overall piping system flexibility.

The procedure to calculate out-of-plane flexibility factors k_x and k_y is similar to the calculation method for in-plane moments except that for out-of-plane bending, a pure out-of-plane moment exists only at one tangent end of the elbow. For all other points throughout the bend of the elbow, there exists a combination of out-of-plane and torsional moments acting on the elbow. Following the procedure of Rodabaugh, et at. (7), it is assumed that the incremental rotation for an elbow section under torsional moment is given by the nominal relationship:

$$d\Theta = M_t(EBR)d /GJ \qquad (2)$$

where   EI = 1.3 GJ
              α = coordinate angle of a curved pipe
              M_t = Torsionial moment
              $d\Theta$ = incremental rotation
             EBR = elbow bend radius=1.5 times nominal pipe diameter

Thus, the rotational flexibility factors for out-of-plane bending (7,9) may be expressed as:

$$k_x = [(\Theta_{ab})_x DE/2000 - 1.021(EBR)]/0.7854(EBR) \qquad (3)$$

$$k_y = [(\Theta_{ab})_y DE/2000 + 0.65(EBR)]/0.5(EBR)$$

where $(\Theta_{ab})_x$ = relative elbow end rotation about the x-axis
              $(\Theta_{ab})_y$ = relative elbow end rotation about the y-axis
                D = outside pipe-elbow diameter
              E = modulus of elasticity

Table 2 shows the resulting flexibility factors for the ten models. The rotational flexibility factors are curved-fitted and given by the

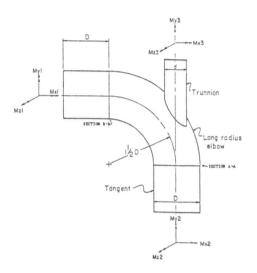

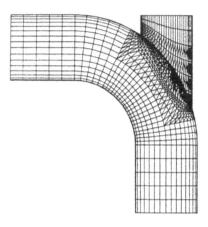

Fig 1. Typical trunnion elbow.

Fig 2. Typical finite element mesh.

Table 1   Dimensional parameters

Model no.	Elbow D (in.)	D (mm)	T (in.)	T (mm)	D/T	d (in.)	d (mm)	Trunnion (in.)	(mm)	d/t	D/d	T/t
1	6.625	168.3	0.280	7.1	23.7	3.500	88.9	0.300	7.6	11.7	1.89	0.93
2	6.625	168.3	0.280	7.1	23.7	4.500	114.3	0.337	8.6	13.4	1.47	0.83
3	12.750	323.9	0.375	9.5	34.0	6.625	168.3	0.432	11.0	15.3	1.93	0.87
4	12.750	323.9	0.500	12.7	25.5	10.750	273.1	0.500	12.7	21.5	1.19	1.00
5	16.000	406.4	0.375	9.5	42.7	8.625	219.1	0.500	12.7	17.3	1.86	0.75
6	16.000	406.4	0.500	12.7	32.0	12.750	323.9	1.000	25.4	12.8	1.26	0.50
7	24.000	609.6	0.375	9.5	64.0	12.750	323.9	0.375	9.5	34.0	1.88	1.00
8	24.000	609.6	0.500	12.7	48.0	16.000	406.4	0.500	12.7	32.0	1.50	1.00
9	36.000	914.4	0.625	15.9	57.6	20.000	508.0	0.625	15.9	32.0	1.80	1.00
10	36.000	914.4	0.750	19.1	48.0	24.000	609.6	1.218	30.9	19.7	1.50	0.62

Table 2   Flexibility Factors for Trunnion-Elbows
and Elbows with Flanged End Constraints

MODEL	h	k_{elbow}	k_{1F}	k_{2F}	k_x	k_y	k_z
1	0.25037	6.59	5.23	4.15	5.32	6.40	5.14
2	0.25037	6.59	5.23	4.15	5.29	6.42	4.70
3	0.17631	9.36	7.01	5.25	7.67	8.85	7.36
4	0.23990	6.88	5.42	4.27	5.62	6.82	4.81
5	0.14746	11.19	8.13	5.91	9.34	10.41	8.78
6	0.19979	8.26	6.31	4.83	6.83	8.25	5.36
7	0.09674	17.06	11.56	7.83	14.49	14.99	13.41
8	0.13038	12.66	9.01	6.42	10.67	11.79	9.83
9	0.10788	15.29	10.55	7.28	12.93	13.68	12.08
10	0.13038	12.66	9.01	6.42	10.64	11.80	9.62

equations:

$$k_x = 0.871 \ k_e - 0.425 \qquad\qquad (5)$$
$$k_y = -0.014 \ k_e^2 + 1.14 \ k_e - 0.509$$
$$k_z = 0.81 \ k_e - (d/D)$$

These equations are subject to the dimensional restrictions $D/2 \leq d \leq D$ and $T \leq t \leq 2T$ and fit the data to within 10 percent.

4. DISCUSSIONS

It has been known for quite some time that bending of an elbow will tend to produce ovalization of the elbow ends. Further, any form of elbow end restraint, such as flanges, reducers, and to a much less degree straight pipe, will tend to inhibit ovalization and thus lower the flexibility of the elbow. In a similar manner, the inclusion of a large trunnion will stiffen a large portion of the elbow that would otherwise contribute to the ovalization and flexibility.

Table 2 shows flexibility characteristics value, h, flexibility of a plain elbow, k_e (or k_{elbow}), flexibility of an elbow with one end flanged, k_{1F}, flexibility of an elbow with two ends flanged, k_{2F}, based on the ASME Code, and the flexibility factors for the trunnion-elbow. To illustrate the effect of the trunnion, Model 1 shows an in-plane flexibility factor of 6.59 for a plane elbow and an in-plane flexibility of 5.14 with the inclusion of a trunnion. Model 2 shown that the same size elbow with a larger and slightly thicker trunnion results in a further reduction in the in-plane flexibility to a value of 4.70. As shown in Table 2, the flexibility reduction due to the trunnion may be greater than for an elbow with one end flanged but is usually not as great as an elbow with both ends flanged. Since the ASME Code requires the flexibility factor of a flanged elbow to be modified from that of a plain elbow, clearly a different flexibility factor should be required for elbows with trunnion attachments. Therefore, the flexibility factor of primary importance, k_z from the current study, is recommended for use in piping analyses involving the use of trunnion-elbows.

5. REFERENCES

[1] Williams, D.K. and Lewis, G.D., "Development of B_1 and C_1 Stress Indices for a Trunnion Elbow Support", Journal of Pressure Vessel Technology, Vol. 106, No. 2, May 1984, pp. 166-171.
[2] ASME Boiler and Pressure Vessel Code, Section III, "Nuclear Power Plane Components", American Society of Mechanical Engineers, New York, 1980 Edition.
[3] Hanger Standards, ITT Grinnel Corporation, Pipe Hanger Division, January 1982.
[4] DeSalvo, G.J. and Swanson, J.A., ANSYS Engineering Analysis System User's Manual, Swanson Analysis Systems, Houston, PA, 1982.
[5] Tso, F.K.W. and Weed, R.A., "Stress Analysis of Cylindrical Pressure Vessels with Closely Spaced Nozzles by the Finite Element Method", NUREG/CR-0507, May 1979.
[6] Lewis, G.D., "Development of Piping Flexibility Factors for Trunnion Elbows", M.S. Thesis, Dept. of Mechanical Engineering, University of South Carolina, 1986.

[7] Rodabaugh, E.C., Iskander, S.K., "End Effects on Elbows Subjected to Moment Loadings", ORNL/Sub-2913/7, March 1978.
[8] Vigness, I., "Elastic Properties of Curved Tubes", Trans. ASME., Vol. 65, 1943.
[9] Crocker, S., and McCutchan, A., "Frictional Resistance and Flexibility of Seamless-Tube Fittings Used in Pipe Welding", Trans. ASME., Fuels and Steam Power, FSP-53-17.

6. NOMENCLATURE

d = outside trunnion pipe diameter
G = shear modulus of elasticity
h = flexibility characteristic (= $T(EBR)/r^2$)
I = moment of inertia
J = polar moment of inertia
k_e = flexibility factor for a plain elbow (= $1.65/h$)
k_{1F} = $1.65/h^{0.833}$
k_{2F} = $1.65/h^{0.666}$
r = elbow mid-wall cross section radius
t = nominal trunnion pipe wall thickness
T = nominal pipe/elbow wall thickness
d_x, d_y, d_z = displacements in directions x, y, z
σ = normal stress
$\Theta_x, \Theta_y, \Theta_z$ = rotations in directions x, y, z

Elastic-plastic analysis of a full-size elbow – Experimental and numerical investigations

K.Kussmaul & H.Diem

Staatliche Materialprüfungsanstalt (MPA), University of Stuttgart, FR Germany

H.J.Rensch & H.Hofmann

Ingenieurunternehmen für spezielle Statik, Dynamik und Konstruktion (SDK), Lörrach, FR Germany

1 INTRODUCTION

Within piping systems, the parts which display the highest flexibility are the pipe bends. In normal operation, the bends are designed on one hand to change the direction of the piping run, while on the other to dissipate reaction forces and moments within the system by elastic deformation, a function to which they are well suited by their greater flexibility as opposed to adjoining straight sections of piping. Because of their specific deformation behaviour when exposed beyond the elastic limit, pipe bends are capable of plasticizing over large areas, thus considerably cushioning transiently loaded systems by energy dissipation as the result of plastic material flow. In this context, considerable importance is assigned to knowledge of the safety margin between incipient local yield and the load at which the pipe bend collapses.

As part of a research project focussed on elastic-plastic bend analysis sponsored by the Federal Minister for Research and Technology of the Federal Republic of Germany, investigations were carried out into the behaviour of pipe bends subjected to large, elastic-plastic deformations. This task also entails determination of load-bearing limits and possible failure mechanisms. An important goal in these investigations is the question as to the precision with which the behaviour of parts of this nature can be represented in a numerical calculation.

Loads are assigned to three different categories – in-plane opening mode, in-plane closing mode and out-of-plane bending – which influence the load-bearing behaviour of pipe bends, as do the geometrical dimensions, the material behaviour and any additional internal pressure which may be applied.

This paper concentrates on the loading condition "in-plane opening-mode bending without internal pressure".

2 SPECIMEN

The elbow examined was manufactured from a seamless drawn pipe by inductive bending; the material was heat-resistant low-alloy ferritic steel 15 MnNi 6 3. With the dimensions $D_0 = 470$ mm, $D_i = 390$ mm and $R = 950$ mm, the 90° bend has a diameter ratio of $D_0/D_i = 1,21$, and may thus be ranked with the transitional range between thin-walled and thick-walled pipes.

Before testing commenced, the actual gemometry, the material parameters and the residual stresses (local) were recorded in detail. With due consideration to symmetrical characteristics, a total of 256 high elongation strain gauges were mounted inside and outside the bend to register local deformation behaviour. An out-of-roundness measuring ring in the crown of the bend helped to establish cross-sectional deformations. The global behaviour of the part was recorded by three deflection transducers. The load was imposed by a 15 MN horizontal tensile-testing machine(Fig. 1), which exerted a member-force combination of normal and transverse force on the pipe, together with a bending moment. However, the long pipe leg (lever) $l = 2900$ mm meant that the bending moment imposed the dominant stress in the elbow.

3 COMPUTATION MODEL

The numerical analysis took the form of a preliminary calculation with the SAN /1/ program system conducted prior to the commencement of testing. The computational model used is a 180° half model of the pipe bend including the adjoining straight pipe leg. It consists of a total of 256 3D degenerated shell elements with 16 nodes per element and 5 interpolation nodes through the wall thickness to determine plastification.

The geometrical data were based on the measured wall thicknesses considering the relatively large differences between maximum (52 mm) and minimum (35 mm) values. The other non-conformances (elliptical ovality and outside diameter-variation) with the nominal dimensions were negligible. The description of the nonlinear load-bearing behaviour is based on an extrained Lagrange formulation. The law of elastic-plastic material behaviour used is based on the Prandtl/Reuss-equations /2/ (J_2-yield theory, cf. v. Mises yield condition, isotrope hardening). Multi-linear approximation of the measured stress/strain curve was used to determine the plastic tangent modulus.

4 EXPERIMENTAL RESULTS

A preliminary test in the elastic range had already demonstrated that two areas of virtually equal stress appear in a bend of this geometry. In contrast to thin-walled bends ($D_0/D_i < 1,2$), high stressing occurs both on the flank and on the intrados at the middle of the pipe bend /3/.

The formation of plasticized zones as a function of load is shown in Figs. 3 and 4, which observe the angular convention of Fig. 2. The latter is a single-plane orthomorphic projection of one half of the pipe surface. In the shaded areas, the material yield strength has been exceeded.

At the load of F = 0,52 MN, pipe behaviour is still completely elastic. Thereafter, initial plastification first appears at the wall inner fibre of the bend flank. As the load increases, plastification also commences on the outer surface at the intrados, extrados and pipe flank. At a load F = 1.03 MN, the bend is plasticized over practically its entire length. At this instant, elongations were still beneath 2.4 %, in other words not yet in the range at which the material investigated begins to harden.

At the end of the test at the tensile force of 2.2 MN and local overall elongation values of up to 6 % the pipe bend had not collapsed. On account of the strain hardening effects of the material and a steady increase in the bending section modulus, the bend retained a load-bearing reserve even beyond the fully plastic state.

5 COMPARISON OF MEASUREMENT AND CALCULATION

Figs. 5 through 7 illustrate a number of characteristic measurements and related computational results. The result of the comparison may be summarized as follows:

1. The results of calculation and measurement evince good correspondence extending into the partially-plastic range.
2. At the transition to the fully plastic range, the calculation predicts a more rigid behaviour than was observed in the test. On the basis of the load, the resultant deviations are approx. 15 %. The deviations between calculation and test at a given load are larger as a result of the rapidly increasing deformations in this range.
3. At roughly twice the initial yield load, a horizontal plateau becomes apparent at which deformations extend without any increase in load. The calculation was terminated at this load, although it subsequently proved possible to increase the test load past this point due to the strain hardening capacity of the material and due to increasing ovalization.
4. In qualitative terms, the formation of the plasticized zones was reflected equally well by calculation and measurement.

6 CONCLUSIONS

In general terms, the bend examined displayed considerable reserves of load-bearing capacity, without any visible sign of damage or plastic collapse. However, this applies only to in-plane opening-mode bending, the loading condition examined. A broadly different load-bearing behaviour with the tendency to collapse may be anticipated in bends subjected to in-plane closing-mode bending.

7 REFERENCES

1   Programm Konzeption SAN:
    Förderungsvorhaben BMI SR 79-01, Juni 1980

2   Reckling, K.A.: Plastizitätstheorie und ihre Anwendung
    auf Festigkeitsprobleme,
    Springer-Verlag, Berlin, 1967

3   Kussmaul, K., Diem, H., Blind, D.: Investigations on the
    plastic behaviour of pipe bends. To be presented at the
    1987 ASME PVP-Conference, June 28 - July 2, SAN DIEGO

Fig. 1. Test rig showing 90° full size bend in position

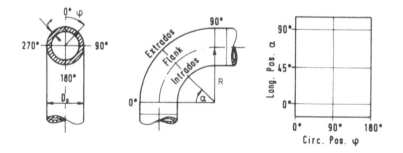

Fig. 2. Angle convention and single plane orthomorphic
         projection of the elbow surface

328

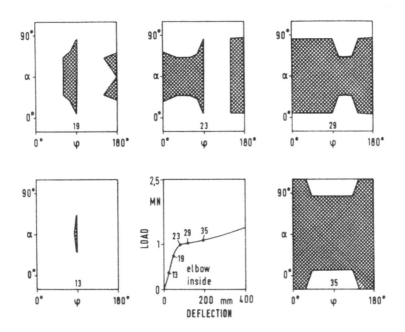

Fig. 3. Formation of plasticized zones (inside surface)

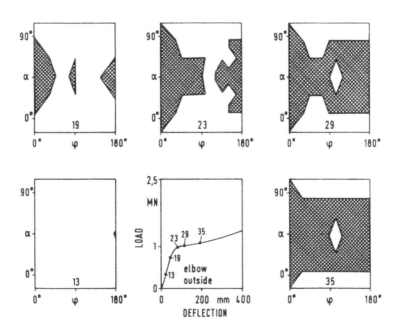

Fig. 4. Formation of plasticized zones (outside surface)

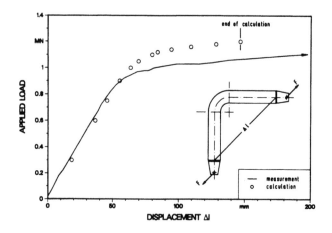

Fig. 5. End deflection of the bend specimen

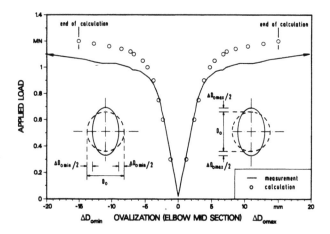

Fig. 6. Ovalization due to increasing loading

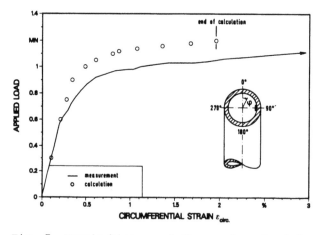

Fig. 7. Strain history at the maximum loaded position

330

Experimental study of behavior of austenitic stainless steel thin-walled elbows – Functional capability

P.Hilsenkopf, B.Boneh & P.Sollogoub
Framatome, Paris la Défense, France
C.Berriaud & P.Jamet
CEA-CEN Saclay, DEDR/DEMT, Gif-sur-Yvette, France

ABSTRACT

Results of a serie of 15 tests on 90 large radius elbows are presented. These elbows were made from Z2 CN 18 10 steel (corresponding to ASME TP 304 L) and had an outside diameter-to-wall thickness ratio of 89.5. They were subjected to in-plane (opening and closing) and out-of-plane bending moments. Changes in elbow angular deflection and ovalization of the mid section were recorded as a function of applied moment. Measurements were made well into the plastic region. Influence of pressure, temperature and cyclic loading were also studied.

The tests therefore supplied extensive data on the behavior of an austenitic stainless steel thin-walled elbow when subjected to large displacements, including its ability to carry the required flow under high loadings. Analysis per the RCC-M (1983) was also performed to quantify flow area reductions at stress limits allowed by these rules. Main results were as follow:

- Tested elbows exhibited satisfactory behavior in deformation and withstood large angular displacements without excessive damage. Althought tests were continued well into the plastic region, it was not possible to take these components to failure.

- The least favourable loading mode was in-plane bending closing the elbow.

- Internal pressure increased component stiffness and therefore favourably affected overall elbow behavior.

- Temperature and preliminary cyclic loading both had an unfavourable effect significantly lowering the experimental limit moment.

- Flow area reductions recorded at the experimental limit moment were low.

Analysis per the RCC-M of the component under study demonstrated substantial safety margins on the experimental limit moment. If class 1 level D service limits are met, functional capability of the tested elbow is assured.

1 NOMENCLATURE

Do   : outside diameter
t   : nominal wall thickness
ρ : radius of curvature
Sy   : code minimum yield stress of material
Su   : code minimum ultimate stress of material
Me   : experimental moment
Ml   : experimental limit moment
$\Delta\beta$: angular deflection of elbow under test
ψ : ovality ratio of elbow mid section
ΔS : reduction in flow area
a : angular rotation applied by jack

2 INTRODUCTION

Functional capability of piping components is closely related to their
capacity for deformation. For this reason, special attention has always
been paid to elbows and bends, which are particularly flexible, and
they have been used as the base component in the various studies on this
question. The results of Ibrahim & Kitz (1978 ; 1983) and Liu, Johnson &
Chang show that ASME code level D service limits suffice to guarantee
functional capability of the tested components. These studies were based
on results of tests on ferritic and austenitic steel elbows with a Do/t
ratio less than 50. They did not cover deformation of thin-walled bends
and elbows defined by 100 > Do/t > 60, about which little is known.The
question to be answered is whether functional capability of such components is assured under accident conditions if service limits only are
met. This paper contributes to answering this question by presenting
results from bending tests on austenitic stainless steel elbows having
an outside diameter-to-thickness ratio of 89.5.

3 TESTS AND MEASUREMENTS

3.1 Description of tested components

The tested components consisted of 90 deg. large radius elbows.
Nominal dimensions were : Do = 179 mm   t = 2 mm ρ = 275 mm
Do/t = 89.5. The elbows were made from two cold-formed shells welded
together along the inner and outer curve lines. The material was
Z2 CN 18 10 austenitic stainless steel (equivalent to ASME TP 304 L) as
specified in the RCC-M. Steel grade, fabrication method and maximum operating temperature and pressure are typical of thin-walled elbows (with
96 > Do/t >58.2) in French nuclear power plant auxiliary systems.
Structural uniformity of tested fittings was verified by macrography and
grain size measurements on a specimen elbow. Tension tests were performed at 20°C and 120°C with the following results:

Table 1. Mean yield and ultimate tensile stress values.

Temperature (°C)	Yield stress (Mpa)	Ultimate tensile stress (Mpa)
20	372	632
120	317	511

3.2 Description of tests

Fifteen tests were carried out at Saclay by the Mechanical and Thermical Studies Department of C.E.A. In-plane opening and closing bending moments were applied to the elbows, which were also subjected to out-of-plane bending moments. The influence of internal pressure (9 bar) and temperature (120°C) on elbow behavior was also studied. In tests 11 and 12, elbows were subjected to preliminary cyclic loading (20 cycles) with the following angular displacements : α = 1.5 deg. for test 11 and α = 2.4 deg. for test 12. Closing bending moments were then applied. Tests were continued deep into the plastic region, beyond the excessive deformation moment and also beyond the instability (collapse) moment whenever possible. However, none of the components failed.

During each test the following values were measured :
Me   : moment applied to elbow mid section
$\Delta\beta$: angular deflection of one end of elbow relative to other
D1   : major diameter of elbow mid section
D2   : minor diameter of elbow mid section
α : rotation of vertical straight portion of the setup
Each test therefore generated plots of Me = $f(\Delta\beta)$ and Me = $f(\psi)$, where ψ is defined by ψ = D1/D2. So we can easily determine the ovality of the most heavily loaded cross-section as a function of the loading level. Plots of Me = $f(\Delta\beta)$ and Me = $f(\psi)$ for tests number 9, 13 and 15 are shown as sample in appendix A.

4 TEST RESULTS

4.1 Experimental limit moment (Ml) and limit rotations.

The experimental limit moment was determined from the Me = $f(\Delta\beta)$ curves. This moment is either the plastic instability moment (Mlip) or the excessive deformation moment (Mlde).
  Mlip : maximum moment reached on moment-rotation curve;
  Mlde : moment for which rotation $\Delta\beta$ is twice the extrapolated elastic rotation.
For each test we consider the following limit rotations :
$\Delta\beta1$: angular deflection of elbow at experimental limit moment.
$\alpha1$: rotation of vertical straigth portion of test assembly at experimental limit moment.
$\Delta\beta M$: maximum angular deflection of elbow in particular test.
αM : maximum vertical straigth portion rotation.
Results are summarized in table 2.

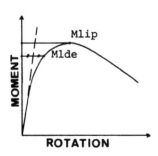

Figure 1. Experimental limit moment

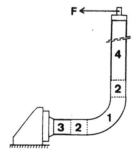

Figure 2. Test setup

333

Table 2. Mechanical behavior of tested components
(except loss of section)

Item of test	Loading	Ml (*) (daN.m)	$\Delta\beta 1$ (deg.)	$\alpha 1$ (deg.)	$\alpha M/\Delta\beta M$
1	clos. ip	456	-	7.3	-
2	clos. ip	413	6.9	7.5	1.01
3	out-of-plane	651	9.0	9.0	1.04
4	clos. ip + pressure	649	10.7	11.2	1.02
5	open. ip + pressure	(1336)	7.7	9.8	2.55
6	open. ip	(952)	8.0	9.3	1.48
7	out-of-plane	649	10.9	10.0	0.98
8	open. ip	(1005)	8.0	9.5	1.59
9	clos. ip	375	7.2	7.5	1.02
10	out-of-plane + pressure	949	10.2	9.2	0.94
11	clos. ip + cycl 1	371	6.7	7.1	1.00
12	clos. ip + cycl 2	367	6.2	6.2	1.00
13	open. ip + temperature	(921)	11.4	13.4	1.61
14	clos. ip + temperature	331	6.0	7.5	1.02
15	out-of-plane + temperature	542	10.2	9.0	1.05

(*) brackets indicate that Ml = Mlde

4.2 Analysis of behavior of tested components

The overall test results show that a bending moment ovalizes the elbow.
Cross-section start to ovalize as soon as elbows are loaded and ovality
increases until the component fails. Failure was never actually attained
but tests were extended sufficiently, (to αM # 30) to reveal the failure mode. Figure 3 illustrates the various ovalization modes recorded in
the tests. From $\alpha M/\Delta\beta M$ values it can be seen that virtually all the
applied displacement is absorbed by the elbow under test, except in the
case of in-plane bending to open the elbow.

In-plane bending to close elbow : ovalization decreases the cross-sectional moment of inertia, hence reducing stiffness of the elbow.The component fails rapidly due to excessive ovalization, flattening the cross-section. This is the most penalizing loading mode (Ml # 400 daN.m)

Out-of plane bending : the phenomenon is identical, though less marked than in the in-plane closing tests, here Ml # 650 daN.m. Cross-sectional deformation becomes irregular above the experimental limit moment when α exceeds 10 deg. The elbow twists back on itself and a collapse region is formed as a re-entrant crease, though without sudden failure of the component.

In-plane bending to open elbow : ovalization increases the elbow's moment of inertia and displaces the weak spot in the test assembly.At a loading level equivalent to α = 12 deg. in the test 5 and 14.6 deg. in test 6, deformation of the elbow stops and a crease occurs at the bottom of the junction between straight portions 2 and 3 .

All subsequent displacements applied by the jack are absorbed by this local unstable area and there is no further deformation of the elbow. This explains why $\alpha M/\Delta \beta M$ is much higher than 1. However, the experimental limit moment of around 1000 daN.m is reached before the crease occurs.

Internal pressure stiffens the elbow and opposes ovalization. The experimental limit moment rises significantly, the gain ranging from 55 % in closing tests to 35 % in opening tests. Internal pressure stiffens the elbow during in-plane opening tests and causes premature local instability, the crease occuring at α = 12   instead of 14.6 .

In closing tests xl varies as the experimental limit moment. This is not the case for the other two types of loadings which produce only slight changes in the limit applied displacement αl.

Preliminary cyclic loading has an unfavourable effect. The elasto-plastic instability moment falls by 11% and there is a more rapid transition from the elastic to the plastic region. Furthermore, αl is 5% lower in test 11 and 17% lower in test 12.

Elevated temperature has an adverse effect. The instability moment drops by about 20% for in-plane loading in the closing direction and by 17% for out-of-plane bending. This influence is less marked for in-plane bending in the opening direction, the excessive deformation moment drop -ping only by 6%. Although it changes significantly in in-plane opening test (increasing by about 40%), there is little change in the value of αl in closing and out-of-plane bending.

M < Ml   end of test

Figure 3. Description of the deformation modes.

4.3 Functional capability

Ovality (ψ) and loss of section (ΔS) : elbows are distorted by the applied loads and become oval in cross-section. This ovalization is uniform for thick-walled elbows but less so for austenitic stainless steel thin-walled fittings. However, the deformed mid section can be considered elliptical up to the experimental limit moment.
Let us consider $\Delta S = 1-\pi/4\psi$ E where E = elliptical cross-section area
Ovality ratio and loss of section were calculed for each test and for each of the following loading levels : 1.9 Sm, 3 Sm, 1.8 Sh, 2.4 Sh, Ml. Sm (class 1) and Sh (class 2 or 3) are the material allowable stress intensities per the RCC-M, based on the specified mechanical properties. Sy and Su values specified in the RCC-M give the following stress intensities : Sm (20 °C) = Sm (120 °C) = 115 Mpa
Sh (20 °C) = 108 Mpa   Sh (120 °C) = 106.8 Mpa
These stress limits give code limit moments calculated from equation 9 in RCC-M B 3652 for class 1 analyses and equation 10 in RCC-M C 3654 for class 2 or 3 analyses.
The flow area reduction recorded for these loading levels are shown in table 3. For the limit moment level table 3 give also the ovality ratio.

Table 3. FLow area reduction ΔS (in %).

Test	Loading level -->	1.9 Sm	3 Sm	1.8 Sh	2.4 Sh	Ml ΔS	ψ
1		*	*	*	0.17	1.69	1.235
2		*	*	*	0.17	1.22	1.196
3		*	*	*	*	2.67	1.304
4		*	*	*	*	2.93	1.319
5	**	*	*	*	*	3.02	1.324
6	**	*	*	*	0.10	4.32	1.401
7		*	*	*	*	3.17	1.333
8	**	*	*	*	0.11	4.44	1.406
9		*	*	0.12	0.31	1.55	1.224
10		*	*	*	*	3.03	1.326
11		*	*	*	0.26	1.53	1.222
12		*	*	0.10	0.21	1.20	1.195
13	**	*	*	*	0.17	8.90	1.626
14		*	0.10	0.18	0.86	1.69	1.235
15		*	*	*	*	3.96	1.379

* Lower than 0.1 %   ** Ml = Mlde

336

4.4 Analysis of results

Level D service limits guarantee that flow area is not significantly reduced. Loss of section are below 0.1 % in class 1 and less than 1 % in classes 2 and 3.

In-plane closing and out-of-plane bending moments produce less than 4 % reduction inf low area at the point of plastic instability.

In the opening direction, elbows did not reach the instability moment. Flow area reduction for the excessive deformation moment is significantly greater than with other types of loading, reaching nearly 9 % in some cases. However, it should be noted that the safety margin between level D service limits and the experimental limit moment is considerable for this type of loading ; a flow area reduction of 0,17 % only was recorded for class 2 or 3 level D service limits.

For in-plane closing and out-of-plane bending, the angular displacement margin (ΔβR) between level D service limits and plastic instability is higher than 4.61 deg. for class 1 and 3.0 deg. for class 2 or 3. In opening bending, this margin with respect to excessive deformation rises to 7.0 deg. in class 1 and 6.5 deg. in class 2 or 3, confirming the great ductility of these components.

In-plane and out-of-plane bending tests were taken to αM = 30 deg., the test stand limit. At this value the mid cross-section of bends subjected to out-of-plane bending is too distorted to be considered elliptical. The deformed shape obtained in closing bending tests is more regular and the flow area reduction formula remains valid. Calculations for test 4, 9, 11, 12 and 14 for rotation αM = 30 deg. provide loss of section values between 13 % and 17 %. The flow area reduction in out-of-plane bending is certainly much greater. Opening bending tests were discontinued at Δβ values approaching 15 deg. because local instability occured on a straight portion next actual elbow.

5 CONCLUSION

- The tested thin-walled elbows demonstrated satisfactory deformation behavior and withstood large displacements without excessive damage. The experimental limit moment, at which flow area reduction did not adversely affect flow capacity, was achieved for displacement values Δβ of 6 deg. to 11 deg., which is considerable.
- RCC-M analysis based on the specified characteristics of tested component demonstrated large safety margins before the experimental limit moment is reached. If class 1 lebel D service limits are met, functional capability of these elbows is guaranted.
- The load involved are considered to be static and primary. The high ductility recorded for thin-walled elbows implies avery substantial margin to failure under seismic conditions.

REFERENCES

RCC-M, January 1983 edition and July 1983 addendum : Design and construction rules for mechanical components of PWR nuclear islands.
Ibrahim, Z.M. & G.T.Kitz 1978. Evaluation of the functional capability of ASME section III class 1, 2 and 3 piping components. ASME paper 78-PVP-83
Ibrahim, Z.N. & G.T.Kitz 1985. Piping functional capability substance or fiction. ASME conference NEW-ORLEANS.

Liu,T.H.,E.R.Johnson & K.C.Chang. Functional capability of ASME class 2/3
   stainless steel bends and elbows, Nuclear Technology Division.
   Westinghouse Electric corporation, Pittsburg, Pennsylvania.

Appendix A : Samples of experimental Me = f (Δβ) and Me = f (Ψ) curves

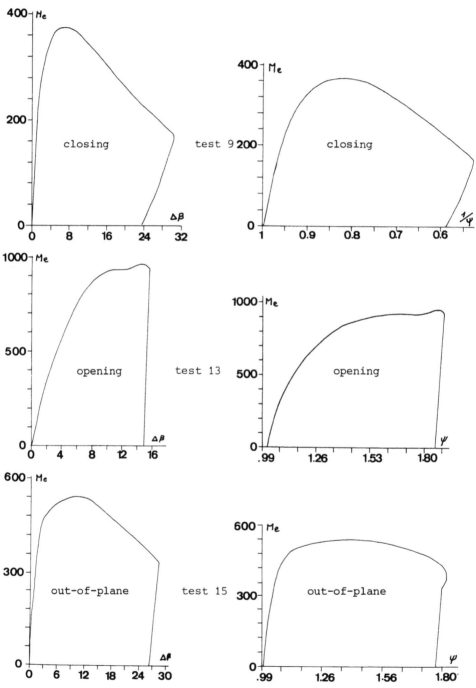

Stress component indices for elbow-elbow connection

R.S.Yadav & V.K.Mehra

Reactor Projects Division, Bhabha Atomic Research Centre, Bombay, India

1 INTRODUCTION

Nuclear piping often has two adjacent elbows lying in different planes and having different elbow angles. Piping stress analysis is done using flexibility factors and stress indices. Detailed stress analysis of elbow connected to straight pipe and subjected to inplane and out of plane loading has been reported in literature (Karabin, Thomson, Natarajan). However, results of stress analysis for adjacent elbows in different planes are not generally available.

In this paper an elbow-elbow connection with one elbow in vertical plane and other in horizontal plane are analysed using three dimensional Finite Element method for different loadings, like axial pull, inplane bending and internal pressure. A parametric study is done for different piping ratios. Results are presented for the run of piping and circumferential distribution along mid sections of the elbows and also at their junction. An attempt is made to estimate the stress component indices for different sections.

2 DESCRIPTION

The elbow in vertical plane is a 90° elbow while the one in horizontal plane is a trimmed 60° elbow. Straight pipes of infinite length ($> 6/\beta$) are connected to each of the elbows to eliminate the end effects. Elbows of three piping ratios (h) of 0.2423, 0.375 and 0.6795 are analysed, where piping ratio is tR/r^2.

Axial pull and bending moments are applied at far end of vertical pipe. The far end of horizontal pipe is fixed. For internal pressure, a force equivalent to $pD/4t$ is applied on far end face of the vertical straight pipe.

3 FINITE ELEMENT ANALYSIS

Three dimensional 8-noded isoparametric brick element available in program SOLID (Buragohain) is used for analysis. A total number of 216 elements bounded by 448 number of nodes are used. Each node has 3 displacements in mutually perpendicular directions as degrees of freedom. Fig.1 shows the typical finite element mesh used. All the three displacements of the nodes lying on far end of the pipe in horizontal plane are restrained. The load is applied on the nodes on far end of the vertical pipe.

It may be noted that load at an end may give rise to a combination of axial force, bending moment and twisting moments.

4 RESULTS AND DISCUSSIONS

Detailed stress distribution for the piping rato of 0.375 are presented here. Fig.2(a) shows the variation of normalised meriodinal and tangential stresses along the runs of the elbows for inplane bending moment applied at straight pipe end for different θ planes. A θ plane is defined as section of elbows cut by a plane making angle θ from the diametral plane containing centre of vertical elbow. Fig. 2(b) shows the variation of stresses along the circumference at ends and at middle length of both the vertical and horizontal elbows.

Figs.3(a) and 3(b) show the variation of normalised stresses along the pipe run and along circumference for the internal pressure. The stresses are quite uniform in elbows whereas end effects are seen near junctions.

Figs.4(a) and 4(b) show the variation of stresses for axial pull. The stresses tend to peak near centre of horizontal elbow while stresses near junction is maximum for vertical elbow.

Fig.5 shows a typical deformed mesh for h equal to 0.375 and subjected to internal pressure.

Table 1 shows the maximum stresses that occur in the elbow system under various loading conditions and for all three piping ratio values studied. Though in general the stresses are higher for centres of elbows, values of stresses are higher at junction compared to those at centre for some load dcases and piping ratios. There is also a change of nature of maximum stress from centre to junction. It may be noted that normalisation is done with respect to nominal stress values at the end where load is applied.

5 CONCLUSIONS

Elbow-elbow connections for three piping ratio values of 0.2423, 0.375 and 0.6795 are analysed using three dimensional finite element method. The stresses for junction and elbow runs are presented. Though in general stresses are maximum at centre of elbow, peak stress occurs at junction of elbows for some load cases and piping ratios. A table of maximum normalised stresses is presented.

REFERENCES

Karabin, M.E. etal. 1986 February. Stress component indices for elbow-straight pipe junction subjected to inplane bending. Jl of PV.Tech,Vol.108.

Thomson G, Spence J.1983 Nov. Maximum stress and flexibility factors of smooth pipe bends with Tangent pipe termination under inplane bending Jl Pr. Vessel tech.

Natarajan, R. etal. 1975. Stress analysis of curved pipes with ends restraints Computer & Structures. Vol.5.

Buragohain, D.N. SOLID-Computer code for finite element analysis of solids. IIT Bombay, India.

Table 1.

			Vertical Elbow			Horizontal Elbow		
	Piping Ratio		0.2423	0.375	0.6795	0.2423	0.375	0.6795
Inplane Bending $\sigma_n = \dfrac{MD_o}{2I}$	$\dfrac{\sigma_\phi}{\sigma_n}$	Junction(J)	1.35	1.35	1.20	1.35	1.35	1.20
		Centre (C)	1.62	1.65	1.57	1.15	-1.23	-1.11
	$\dfrac{\sigma_\theta}{\sigma_n}$	(J)	0.46	0.27	0.26	0.46	0.27	0.26
		(C)	0.68	0.60	0.43	0.28	0.36	0.15
Axial pull $\sigma_n = \dfrac{FRD_o}{2I}$	$\dfrac{\sigma_\phi}{\sigma_n}$	(J)	-1.33	-1.44	-1.29	-1.33	-1.44	-1.29
		(C)	-0.67	-0.69	-0.64	-1.86	1.90	1.76
	$\dfrac{\sigma_\theta}{\sigma_n}$	(J)	-0.42	-0.37	0.26	-0.42	-0.37	0.26
		(C)	-0.30	-0.27	0.35	-0.44	-0.38	0.24
Internal Pressure $\sigma_n = \dfrac{PD_o}{2t}$	$\dfrac{\sigma_\phi}{\sigma_n}$	(J)		0.44	0.33		0.44	0.33
		(C)		0.34	0.35		0.34	0.35
	$\dfrac{\sigma_\theta}{\sigma_n}$	(J)		0.88	0.72		0.82	0.72
		(C)		0.93	0.77		0.93	0.77

WALL THICKNESS . . . . . t
PIPE MEAN RADIUS . . . r
ELBOW MEAN RADIUS . R
O. D. OF PIPE . . . . Do
MOMENT OF INERTIA . . . I

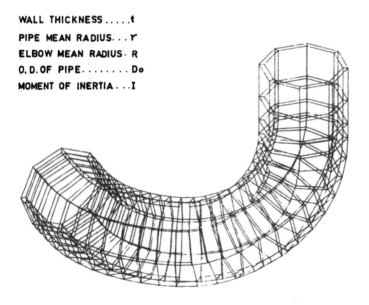

FIG.1 FINITE ELEMENT MESH OF ELBOW-ELBOW CONNECTION

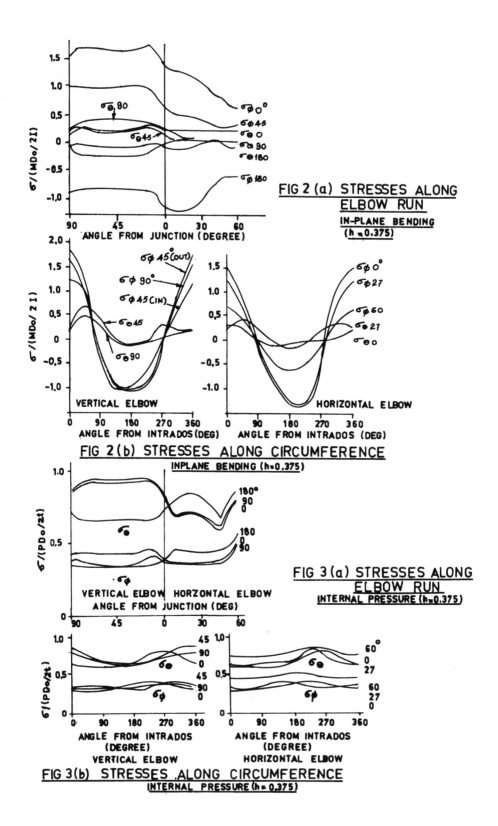

FIG 2 (a) STRESSES ALONG ELBOW RUN
IN-PLANE BENDING (h = 0.375)

FIG 2 (b) STRESSES ALONG CIRCUMFERENCE
INPLANE BENDING (h = 0.375)

FIG 3 (a) STRESSES ALONG ELBOW RUN
INTERNAL PRESSURE (h = 0.375)

FIG 3 (b) STRESSES ALONG CIRCUMFERENCE
INTERNAL PRESSURE (h = 0.375)

342

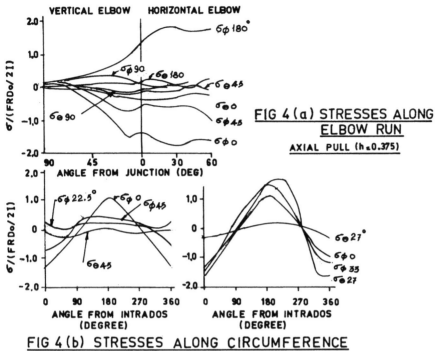

VERTICAL ELBOW   HORIZONTAL ELBOW

σ_ϕ 180°

σ_ϕ 90 σ_θ 180 σ_θ 45

σ_θ 90 σ_θ 0 σ_ϕ 45 σ_ϕ 0

$\sigma/(FRD_0/2I)$

ANGLE FROM JUNCTION (DEG)

FIG 4 (a) STRESSES ALONG
ELBOW RUN
AXIAL PULL (h=0.375)

σ_ϕ 22.5° σ_ϕ 0 σ_ϕ 45

σ_θ 45

$\sigma/(FRD_0/2I)$

σ_θ 27° σ_ϕ 0 σ_ϕ 35 σ_θ 27

ANGLE FROM INTRADOS
(DEGREE)

ANGLE FROM INTRADOS
(DEGREE)

FIG 4 (b) STRESSES ALONG CIRCUMFERENCE
AXIAL PULL (h=0.375)

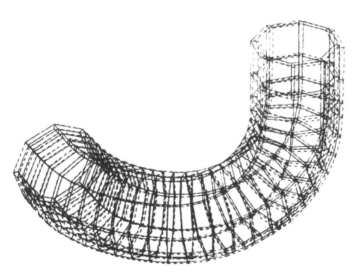

FIG. 5 DEFORMED SHAPE FOR INTERNAL PRESSURE
(h=0.375)

Components II: Materials selection and stress analysis

Effect of copper and phosphorus on the radiation stability of CrNiMoV steel

P.Novosad, J.Koutský & E.Keilová
Nuclear Research Institute, Rez, Czechoslovakia

1 INTRODUCTION

The radiation damage of reactor pressure vessel steels is influenced by both the working condition (temperature, neutron fluence), and chemical composition. Because there is only a little possibility to change the working conditions of pressure reactors, the most effective means for radiation embrittlement reduction are changes of chemical composition of steels. Considerable influence of impurities, first of all copper, phosphorus, arsenic, tin and antimony on the radiation embrittlement was found in /1/, /2/, /3/. The outstanding synergic effect of both nickel and copper were proved in some RPV steels. Special attention is paid to copper and phosphorus at the CrNiMoV and CrMoV steels. The evaluation of the influence of these both elements on the radiation stability is object of this study.

2 EXPERIMENTAL MATERIAL AND METHODS

The melt of CrNiMoV steel, produced in ŠKODA Plzeň Works, was used for the investigation. Remelting was carried out in the Iron and Steel Institute, Karlštejn, and several small melts were produced from it. Their production was made in an induction crucible furnace with basic lining of 100 kg capacity. After copper and phosphorus addition on the required content, the ingots were produced from the melt. The 50 kg ingots were overforged into plates of dimensions 28 x 200 x 600 mm. The forging temperatures were kept in the range 1180 - 950°C, the homogenization at 1200 °C/10 hours was made before forging.

The heat treatment followed: 920 °C/2 hrs./air + 650 °C/10 hrs./air + 620 °C/25 hrs. + 650 °C/20 hrs./furnace to 300 °C/air. Chemical composition of the melts is given in table 1.

The following specimens were produced of each melt: the tensile strength specimens with diameter 4 mm, specimens for Ch-V dynamic fracture toughness test with fatigue crack, and foils for electron microscope examination.

Tension tests were performed on the electro-hydraulic testing machine – Instron 1251 at the cross-head speed 8,3.10⁻⁶ m/s. The testing temperature was 20 °C, the magnification of deformation on the record was 100 x.

For the dynamic fracture toughness test, a fatigue crack under notch was made on the specimens by a Dynatup Precracker B device. Starting load was 3,5 kN, which was equivalent to the value K_f = 10 MPa m$^{-1/2}$ for the start of cycling. The proper tests were performed on an instrumented Tinius-Olsen pendulum impact testing machine, controlled by Eti 300 computer, equipped by a printer, and a recording unit with a pair of floppy disks. The ram impact rate was changed in the range from 5,6 m/s to 1 m/s. The test temperatures were used in the range from -100 $^{\circ}$C to + 170 $^{\circ}$C.

The all equipments were installed in special boxes, equipped with closed TV circuits.

Electron microscope examinations were realized in a Tesla Bs 540 electron microscope operated at 120 kV. The thin foils were prepared by an electrolytic thinning in a Tenupol device.

The irradiation experiments were realized in Chouca Mt rigs, placed in the VVR-S research nuclear reactor at the Nuclear Research Institute, Rez. In the rigs filled with helium atmosphere, the temperature was kept in the range of 280 - 290 $^{\circ}$C. The fluence of fast neutrons were determined from ^{63}Cu (n,α) ^{60}Co reaction and thermal neutrons from a ^{59}Co (n,γ) ^{60}Co reaction.

3 RESULTS

The microstructures of all unirradiated melts examined in a light microscope are very similar, and appear as a mixture of tempered bainite and martensite. The martensite and bainite laths were determined by a transmission electron microscopy, the granular bainite is characterized by a polyedric shape of grain. The decrease of dislocation density, the creation of dislocation network as well as subgrain formation, take place during a recovery. The differences in dislocation densities inside of the laths ($\approx 10^{14}$ m^{-2}) evaluated on both unirradiated and irradiated material are not conclusive. The dislocation loops are hardly recognizable within dislocation structure, their amount is probable a little higher after an irradiation. Carbides, precipitated both inter and intragranular, were identified by a selective electron difraction. The presence of M_7C_3, $M_{23}C_6$ and M_3C particles was found out. Neither very fine particles (≤ 20 nm) segregated on the dislocations, nor the needles within material was possible to determine by this method; carbides type of V_4C_3 and Mo_2C are supposed.

The results of tensile tests are given in table 2. The mechanical properties of all melts in starting state are practical identical. The outstanding changes after an irradiation appeared in the yield point Rp0,2, the contraction Z and tensile ductility A5. For a sensibility of individual melts evaluation to a radiation hardening appreciated by a yield point change , the values of B-constant were calculated from an equation Rp0,2 = B $(\phi . 10^{-22})^{1/3}$ (table 2). It is evident that both the radiation hardening and B value increase with an increasing of copper content, tensile ductility and a contraction decrease on the contrary. From a point of view of the radiation hardening, the worst melt contains 0,52 % Cu and 0,021 % P.

Radiation embrittlement was appreciated according to the change of transition temperature at the value 100 MPa m$^{1/2}$ of dynamic fracture toughness. The results of these tests are given for both the irradiated and unirradiated state in table 3. For the better irradiation effect appreciation, a value of A-constant was calculated from the equation ΔTT_{100} = A.$(\phi.10^{-22})^{1/3}$.

The A-constant values for the average notch toughness transition temperature shifts, determined at the identical melts

are given in this table, too. In all specimens, area of toughness
fracture plane was determined. The transition temperatures, as well
as their shifts induced by the irradiation, were determined according
to the criterion 50 % of ductile fracture in the fracture plane. The
values are also given in table 3.

The values of the transition temperatures in starting state show
considerable scatter without any dependance on the composition. The
smallest transition temperature T_{100} MPa m$^{1/2}$ shift in an irradiated
state was determined at the melts with the smallest copper content.
Radiation embrittlement shows a clear dependance of copper content,
however the effect of phosphorus is weak.

On the selected specimens, a fractographic analysis in both the unirradiated and irradiated state was performed.

At the fracture planes of all specimens, both irradiated and unirradiated, two types of fracture were found: a ductile fracture presented by the dimple fracture, and a brittle fracture with largely trans-
crystalline cleavage. No differences in fracture morfology were found
between irradiated and unirradiated states. Neither the effect of
both copper and phosphorus on the fracture morfology appeared, nor an
anomaly was found in the inclusions occurence on the fracture planes.

CONCLUSION

The following conclusions can be drawn from this study:

1/ Radiation hardening, represented by differences of yield point before and after a neutron irradiation, is negligible for copper content less than 0,08 %. A radiation hardening increases with the increasing copper content.

2/ Radiation embrittlement defined by the transition temperature shift
   on a transition curve of dynamic fracture toughness is minimum at
   the copper content less than 0,08 %.

3/ Phosphorus effect on the mechanical properties after irradiation
   is very weak - quite overlapped by a copper influence.

4/ The electron microscope examinations and the dislocations density
   measurements allow to assume that an increased hardening and embrittlement after an irradiation of steels with different copper
   and phosphorus content, is caused first of all by copper vacancy
   complexes, or by finely distributed copper-rich precipitates, formed due to a neutron irradiation.

LITERATURE

Potapovs U., Hawthorne J.R. 1967. Nucl. Appl., 6, 27.
Hawthorne J.R. et.a. 1975.
  ASTM STP-570, 83.
Smidt P.A.Jr., Sprague J.A. 1973.
  ASTM STP-529,78.

Table 1. Chemical composition of the experimental melts of CrNiMoV Steels

Melt	C	Mn	Si	P	S	Ni	Cr	Mo	Cu	V
1306/1	0,15	0,48	0,17	0,012	0,010	1,28	2,06	0,56	0,30	0,10
1306/2	0,15	0,48	0,17	0,018	0,010	1,28	2,06	0,56	0,30	0,10
1308/1	0,15	0,34	0,21	0,015	0,009	1,28	1,89	0,55	0,52	0,09
1308/2	0,15	0,34	0,21	0,021	0,009	1,28	1,89	0,55	0,52	0,09
1310/1	0,16	0,34	0,20	0,014	0,009	1,27	2,14	0,58	0,08	0,10
1310/2	0,16	0,34	0,20	0,021	0,009	1,27	2,14	0,58	0,08	0,10

Table 2. Results of static tensile tests

Melt	$\varnothing \cdot 10^{23}$ /n/m²/ E 1MeV	RpO,2 /MPa/	Rm /MPa/	RA /MPa/	A_5 /%/	Am /%/	Z /%/	$B = \dfrac{R_{p0,2}}{(\varnothing \cdot 10^{-22})^{1/2}}$	B from /4/
1306/1	–	540	648	1276	19,5	6,7	70	–	–
	2,2	701	799	1329	17,5	7,5	61	57	90
1306/2	–	534	670	1325	19.9	6,9	71	–	–
	2,2	756	833	1361	16,3	6,9	59	79	84
1308/1	–	520	631	1357	20,3	7,1	72	–	–
	2,2	731	802	1406	16,7	6,6	63	75	96
1308/2	–	528	635	1261	19,3	7,0	70	–	–
	2,6	726	797	1414	16,9	7,4	62	67	85
1310/1	–	538	642	1380	18,8	6,4	72	–	–
	2,6	632	718	1401	17,5	6,6	69	32	33
1310/2	–	542	648	1293	17,4	6,4	65	–	–
	2,6	652	735	1313	16,7	7,0	64	37	44

Table 3. Results of dynamic fracture toughness

Melt	$\emptyset.10^{23}$ /n/m²/ E 1MeV	TT 100MPa 1/2	TT 50%	TT 100MPa m 1/2	TT 50%	TT(°C) average	$A=\dfrac{TT}{(\emptyset.10^{-23})^{1/3}}$	A from /4/
1306/1	–	+2	-10	–	–	–	–	–
	2,2	+116	+95	114	105	110	41	43
1306/2	–	+23	+ 5	–	–	–	–	–
	2,2	+131	+95	108	90	100	39	44
1308/1	–	- 35	-35	–	–	–	–	–
	2,2	+ 85	+85	121	120	120	43	48
1308/2	–	- 44	-42	–	–	–	–	–
	2,6	+101	+103	148	145	147	50	43
1310/1	–	- 32	- 15	–	–	–	–	–
	2,6	+ 10	+ 21	42	36	39	14	14
1310/2	–	+ 8	- 12	–	–	–	–	–
	2,6	+ 54	+ 58	46	70	58	16	18

Influence of the chemical composition of RPV-steels on radiation embrittlement

J.Ahlf, D.Bellmann & F.J.Schmitt
Institut für Werkstofforschung, GKSS-Forschungszentrum Geesthacht, FR Germany
J.Föhl
Staatliche Materialprüfungsanstalt, Universität Stuttgart, FR Germany

1 INTRODUCTION

As a contribution to the research program "Integrity of Components" a number of special melts of steel 22NiMoCr37 (corresponding to ASTM A508 cl.3) have been fabricated and investigated.

These melts were originally tailored to investigate effects of chemical composition on the sensitivity to stress relief cracking during post weld heat treatment, see Kussmaul et al. (1986). For this reason the content of some of the residual elements was intentionally varied, and did not always meet the stringent demands of the appropriate regulatory requirements.

It was considered worthwile to investigate the irradiation response of these melts, too, since the influence of phosphorus content on irradiation embrittlement is not yet well established. Potapovs and Hawthorne (1969) found a significant post-irradiation transition temperature increase when phosphorus (and sulfur) were added to a steel with low content of residual elements, and more recently Hawthorne (1985) demonstrated a significant enhancement of the embrittlement when the phosphorus content is varied from .003 to .025% in a low copper (<.03%) steel, whereas this effect was small in a high copper (.30%) steel. On the other hand, the Metal Properties Council (1983) concluded from its study that "an independent effect of phosphorus on irradiation behavior could not be supported". Whereas phosphorus is explicitly taken into account in USNRC Reg. Guide 1.99 rev. 1, this is no longer the case however in the proposed revision 2.

As for the influence of sulfur on irradiation embrittlement, Potapovs and Hawthorne (1969) found that .02% sulfur added to a low copper melt (<.005%) led to the known reduction of upper shelf energy in the unirradiated state but did not lead to a significant post-irradiation transition temperature shift or upper shelf reduction. More recently, Hawthorne (1981) stated that a high sulfur content (.029%) appears to reduce the detrimental effect of a high copper content. From the investigation of welds, Fisher et al. (1984) concluded that this could be due to the fact that copper bound in copper sulfide is not available for forming irradiation induced copper precipitates.

2 MATERIAL CHARACTERIZATION

Melting and casting procedure of the materials correspond to production conditions for heavy ingots. Each of the selected 8 heats was taken from a basic melt of about 20 t weight (basically 22NiMoCr37 steel) and was adjusted in chemical composition according to the specified values. Each ingot was cast with a weight of about 5000 kg without degassing.

After casting, the ingots were forged to bars of 300 mm thickness (forging ratio 2.5), corresponding to the initial thickness of forged rings for large reactor pressure vessels. The material was forged main-ly in one direction leading to strong anisotropy in toughness depending on the impurities and segregations in the steel. From this point of view the materials represent a lower bound material state with respect to real reactor pressure components for which the chemical composition, the melting and forging procedure have been optimized to achieve high toughness combined with low anisotropy. After a post forging heat treatment the bars were quenched in water, tempered for 12 h at 640 °C, and then air cooled so that a fine grained bainitic microstructure was achieved.

All specimens for irradiation and unirradiated controls were taken from 1/4 T to exclude near surface inhomogeneities resulting from quen-ching.

The chemical composition of the material was determined adjacent to the specimens at six positions by means of optical spectroscopy. The mean values are compiled in table 1, together with the requirements ac-cording to VdTÜV 365(4.72) and a restricted analysis which represents an updated requirement for reactor components.

Table 1: Chemical composition in wt% (mean values from 6 locations) in comparison with requirements for nuclear grade steel 22NiMoCr37

heat	C	Si	Mn	P	S	Cr	Ni	Mo	V	Al	Cu	W	Co	Nb	Sn	Ti	As	Pb	Sb
Specified																			
VdTÜV	.17	...	.50	...	...	.30	.60	.50	...	...	...	...	...	...	...	...	...	...	...
365(4.72)	.25	.35	1.00	.025	.025	.50	1.00	.80	.05	.05	.20	...	...	...	...	...	...	...	...
Restricted	≤.20	.20	.85	≤.008	≤.008	≤.40	1.20	≤.55	≤.01	.01	≤.10	...	≤.003	...	≤.01	...	≤.015	...	≤.005
KS16S	.21	.28	.72	.006	.006	.48	.74	.52	.04	.01	.11	.01	.01	<.01	.01	<.01	.01	<.01	<.01
KS16K	.17	.37	.76	.005	.004	.41	.76	.58	.01	"	.20	"	"	"	.01	"	"	"	"
KS16H	.19	.25	.77	.004	.007	.41	.76	.58	.01	"	.20	"	"	"	.02	"	"	"	"
KS16M	.20	.21	.71	.004	.015	.41	.76	.58	.01	"	.20	"	"	"	.01	"	"	"	"
KS16G	.22	.27	.69	.015	.005	.42	.74	.55	.02	"	.19	"	"	"	.02	"	"	"	"
KS16E	.20	.27	.74	.022	.012	.43	.75	.58	.01	"	.21	"	"	"	.02	"	"	"	"
KS16D	.21	.24	.71	.018	.013	.44	.75	.66	.01	"	.21	"	"	"	.02	"	"	"	"
KS16C	.20	.18	.73	.020	.012	.44	.75	.72	.01	"	.21	"	"	"	.02	"	"	"	"

To assess the irradiation response of the materials, tensile and Charpy-V-notch specimens were used. For two of the eight heats (C and G) specimens were taken from the transverse (T-L) and longitudinal (L-T) direction, whereas from the other heats only longitudinal speci-mens were used.

As can be seen from the data in table 2, the 41 J transition temperature for the unirradiated state ranges from -20°C to +30°C and the
upper shelf energy ranges from 80 J to 170 J. As a consequence of production conditions and residual element content not all melts meet
nuclear grade requirements.

3 EXPERIMENTAL PROCEDURES

The irradiation experiment was performed in a capsule with large specimen volume, see Ahlf et al. (1983), in a border position of the 15 MW
swimming pool type research reactor FRG-2. To reduce gamma heating in
the specimens a tungsten gamma shield 47 mm thick was placed between
the reactor and the capsule. The irradiation temperature was measured
with chromel-alumel thermocouples and controlled with electric heaters.
The average irradiation temperature was 290 ± 3°C.
    The neutron exposition was determined from the reaction Fe54(n,p)Mn54
by evaluating the activation of monitor wires spanned over the specimen
volume inside the irradiation capsule, see Bellmann et al. (1982). The
evaluation was based on neutron spectra obtained from two-dimensional
transport calculations with DOT IV. The dosimetry cross sections were
taken from ENDF/B-V. The spectrum averaged cross section for the expo-
sition in units of dpa was calculated using the differential cross
sections given in ASTM E693-79. Fluence variation within a specimen set
was less than 5%. The average flux density was $3 \cdot 10^{16} m^{2} s^{1}$. The dpa to
fast fluence (E>1MeV), ratio is dpa $[\%]$/fluence$[10^{23} m^{2}]$= 2,14.
    The tensile tests were performed on a 250 kN Instron testing machine,
the Charpy tests on an instrumented 300 J pendulum. The transition
curves for the impact energy and the lateral expansion were approximated with the Gaussian integral.

4 RESULTS AND DISCUSSION

The initial and post-irradiation yield strength and hardness at room
temperature are plotted in figure 1. The yield strength increase is
less than 15% for steel S (low Cu, P and S)and nearly double as high

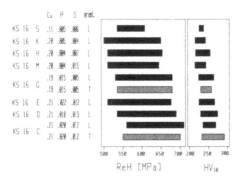

Figure 1. Increase of yield
strength (R_{eH}) and hardness
(HV_{10}).

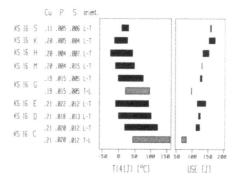

Figure 2. Increase of 41 J transition temperature and change of
upper shelf energy.

(30 ± 3 %) for all the other steels. Also the hardness increase is
significantly lower (only 5%) for steel S than for the other steels (14
to 25%). This strongly indicates that yield strength and hardness increase on irradiation are predominantly influenced by the copper content (.11% for steel S; .19 to 21% for the other steels) whereas an
influence of phosphorus and sulfur content is not discernible for the
steel compositions investigated.

With respect to the Charpy impact properties given in table 2 and figures 2 and 3, the melts can be subdivided into groups. The clean low
copper melt S shows little 41 J shift and no upper shelf reduction.
Doubling the copper content with phosphorus and sulfur kept low (melts
K and H) leads to enhanced embrittlement which is more pronounced for

Table 2: Results of the Charpy impact tests

heat	orientation	fluence (10²³m⁻²)	T(41J) (°C)	T(68J) (°C)	T(0,9mm) (°C)	U.S.E. (J)
KS16S	L-T	0	8	19	16	156
		2,54	29	40	40	159
KS16K	L-T	0	-14	-3	-9	171
		2,54	26	45	46	152
KS16H	L-T	0	-26	-8	-14	152
		2,55	42	62	63	133
KS16M	L-T	0	-11	-6	-7	130
		2,50	48	72	71	131
KS16G	L-T	0	-2	9	2	131
		2,47	76	108	108	125
KS16G	T-L	0	20	39	30	98
		2,23	96	149	131	97
KS16E	L-T	0	-10	9	2	143
		2,50	93	130	128	116
KS16D	L-T	0	-1	16	11	126
		2,47	102	147	139	120
KS16C	L-T	0	18	38	32	124
		2,41	121	162	152	112
KS16C	T L	0	43	81	56	82
		2,32	161	-	200	67

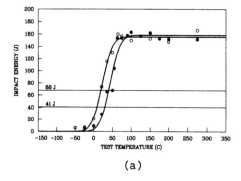
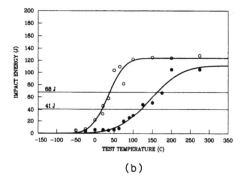

(a)   (b)

Figure 3. Impact energy curves for the unirradiated and the
irradiated state for (a) KS16S (L-T) and (b) KS16C (L-T).

steel H than for steel K for a yet unknown reason. Addition of sulfur
(melt M) does not further increase the 41 J shift but leads to a significant decrease of the slope of the transition curve as reflected by
the high 68 J shift. Addition of phosphorus with sulfur kept low (melt
G) leads to an increase in the 41 J shift. Melts C, D and E which have
similar chemical composition with a further increased phosphorus content and rather high sulfur content show the highest 41 J shifts and a
significant decrease of the slope of the transition curves. It is
clearly shown that an increased concentration of phosphorus and sulfur
leads to significantly enhanced embrittlement. Although it is difficult
to separate the individual contributions of phosphorus and sulfur, the
former seems to play the dominant role.

It is noted that the transition curves of lateral expansion are
shifted much more than the impact energy curves at 41 J. In most cases
the .9 mm lateral expansion shift is nearly equal to the 68 J shift.
There is not much difference in the 41 J shift and shape of the transition curves between T-L and L-T orientation for melts G and C apart
from a lower upper shelf energy in the transverse orientation. The
reduction of upper shelf energy varies from 0 to 19% with no apparent
correlation with the chemical composition.

Following the analysis of Hiser (1985) the measured 41 J shifts were
compared to the predictions of different correlation models and the
trend lines of US NRC Reg. Guide 1.99 rev. 1 and prop. rev. 2 the
latter based only on surveillance results from power reactors. The
result is shown in figure 4. Since most correlation models attribute
the embrittlement almost entirely to the action of copper (and more
recently additionally to nickel which was constant in the heats
investigated) they do not reflect the contribution of phosphorus and
perhaps sulfur and give rather poor correlation. Remarkably the only
exception is Reg. Guide 1.99 rev. 1 which explicitly presumes a
phosphorus effect. It gives a very conservative prediction of the shift
but it shows the correct trend.

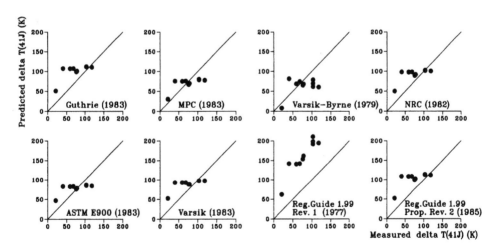

Figure 4. Comparison of the measured 41 J transition temperature
shifts with different correlation models and trend curves.

5 CONCLUSIONS

Eight special melts similar to the German RPV steel 22NiMoCr37 (corresponding to ASTM A508 cl.3), seven of which had an intentionally high copper content (.2%) and varying phosphorus and sulfur contents, were irradiated at 290°C to a fluence of $2,5 \cdot 10^{23} \text{m}^2$, E>1MeV. It was shown that the irradiation induced embrittlement as determined from the 41 J Charpy shift was strongly enhanced by increasing the phosphorus and sulfur concentration, phosphorus presumably playing the main role. Comparing different correlation models for the transition temperature shift the trend curves of USNRC Reg.Guide 1.99 rev. 1 give the best correlation, however on a very conservative level.

REFERENCES

Ahlf, J., D. Bellmann and H. Martens. 1983. Operating Experience with Capsules for RPV Steel Irradiations with Large Specimen Volume. P.v.d.Hardt and H. Röttger (eds), Irradiation Technologie, Brüssel, 593-603.

Bellmann, D., J. Ahlf, P. Wille and G. Prillinger. 1982. Neutron Dosimetry in Irradiation Capsules for Large RPV Steel Specimens. NUREG/CP-0029, Vol. 1, 579-585.

Fisher, S.B., J.E. Harbottle and N.B. Aldridge. 1984. Microstructure Related to Irradiation Hardening in Pressure Vessel Steels. Proc. of the Conf. on Dimensional Stability and Mechanical Behaviour of Irradiated Metals and Alloys. Brighton, 11-13 April 1983.

Hawthorne, J.R. 1981. Significance of Selected Residual Elements to the Radiation Sensitivity of A302B Steel. Trans. Amer. Nucl. Soc. 38: 304-305.

Hawthorne, J.R. 1985. Composition Influences and Interactions in Radiation Sensitivity of Reactor Vessel Steels. Nucl. Eng. and Design 89: 223-232.

Hiser, A.L. 1985. Correlation of C_v and K_{Ic}/K_{Jc} Transition Temperature Increases Due to Irradiation. NUREG/CR-4395.

Kussmaul, K., W. Schellhammer. 1986. Rißempfindlichkeit, Verformungs- und Bruchverhalten der wärmebeeinflußten Zone von dickwandigen Schweißverbindungen hochfester Feinkornbaustähle. Kom. d. Europ. Gemeinsch.: Techn.Forschg. Stahl, Bericht EUR 9890 DE.

Metal Properties Council Subcommittee 6 on Nuclear Materials. 1983. Prediction of the Shift in the Brittle-Ductile Transition Temperature of Light-Water Reactor (LWR) Pressure Vessel Materials. J. of Testing and Eval. 11: 237-260.

Potapovs, U. and J.R. Hawthorne. 1969. The Effect of Residual Elements on the Response of Selected Pressure-Vessel Steels and Weldments to Irradiation at 550°F. Nucl. Appl. 6: 27-46.

Improvement of valve lift characteristic of a main steam isolation valve to reduce pipe loading

E.Brehm

Kraftwerk Union AG, Offenbach, FR Germany

Introduction

Safe design of piping systems in nuclear power plants requires the demonstration that the piping will withstand all possible static and dynamic loads during normal operation or accidents. Among the forces causing
such loads are the fluiddynamic forces due to operational or accidental
transients. They result from rapid changes of the flow conditions in the
piping system. Especially fast valve opening or closing can create
large pressure changes leading to significant shock loadings. Not only
should these loads be kept below the specified values, aspects of care-
fuel treatment of plant during longterm operation demand loads as low
as possible too.
Such aims call for a precise tuning of the valve lift characteristic to
the requirements. Often this means an inprovement of an already existing
valve construction.
For valves which are installed in the plant, such an improvement or
tuning has to be done without extensive testing. With only few tests one
can get certainty about the functionability of the valve and the effectiv
ness of improvement measures only by mathematical modelling of the
piping system and the valve.
Numerical calculations then show the influence of possible improvements
on the valve lift characteristic and the system's response.
Such a procedure for reducing fluiddynamic pipe loads caused by closure
of self-controlled isolation valve in the main steam line of a KWU BWR is
presented in the following.

Fluiddynamic Loads Generation by Valve Closing

Every closure of a Main Steam Isolation Valve (MSIV) causes fluiddynamic
forces. These forces originate from pressure and mass flow changes in the
piping system. They depend on the reactor power and rise with initial
steady state flow rate through the steam line, i.e. the flow to the turbine.
Fig. 1 shows schematically the main steam line from the reactor pressure
vessel to the turbine valves. Upon demand, the MSIVs isolate the main
steam path out of the containment by valve closure.
During the initial stage of valve closure the flow in the pipeline is
only influenced by a slightly growing flow resistance of the valve. This
changes, when the flow area in the MSIV becomes flow-limiting. Then

359

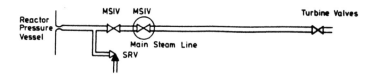

Figure 1. System model of a main steam line

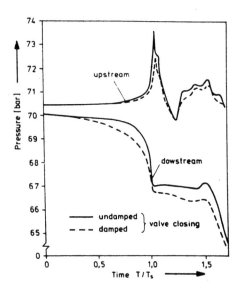

Figure 2.
Pressure histories in main steam
flow upstream and downstream
MSIV during valve closing

flow rate: 60% of nominal flow,
time normalized by valve closing
time

significant throttling of the flow in the valves reduces the mass flow
through the valve to below the value of the flow into the turbine.

As a consequence, pressure downstream decreases while upstream it grows
(see Fig. 2). This pressure difference is intensified when the pressures
reach the critical pressure ratio and the flow velocity in the valve
attains the speed of sound. Flow rate through the valve then is no
longer influenced by decreasing pressure downstream, meaning that the
growing pressure difference accross the valve cone cannot even partly
be compensated by higher flow velocity. This pressure difference
creates an additional dominant closing force accelerating the valve
cone rapidly during the final time span of cone motion. As this force
grows with reactor power, the closing time of the valve decreases
with power (Fig. 4) while impact velocity increases (Fig. 5), producing an unnecessary mechanical impact load at the valve itself.

Load Reduction

Reducing fluiddynamic loads requires to modify the valve lift characteristic during the final stage of cone motion by diminishing the resulting closing force. This can be done by an additional hydraulic damping
of the valve lift which is easily installed at the MSIVs in use.
Figure 3b shows the principle design of such a measure.

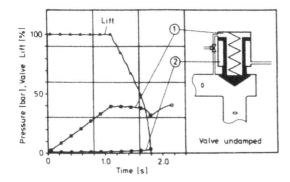

Figure 3a.
Calculated valve lift history and pressure histories in the control chamber ① and the lower piston chamber ② for undamped valve closing at nominal flow rate

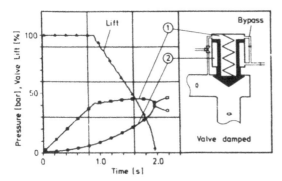

Figure 3b.
Calculated valve lift history and pressure histories in the control chamber ① and the lower piston chamber ② for damped valve closing at nominal flow rate

As the MSIV works according to the loading principle in case of valve closing a defined bypass allows steam to intrude from the upper control chamber to the lower valve piston chamber which normally is exhausted.
If steam outflow of this chamber is throttled, valve closing causes compression of the steam inducing a force directed against cone motion and thus lowering valve lift velocity.
The effectivness of this measure can be optimized by tuning of the three throttles or orifices - pilot valve, bypass throttling and exhaust orifice.
Optimizing the damping function requires a great number of calculations with different parameter combinations and boundary conditions. Thereby one has to obey restrictions like specified maximum closing times in case of pipe breakes and sufficient fast closing even at low flow rates.

Numercial Modelling

Isolation valve closure stops steam flow in the main steam line, leading to rapidly changing flow conditions which influence valve closing itself. So valve cone motion and fluid flow interact with each other giving a strong coupling between MSIVs and their system environment. Mathematically this means that the differential equations describing cone motion and fluid flow form a nonlinear coupled system of equations. Therefore numerical calculation of the valve lift history requires not only modelling the MSIV itself (see Fig. 3), but the whole main steam line from the reactor pressure vessel to the turbine (Fig. 1) has to be described sufficiently precise in a computer model.

361

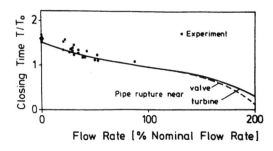

Figure 4.
Dependence of valve closing
time on initial flow rate in
the main steam line for
undamped valve closure
Comparision of calculation
and experiment

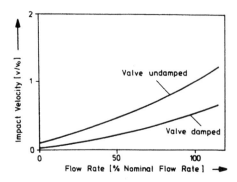

Figure 5.
Dependence of valve cone impact
velocity on initial flow rate in
the main steam line

Calculations were performed with the thermalhydraulic computer program
2PHI1K for one-dimensional flow in pipe line networks using SUPVERA,
a program for calculation of the lift history of selfcontrolled valves,
as a submodel.

This code combination simultaneously solves the equations of homogenous
two-pase flow including pure steam for the flow in the piping, determines all forces influencing valve lift and calculates valve lift
history by solving the equation of motion for the valve cone.
Some results of such a calculation for an undamped valve are shown in
Fig. 3 a. They clearly demonstrate the increase of lift velocity during closure.
Calculated closing time dependence on flow rate was verified by
experimental data (see Fig. 4) in the range of operational mass
flow, i.e. up to 100 % nominal flow rate. In case of pipe rupture,
the mass flow through the main steam line is limited by the flow limi-
ter at the pressure vessel to a value of 200 % nominal flow. Valve closing characteristics then depend on the location where rupture occurs.
The calculated lift histories for such accidents agree well with the
HDR-blow-down experiments /1/.

Results

Calculation of an improved valve shows a significant pressure rise in
the lower piston chamber of the valve during the final stage of cone
motion (cf Fig. 3a and 3b) which gives a sufficient damping force reducing the impact velocity to nearly 50 % of the undamped value
(see Fig. 5). In comparision to undamped closure (Fig. 3a), the gra-
dient of the cone velocity is less steep (Fig. 3b).

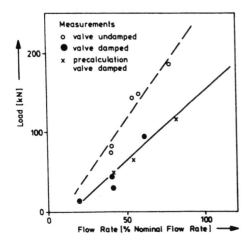

Figure 6.
Measured and calculated maximum
pipe loads at a shock absorber
during undamped and damped MSIV
closing
(straight lines by linear
  regression analysis)

Lower velocities during the final stage of cone motion can be reached without significant increase in total closing time by a precise
tuning of the throttles and orifices (cf Fig. 3a and 3b).
As a consequence of moderate cone motion, the pressure peak ahead and
the steepness of pressure decrease behind the valve are reduced too as
experiments show (see Fig. 2, where the times for damped and undamped valve closing are normalized so that time 1.0 in both cases represents complete valve closure).
During undamped valve closure the pressure peak upstream reaches about
70 % of its theroretical maximum possible value due to the Joukovksi
formula, while for an improved valve this value stays below 46 %.
Calculated load reductions were verified by force measurements at the
piping. One of the control points was a shock absorber of the main
steam line downstream near the MSIVs. The maximum measured values of
forces at this point for undamped and damped valve closing is shown
in figure 6, where in addition the precalculated load values for
damped closure are marked. They agree very well with the measurements. As figure 6 demonstrates, this simple valve improvement reduces
loads of the piping up to about 40 %.

Conclusion

In operating BWR's the closure of stream valves results in loads on
piping and components. These loads often can drastically be reduced
by slightly modifying and/or fine tuning these valves.
A computer program has been development and tested to optimize the
valve design and performance. Applying the results of precalculations valves can easily be adjusted so to reduce testing time and
reactor outage.

References

/1/   HDR Sicherheitsprogramm, Blowdown Versuche,
         Versuchsgruppe DIV 1
         GfK, 1977

Modelling methods in stress analysis of pipe coupling clamps

B.K.Dutta, H.S.Kushwaha & A.Kakodkar
Reactor Engineering Division, BARC, Trombay, Bombay, India

1 INTRODUCTION

Pipe coupling using clamps are becoming more and more popular components in nuclear power plants and are used in place of conventional flanges because of their compactness, easy main-tenance & more reliability. They are used in large numbers in Pressurised Heavy Water Reactors (PHWR) such as at the joint between feeder pipe and End-fitting, in F/M housing etc. Integrity of these clamps have direct effect on over-all safety of the nuclear power plants. This necessiates proper design, fabrication, installation & maintenance of these components.

Proper design of these clamps is a challenge to the design-er. This is because of changing boundary conditions at the interface with the hub during various stages of loading. A detail stress analysis of clamps considering changing bound-ary conditions under various loadings can be done using fini-te element technique. In the following sections, two finite element modelling methods to simulate clamps along with hubs are described. Both these methods assumed absence of friction between the clamp and hubs during bolting, whereas absence of relative movement between them was assumed during other stages of loadings.

2 MODELLING METHOD-1

This modelling method is applicable to axisymmetric hubs under different axisymmetric external loads (other than bolt loads). In this method a complete 3-D discretization of hubs and clamps together is avoided though bolt loads on the clamps are asymmetric. Two finite element models are used in this method. In one model one-fourth of the clamp is discretized along with radial springs at the interface to simulate hubs. Whereas in the second model hubs with the clamp is discretized by axisymmetric elements. Using these two models together, the stress picture in hubs and clamp can be computed for bolt and all other axisymmetric loads as explained below.

The 3-D model of the clamp is first analysed for bolt

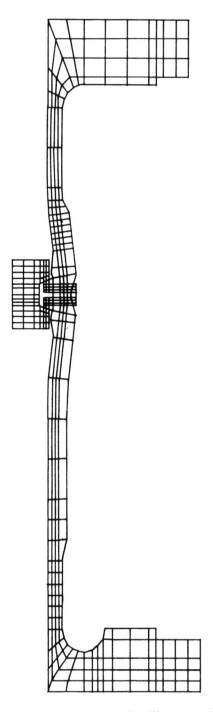

Fig 1.- Deformed shape of the fuelling machine housing
due to bolt load

load alone. The radial deformation of the clamp is generally found to be independent of the angle because of a slot provided to obtain uniform compression on the hubs. This analysis also generates stress picture in the clamp body due to bolt load. The uniform radial displacement computed from this analysis is then used in the axisymmetric model to obtain equivalent axisymmetric radial load, which provides the same radial displacement. This load when applied alone in the axisymmetric model generates stress distribution in the hubs due to bolt load. This load is then used along with other axisymmetric loads to obtain final stress distribution in the hubs. To obtain stress distribution in the clamp due to axisymmetric loadings on the hubs, displacement at the hub surface is calculated due to external loadings only. Equivalent load corresponds to this displacement along with bolt load is then applied to 3-D model of the clamp to compute stress distribution.

This method of analysis has been used to calculate stress distribution in the clamp hub assembly of pressure housing of fuelling machine of 500 MWe pressurised heavy water reactor (Dutta, 1985). The 3-D model of clamp is first analysed for bolt load of 47000 kgf per bolt. The uniform radial deformation of 0.13 mm of hub surface is obtained from this analysis. The equivalent radial load which gives 0.13 mm radial displacement at the hub surface in axisymmetric model is found to be 500 kgf/m.m. of circumference. This load is then used in the axisymmetric model along with other loads to compute stresses in hubs. Fig.1 shows the deformed shape of the pressure housing due to bolt load. Stress intensities obtained by adopting this method of analysis for the present case are quoted in paper F 7/8 of the present conference.

3   MODELLING METHOD-2

This modelling method is applicable to any general shaped hubs under different loadings. The analysis is done on a 3-D model of hub-clamp assembly. To consider zero friction between clamp and hub "fluid elements" are used at the interface. The properties of these elements are so adjusted that as young's modules tends to zero, poison's ratio tends to 0.5 and bulk modules remains same. With this approach any standard solid mechanics computer code can be used for this purpose. While considering the other loads, if assumption of zero friction at the interface is made, results obtained are conservative for the hub, but not so for the clamp. The other extreme assumption of no relative movement between the two surfaces though computes conservative results for the clamp but not for the hub. Hence it is advisable to compute results separately by making both the assumptions. To consider no relative movement between the two surfaces, the fluid elements can be converted to solid elements by restoring their mechanical properties.

This analysis method is used to analyse high pressure pipe couplings (Dutta, 1986). A 3-D model of the pipe coup-

367

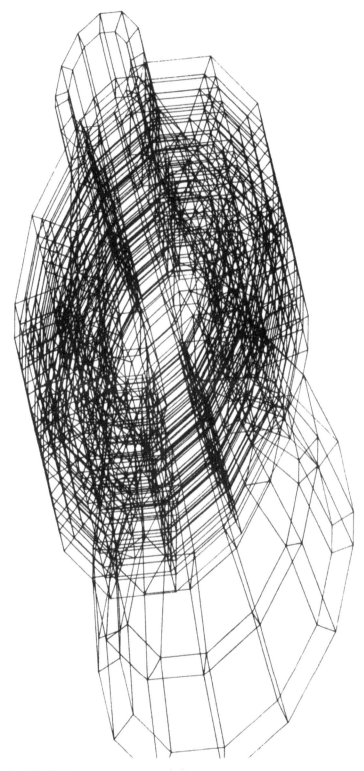

Fig 2. 3D-Finite element model of feeder coupling for KAPP.

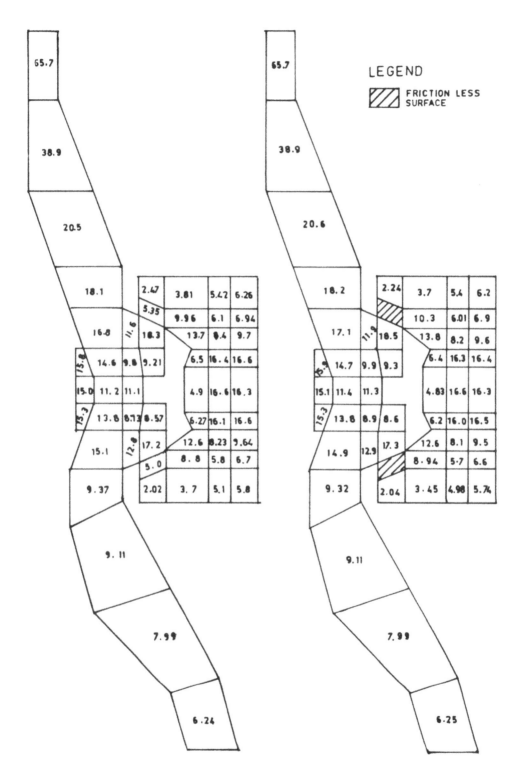

Fig.3 Stress intensities for different interface condition for level- C.

lings is made (Fig 2). The model is analysed for various
loading cases. Results obtained for a typical loading case
for service level C are shown in Fig.3. For this case no
significant difference in stress intensities is obtained
for the assumptions of zero friction and no reletive movement between interfaces.

REFERENCES

Dutta, B.K., Ramana, W.V., Kushwaha, H.S. & Kakodkar, A. 1985
   Stress analysis of pressure retaining components of fuell-
   ing machine of 500 MWe nuclear reactor. Ref.RED/AK/86/257.
Dutta, B.K., Ramana, W.V., Kushwaha, H.S. & Kakodkar A. 1986
   A report on Stress analysis of high pressure pipe couplings
   for KAPP. Ref: RED/AK/1393/86.

Stress indices for nonradial branch connections subjected to out-of-plane moment loadings

P.P.Raju

Sudbury, Mass., USA

1 ABSTRACT

The purpose of the subject paper is to disseminate the results and fin-
dings of the Pressure Vessel Research Committee(PVRC) Phase 2 program
relative to the development of stress indices for 45 degree lateral connections under out-of-plane moment loading applied on the branch pipe.

The objective of the earlier Phase 1 program was to develop stress indices for lateral connections under three load cases namely, internal
pressure, and in-plane moment loadings on the branch and the run pipe
(or the vessel). A complete discussion of the results and conclusions
of the Phase 1 program and the recommended indices for the design of
lateral connections in vessels and piping can be found in WRC Bulletin
301 (Reference 1). The purpose of the Phase 2 program described herein
is to ascertain the relative magnitudes of the in-plane and out-of-plane moment indices through selected parametric studies and recommend
comprehensive C_2K_2 indices for use with equations in NB-3650, Section
III of the ASME code and for the design of nozzles per NB-3338 Stress
Index Method. The Phase 2 parametric studies performed on PVRC Models
2and 4 (d/D=0.5, D/T=10,40) indicate that the out-of-plane moment index
could far exceed the in-plane index and the use of in-plane index to
envelop the resultant moment in piping analysis could be unconservative.

2 INTRODUCTION

At the conclusion of the Phase 1 of the PVRC program on 45 degree lateral connections that resulted in the development of stress indices for
internal pressure and in-plane moment loadings, the representatives of
the fitting manufacturing industry felt that generating stress indices
for out-of-plane loading on the lateral branch would be a worthwhile
PVRC contribution to the technical literature and that piping designers
could greatly benefit by these data.

A subsequent literature search revealed that data, both theoretical
and experimental, on stress indices for out-of-plane moment on lateral
(nonradial) branch connections were scarce and the limited data available were conflicting and inconclusive from the view point of the relative magnitude and the trend of the index in comparison with the in-
plane moment index. As a result, the PVRC launched the Phase 2 parametric

study to develop and recommend stress indices for out-of-plane moment
loading for incorporation into various ASME codes and comparison with
in-plane moment indices so that the relative impact can be assessed in
the structural integrity evaluation of branch connections.

For the Phase 2 study, it was decided to select two of the four models
already analyzed for in-plane moment loading under the Phase 1 program.
Models 2 and 4 with d/D=0.5 and D/T of 10 and 40 respectively were chosen for the reported study. The geometric parameters of these two models
were found to nearly represent the configuration used in nuclear and
petrochemical applications.

The purpose of the subject paper is to present and discuss the relative magnitude of the in-plane and out-of-plane moment indices based on
the selected parametric studies and the proposed piping branch connection
and vessel nozzle indices for use with the Stress Index Method permitted
by Section III of the Boiler and Pressure Vessel Code.

A detailed discussion of (1) the technique and the procedure used to
generate the 3-D finite element model, (2) the simulation of the out-of-
plane moment through weighted nodal forces, and (3) antisymmetric boundary conditions is presented to promote similar parametric studies by
the pressure vessel and piping industry.

The summary of results includes both in-plane and out-of-plane stress
indices for critical regions in order that an accurate and comprehensive
evaluation of nonradial branch connections could be made in the future.

3 MODEL CONFIGURATIONS

The geometry of Model 2, along with coordinate axes and the edge boundary conditions for the in-plane moment loading is shown in Figure 1.
The comparable dimensions for Model 4 are as follows. The inside diameter D of the main vessel (or run pipe) is 30 inches (25.4mm/inch) and
the wall thickness T is 0.75 inch. The lateral branch or nozzle inter-
sects the main vessel or run pipe at an angle of 45 degrees. The inside
diameter d of the nozzle or branch is 15 inches and the wall thickness
t_n at the reinforced section of the nozzle is 2.60 inches. The wall thickness t of the branch pipe is 0.375 inch. The inside corner radius
r_1 and the outside fillet radius r_2 are 0.375 and 0.75 inch respectively. The radius r_3 at the transition between the branch pipe and the
reinforced section of the nozzle is 1.575 inches.

The intersection of the axes of the main vessel and the lateral nozzle
is chosen as the global origin for both the models. The boundaries of
Model 4 are extented to a length L_1 of 45 inches to the right of the
global origin and a length L_2 of 15 inches to the left of the global
origin. These lengths amount to a boundary layer of $4.0(RT)^{\frac{1}{2}}$ or greater
from the acute and obtuse corner regions and are considered adequate
to attenuate the edge effects. A length L_3 of 45 inches is considered
along the lateral axis for Model 4. This provides ample length beyond
Section A-A (Figure 1) to attenuate the edge effects toward the free
end. All the parameters discussed above are shown in Figure 1 for Model
2.

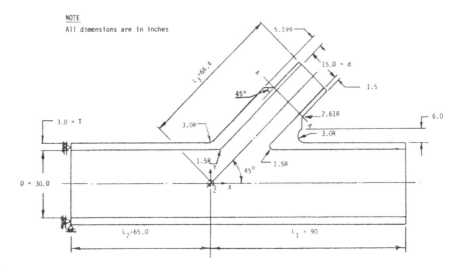

NOTE
All dimensions are in inches

Figure 1 PVRC Model 2. Geometry

4 3-D FINITE ELEMENT MODEL

A detailed discussion of the technique and procedure used to generate the 3-D finite element model is presented in Reference 1. A brief description is, however, presented herein to make this paper a stand alone document.

The finite element model was developed using the 3-D isoparametric solid element (STIF 45) of the general purpose ANSYS computer program (Reference 2). This element is defined by eight nodal points having three degrees of freedom at each node. Since the model is symmetric about the longitudinal X-Y plane (Refer to Fig.1), a half symmetric finite element model is generated to reduce the number of elements to be used in the computation. The ANSYS node and element generator (PREP7) is used in generating the models. This generator is an extended capability version of the node and element generation capability in the ANSYS program. PREP7 calculates points along the intersection of surfaces. Selective plots and printouts may be produced as the model is generated. Model editing features are also availabe in PREP7.

The finite element models are generated in several segments and merged at the end. In all, the half symmetry finite element mesh of Models 2 and 4 consists of 2600 and 2208 isoparametric solid elements respectively. As a minimum, three elements through the thickness are used in all regions except the lateral branch pipe where two elements through the thickness are considered sufficient. In the circumferential direction of the branch, the model consists of one element every 10 degrees and the main vessel is modeled with one element every 10 degrees upto 90 degrees and one element every 15 degrees between 90 and 180 degrees. In the critical acute crotch region, a row of seven elements is used to obtain a smooth stress profile. For the innermost elements around the crotch region (from 0 to 180 degrees) a twenty seven integration points

373

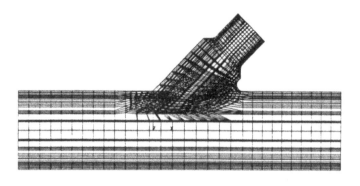

Fig.2 PVRC Model 2. Front View

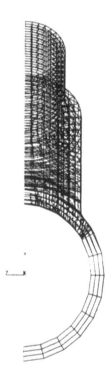

Figure 3 PVRC Model 2 Side View

printout option is selected. A discussion on the basic concepts and formulation of isoparametric elements is included in Reference 3.

Since the wave front size is dependent upon the order in which the elements are input, the elements generated are reordered to decrease the wave front size. Two finite element plots of Model 2 are presented in Figures 2and 3. Different viewing vectors are used in generating these plots. Detailed sectional plots can be found in WRC Bulletin 301.

5 MATERIAL PROPERTIES

For the elastic analysis of the lateral models reported herein, a modulus of elasticity of 30×10^6 psi and a Poisson's ratio of 0.3 are used.

6 APPLIED LOAD

In the current evaluation, an out-of-plane moment that would cause a nominal stress of 1000 psi at the loaded edge is applied over the branch pipe in terms of nodal forces. The equivalent nodal forces are calculated using the following procedure and applied on all 57 nodes.

6.1 Calculation of nodal forces-Model 4

Nodes are spaced at 10 degree interval in the circumferential direction, as previously mentioned. Three nodes are placed in the thickness direction.

   Total number of nodes = 57 (0 degrees-180 degrees)
   Applied moment = 68×10^3 in-lbs
   Accounting for half symmetry of the model, the moment transmitted by 57 nodes is equal to 34×10^3 in-lbs(one half of the applied moment).Nodal forces are also halved at the symmetry boundary nodes.
   $M = 34 \times 10^3$ in-lbs. $= F_0(0.255 \ r_0 + 0.5 \ r_m + 0.245 \ r_i) \measuredangle \ Sin^2 \theta$
   Where:

θ = 0, 10, 20, ...180
r_0 = outer radius of the branch pipe
r_i = inner radius of the branch pipe
r_m = mean radius of the branch pipe
F_0 = nodal force vector
$= \dfrac{34 \times 10^3}{69.2} = 491.3$ lbs.

6.2 Outer nodal forces

F_i = 0.255 x 491.3 x Sin θ
(θ = 0 , 10, 20, ...180)
Substituting θ values in the expression above, the individual nodal forces can be obtained as follows:
Node location: Degrees
0 10 20 30 40 50 60 70 80 90 100 110 120 130 140 150
Nodal Forces: lbs
0 21.8 42.9 62.7 80.6 96.0 108.5 117.8 123.4 125.3 123.4 117.76 108.5
96.0 80.6 62.7
Node location (Continued)
160 170 180
Nodal Forces (Continued)
42.9 21.8 0

6.3 Mid nodal forces

F_i= (0.5 x 491.3 Sin θ)
θ = 0, 10, 20, ...180
Individual nodal forces can be calculated as demonstrated above.

6.4 Inner nodal forces

F_i = (0.245 x 491.3 x Sin θ)

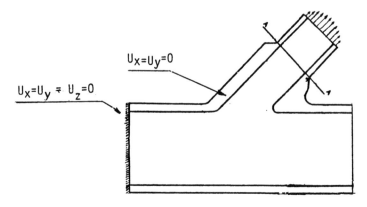

Figure 4 Schematics. Applied Load & B.C's

The nodal forces based on the expression above can be calculated in the same manner as demonstrated above.

A similar procedure is adopted to calculate the nodal forces for Model 2. The applied nominal stress at the loaded branch end is maintained at 1000 psi for this model also.

7 BOUNDARY CONDITIONS

In order to induce an out-of-plane moment loading on lateral half symmtry model, anti-symmetry boundary conditions are imposed on the symmetry plane.
  Symmetry plane costraints
  $U_x=0$, $U_y=0$
  $U_z=$ Unconstrained
  The left end of the model is completely constrained by restraining all three degrees of freedom ; i.e; $U_x = U_y = U_z =0$. This edge condition is different from the in-plane moment loading case for obvious reasons. The schematics of the nodal forces, constrained degrees of freedoms and the boundary conditions are presented in Figure 4.

8 ANALYSIS PROCEDURE

Revised input files were prepared from the data stored on magnetic tapes saved from the Phase 1 program to reflect the changes in the boundary and constraint conditions and the nodal forces from the earlier in-plane moment load evaluation.

   A data check run was made to verify the input. Upon complete verification of the input data, a solution run was made and the results of the analysis were saved on tape for post-processing. The computed results were checked for consistency and accuracy in the key regions.Several stress and displacement plots were also obtained for selected planes and cross sections to detect any anomalies. However, only integration point stresses extrapolated to the surface are used in computing the maximum stress index which for the out-of- plane moment occurs at the transverse plane, 90 degrees from the symmetry plane.

9 SUMMARY OF RESULTS

The stress indices based on the extrapolated Gaussian point values on the inside and outside surfaces of the selected locations are presented below in Table 1 and compared with the in-plane momentloading.

Table 1 Summary of stress indices

Location	Surface	Model	Branch Moment	
			In- Plane	Out-of-Plane
Acute Crotch	Inside	2	0.10	0.05
		4	0.13	0.07
Obtuse Crotch	Inside	2	0.02	0.01
		4	0.11	0.07
SectionA-A Acute	Outside	2	1.40	0.31
		4	1.33	0.28
SectionA-A Obtuse	Outside	2	1.40	0.30
		4	1.32	0.24
Entire Model	---	2	1.40	2.70
		4	1.32	2.50

10 PROPOSED ASME STRESS INDICES

For the design of nonradial nozzles in pressure vessels per NB-3338 of Section III of the ASME B&PV Code, a stress index value of 2.70 is recommended for out-of-plane moment loading applied on the branch end. This index is applicable only if the axis of the nozzle makes an angle less than or equal to 45 degrees with the normal to the vessel wall and if $d/D \leq 0.50$. The indices for internal pressure and in-plane moment loading are available from WRC-Bulletin 301. The dimensional limitations specified in the above bulletin shall apply here also.

For the design of lateral branch connections in piping per NB-3680 of the Code, the recommended C_2K_2 for the out-of-plane moment applied on the branch pipe is 2.70. This value is comparable to the minimum recommended value for the branch moment index in a radial branch connection. Obviously, a reduced value could be justified for radial branch connection based on the similar value calculated here for lateral branch connections. Indices for other loading conditions are available from WRC-Bulletin 301.

11 CONCLUSION

The results and recommendations presented here and in WRC-Bulletin 301

represent a significant contribution from the Pressure Vessel Research
Committee in promoting a safe and economical design of pressure vessel
and piping components. The proposed stress indices fill a major void left
in the Code since its inception. It is, however, recommended that the
industry follow the lead established by PVRC through these benchmark
studies and expand the database to develop comprehensive design guidelines for nonradial branch connections. The details of what the expan-
ded database should include are clearly spelled out in Section 17.0 of
WRC Bulletin 301.

12 ACKNOWLEDGMENT

The author wishes to express his appreciation to the Pressure Vessel
Research Committee for supporting this important investigation for almost a decade. The permission granted to publish this material is also
gratefully acknowledged.

13 REFERENCES

Raju,P.P. A parametric three dimensional finite element study of 45 deg-
    ree lateral connections. WRC Bulletin 301. Welding Research Council
    New York,N.Y.January, 1985.
DeSalvo,G.J & Swanson,J.A. ANSYS Engineering Analysis Systems. User's
    Manual. Swanson Analysis Systems Inc. Houston,PA. February,1982.
Ergatoudis,I; Irons,B.M & Zienkiewics,O.C. Curved, isoparametric, quad-
    rilateral elements for finite element analysis. International Journal
    of Solids and Structures,1968, Volume 4. pp. 31-42.

Stress analysis in piping connection thermal sleeves – Some design criteria

V.Zubizarreta Enríquez & J.Martínez Lao
Empresarios Agrupados, S.A., Madrid, Spain

1. INTRODUCTION

The protection of nozzles and connections against sudden changes in fluid temperature is normally performed by means of thermal sleeves. Through these elements, the component area subjected to higher mechanical stresses (structural discontinuity) is thermally shielded, which reduces the thermal stresses. So consequently, the thermal sleeve itself and its union to the component are the areas most affected by thermal shock.

In the case of cold water injections in the reactor primary circuit, the temperature drop may be very high (over 300ºC), producing extremely high thermal gradients and stresses, which particularly affect fatigue analysis. The situation arising from cold water injection on steam is especially serious, owing to the sudden condensation produced in the mixture, which could also spread to the chamber located between the sleeve and the piping. In these conditions, it may be a problem for the sleeve to support all the transients postulated for plant life if its design is not suitable.

2. TYPES OF PIPING CONNECTION THERMAL SLEEVES

The solutions adopted in thermal sleeve design are most varied. Just as in the pressure vessel nozzles there are various coaxial sleeves, in the case of piping connections there is usually only one sleeve in the injection piping or, in very critical cases, the piping area receiving the injection is shielded with another sleeve. A spray located at the end of the injection sleeve may be provided to improve homogeneization of the mixture. Figure 1 on the next page shows two basic types.

The main parameters that determine the necessity of having thermal sleeves in the connection, as well as choosing the type, are the magnitude and intensity of the thermal drop and the characteristics of the flow in the piping that receives the injection. From this second point of view one can talk about injection in piping with or without flow, as well as injection in liquid or steam phase. Injection in flow implies a permanent mixture regime which, depending on the respective flows, will determine the resulting temperatures in the mixture zone. As the mixture is not homogeneous, the fatigue effect resulting from the high number of cycles on the interior surface will have to be taken into account also, due to uncertain impact of cold drops.

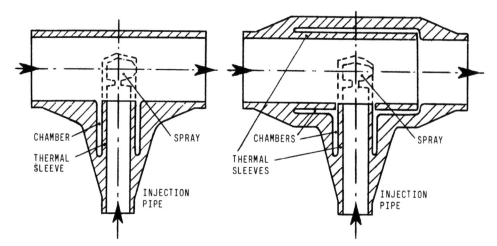

Figure 1. Thermal sleeve types: One thermal sleeve only in the injection piping and double thermal sleeve.

The most critical situation occurs in the injection in dammed steam. In this case the double sleeve type should be chosen. It is convenient to place the injection piping vertically so that it is initially full of condensate in balance with the steam. Thus, a lesser thermal conductance in the sleeve and chamber assembly is obtained.

3. ANALYSIS METHOD

By having the thermal sleeve in the connection thermal shock is avoided in the crotch region of the two pipes, which is where the highest local stresses caused by pressure and mechanical loads are found. Therefore, this region may be studied by means of normal piping analysis formulas, with the suitable stress indices.

The thermal problem, therefore, is transferred to the piping and sleeves, especially in its joint areas, which is where geometric discontinuity exists. Taking advantage of the axial symmetry presented by these portions, an axisymmetric finite element model may be considered for the analysis, which goes from a piping section sufficiently far from the thermal sleeve up to the crotch region. Figure 2 on the next page shows a model of these characteristics.

In the chamber located between the thermal sleeve and the protected piping there is thermal conductance. This is true when the injection is performed in liquid or, in the case of injection in steam, if there is condensate. In the event that there is steam in the chamber, the process may be most complex, as this steam on condensating will allow water to enter the chamber which, in turn, will vaporize as soon as the temperature permits it. In these conditions it is only feasible to perform very conservative hypotheses that envelop real behavior, such as assuming that cold water instantaneously enters the chamber.

The first step will be to perform the heat transfer analysis during the injection transient. Steady state temperature distribution corresponding to the hot fluid in the piping (practically constant temperature if there

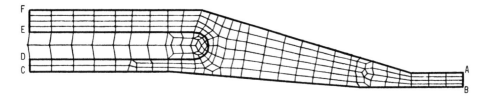

Figure 2. FEM model of thermal sleeve.

is external isolation) is considered as an initial condition. As environmental conditions we will have isolation in the end sections, AB and CDEF, where the heat axial flow is almost nil, forced convection on the internal surface of model BC and free convection or isolation on the external FA surface. As a result of this analysis temperature distributions for each instant of time are obtained.

The following is the stress analysis, both due to temperature and to pressure and mechanical loads transmitted by the piping. A program with capacity for both axisymmetric (temperature, pressure and normal force) and non-axisymmetric (shear forces and bending and twisting moments). The model is the same as in the thermal analysis, eliminating water elements from the chamber and considering fixed the EF section as a boundary condition.

Distribution of the stresses obtained are linearized in the most critical sections to obtain membrane and bending stresses. Logically, different time snapshots have to be taken into account during the transient to choose the most critical one in each section.

Finally, a post-process of all the stresses obtained must be performed both thermal and for pressure and load unit cases. The basic stages of this calculation phase will be: Factorization and load combination; stress categorization; stress range calculation; cyclic analysis (fatigue).

4. RESULTS AND DISCUSSION

Two cases of piping connections protected with double thermal sleeve pertaining to the primary system of a PWR nuclear power plant have been analyzed. Figure 3 shows the sizes (in mm.) and basic characteristics of the sleeve corresponding to the injection piping. Three variants of sleeve thickness have been taken into account for the first case, maintaining the sleeve and chamber thickness sum constant.

The FEM programs used have been DOT for temperature calculation and SAP IV and ASHSD2 for stress calculation. An in-house developed program, specifically adapted to the ASME III Class 1 code procedures, was used for the post-process. When developing this program an effort has been made to maintain a reasonable balance between overall simplifications made in procedures and the conservatism which, as a consequence thereof, is introduced in the analysis.

	Case 1		Case 2
L	195.0		267.0
d	68.9		90.0
D	98.9		122.0
t_s	1.a	7.0	5.0
	1.b	7.5	
	1.c	10.0	
t_p	25.0		8.5
Th	346ºC		188ºC
Tc	30ºC		43ºC
Δt	0.5 sec.		0.3 sec.

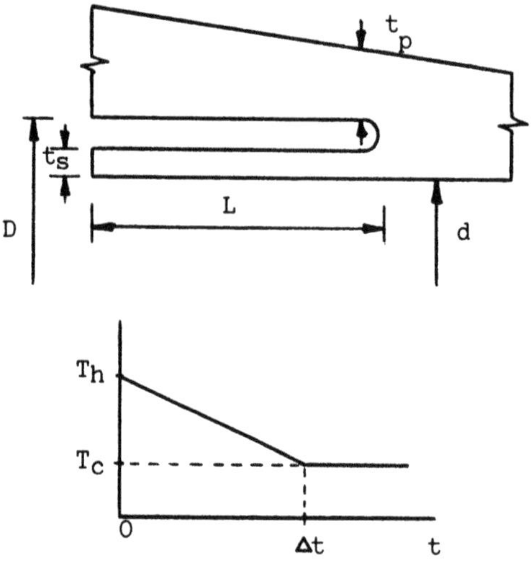

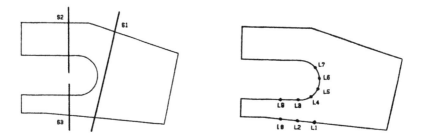

Figure 3. Basic characteristics and sizes of analyzed thermal sleeves.

Figure 4. Critical sections and locations.

Figure 4 shows the most critical three sections and nine locations from the point of view of stresses caused by the temperature. These stresses are considerably higher than those caused by pressure and mechanical loads, and it may be affirmed that these govern the design.

Evolution of the temperature and thermal stress intensiy during the transient corresponding to Case 2 is shown in Figure 5 for the elements comprising sections S1 and S2. The shielding effect of the thermal sleeve on section S2 is appreciable in the curves.

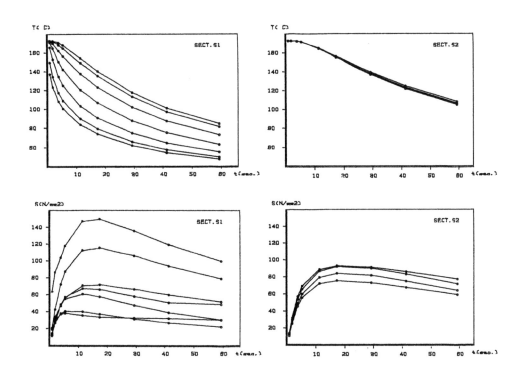

Figure 5. Temperatures and thermal stress intensities for case 2.

Table 1. Membrane and bending stress intensities due to temperature.

Stresses	Sections	Cases			
		1a	1b	1c	2
Membrane (N/mm^2)	S1	70.9	69.2	69.1	19.6
	S2	125.3	121.6	118.5	67.0
	S3	273.5	255.3	231.5	103.3
Membrane + Bending (N/mm^2)	S1	470.6	474.0	478.3	124.6
	S2	233.7	234.4	244.0	94.8
	S3	370.4	397.4	420.4	148.3

As a result of stress linearization in each section, the membrane and bending stresses are obtained and the peak stresses are filtered. Table 1 shows the values corresponding to the above mentioned critical sections. The variation in thickness of the thermal sleeve indicated in case 1 does not have an important effect on the inerized stresses. The values obtained in sections S1 and S3 for total stresses exceed the admissible limit (3 Sm). So, it has been necessary to perform a simplified elasto-

383

plastic analysis. Section S3 is critical from the thermal ratchet, but as there are no primary stresses the thermal stresses are not subject to limitation.

In the fatigue analysis not only the elements constituting the sections are taken into account, but also the elements that surround the bottom of the chamber. Table 2 shows the values of the alternating thermal stress intensities in the nine critical locations shown in Fig. 4.

Table 2. Alternating stress intensities due to temperature.

Stresses	Locations	Cases			
		1a	1b	1c	2
Membrane	L1	375.0	328.3	345.5	69.0
+ Bending	L2	281.9	274.1	298.5	89.0
+ Peak	L3	152.8	99.2	106.2	78.7
one half	L4	229.0	148.1	109.1	105.9
(N/mm^2)	L5	269.8	378.0	474.3	124.7
	L6	654.0	542.1	636.5	149.8
	L7	591.3	439.2	536.2	120.7
	L8	235.4	230.1	256.5	76.0
	L9	174.5	152.1	131.9	57.4

The most critical location in all cases is L6. For the variants of case 1 the lowest value is that of case 1b, that is, when the sleeve and chamber thicknesses are equal (7.5 mm). This behavior may be justified if we consider that the increase in sleeve thickness increases the shield effect but, on the other hand, it decreases the radius at the bottom of the chamber, which concentrates the stresses. Therefore, there is an optimum situation which coincides with case 1b.

As required in the simplified elasto-plastic procedure, the alternating stresses in all elements forming part of sections S1 and S3 were penal-- ized, as well as all locations in those surroundings. The fatigue analy- sis was performed with the resulting values, obtaining cumulative usage factors very close to the unit in all cases.

5. SUMMARY AND CONCLUSIONS

A statement of the problems that present piping connections subjected to sudden cooling has been made, emphasizing the importance of a suitable choise of the design to reduce thermal stresses. The analysis methodology has also been explained by means of finite element programs.

The application to two specific thermal sleeve cases has permitted dis- cussion of the results in the most critical sections and locations. On modifying the dimensions of one of them, the impact on thermal stresses has been verified, proving that it is significant only as far as fatigue is concerned.

The stress values obtained are very high, making the simplified elasto- plastic procedure necessary. Even after this, the results are very near the admissible limits. This means that in even more critical cases than those considered above, design validation would require a stress analysis based on plastic behavior.

Stress and fatigue analysis of fuelling machine housing of 500 MWe PHWR

B.K.Dutta, W.V.Ramana, H.S.Kushwaha & A.Kakodkar
Reactor Engineering Division, BARC, Trombay, Bombay, India

1 INTRODUCTION

One of the most appealing features of the Pressurised Heavy Water Reactors is the online refuelling capability. For this a fuelling machine is used. This machine opens a reactor channel by removing a seal plug and a shield plug and then does the necessary fuelling by pushing fuel bundles from a fuel magazine by rams. After necessary fuelling the machine closes the channel automatically. One of the most important parts of the fuelling machine is its pressure housing which becomes a part of the reactor channel during refuelling operation. It houses the fuel magazine, separators and rams. Beside channel pressure and other mechanical loads, the pressure housing experiences thermal transients during refuelling. The housing consists of two cylindrical shells having one end-closer in each. They are connected with each other by a large sized coupling. There are many holes on both the end-closers to accommodate ram movement, separators and magazine drive mechanisms. Some of these holes intersect with each other in the housing end-closers and hence end-closers are reinforced accordingly. This also makes the end-closers nonsymmetric.

In the following sections the various analysis done to compute general stress distribution, stress concentration factors near to various holes, temperature transients during refuelling and also allowable fatigue cycles for pressure housing of fuelling machine for the proposed 500 MWe are described.

2 STRESS ANALYSIS OF PRESSURE HOUSING FOR BOLT AND PRESSURE LOADS

The general stress distribution in pressure housing is picked up from an approximate axisymmetric model. For this a 2-D axisymmetric model of complete housing along with couplings is prepared. To input equivalent bolt load in this model, it is found necessary to make a 3-D model of one-fourth coupling along with springs at the inner surface to simulate

385

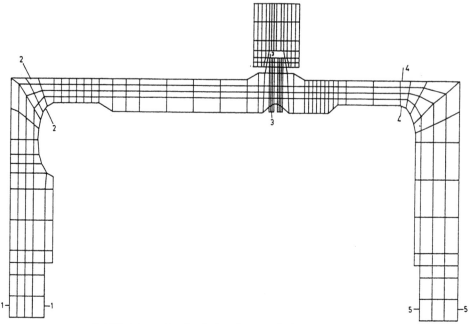

Fig1 2D Axisymmetric discretization of pressure housing

stiffness of hubs. To consider the effect of bolt loads in axisymmetric model, radial loads are applied on hubs which produces the same deflection as obtained in 3-D model due to bolt load alone. This method of analysis is explained in detail in paper F 7/5 of the present conference.

351 isoparametric 8 noded finite elements connected by 1241 nodes having 2 degrees of freedom per node are used in axisymmetric model (fig.1). Whereas in 3-D model, 1226 elements with 1807 nodes having 3 degree of freedom per node are used. These two models are first analysed for bolt loads and then for bolt with pressure loads. Magnitudes of different stresses, such as membrane, bending and peak, are calculated at different cross sections from the finite element results and compared with corresponding codal limits (Table 1).

3   STRESS ANALYSIS FOR COMPUTING STRESS CONCENTRATION FACTORS

As mentioned above, axisymmetric analysis of the pressure housing is done to compute general stress distributions. It is then necessary to modify stresses in the heads of both the shells by multiplying suitable stress concentration factors to take into consideration effect of holes. Though in the head of one shell holes are well-shaped and hence a theoretical calculations is justified, but same is not true for the other head in which three holes intersect with each other. These holes are to accommodate ram movement and separators. This shell which is popularly known as End-cover

TABLE-1   Stresses compared with allowable values at different
cross sections of pressure housing.

For Bolt Loads

Cross Section No.	Stress category	Max.stress intensity	Stress limit	Allowable value	Remarks
1	Pm	0.6×10^{-4}	Sm	20	Safe
	Pm+Pb	0.1×10^{-3}	1.5 sm	30	Safe
	Total	0.1×10^{-3}	-	-	Unclassified
2	Pl	$.4 \times 10^{-3}$	1.5 Sm	30	Safe
	Pl + Q	$.3 \times 10^{-2}$	3.0 Sm	60	Safe
	Total	$.6 \times 10^{-2}$	-	-	Unclassified
3	Pl	7.65	1.5 Sm	30	Safe
	Pl + Q	10.8	3.0 Sm	60	Safe
	Pl + Q+F	14.8	Sa	50	Safe
4	Pl	0.3×10^{-1}	1.5 Sm	30	Safe
	Pl + Q	0.13	3.0 Sm	60	Safe
	Total	0.163	-	-	Unclassified
5	Pm	$.3 \times 10^{-3}$	Sm	20	Safe
	Pm + Pb	$.2 \times 10^{-1}$	1.5 Sm	30	Safe
	Total	$.2 \times 10^{-1}$	-	-	Unclassified

For Pressure and Bolt Loads

Cross Sec.No.	Stress category	Max.Stress intensity	Stress limit	Allowable value	Remarks
1	Pm	0.8	Sm	20	Safe
	Pm+Pb	0.9	1.5 Sm	30	Safe
	Total	0.9	-	-	Unclassified
2	Pl	3.0	1.5 Sm	30	Safe
	Pl + Q	6.2	3.0 Sm	60	Safe
	Total	9.5	-	-	Unclassified
3	Pl	5.9	1.5 Sm	30	Safe
	Pl + Q	13.2	3.0 Sm	60	Safe
	Pl+Q+F	24.8	Sa	50	Safe
4	Pl	3.7	1.5 Sm	30	Safe
	Pl + Q	12.4	3.0 Sm	60	Safe
	Total	15.7	-	-	Unclassified
5	Pm	1.90	Sm	20	Safe
	Pm + Pb	5.66	1.5 Sm	30	Safe
	Total	5.66	-	-	Unclassified

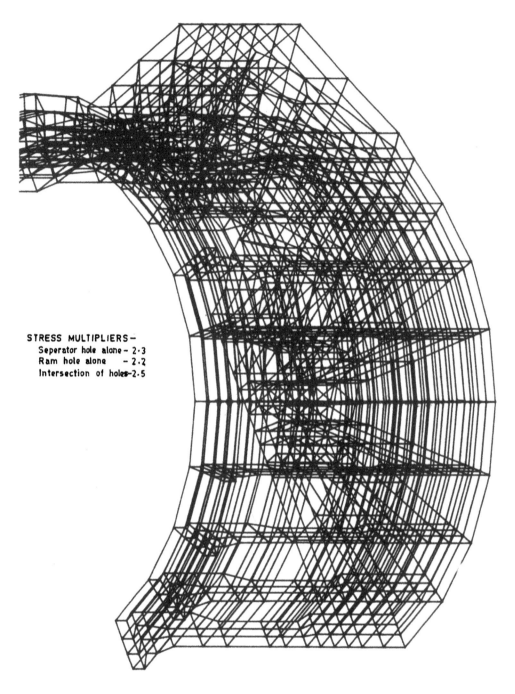

STRESS MULTIPLIERS—
Seperator hole alone – 2·3
Ram hole alone   – 2·2
Intersection of holes–2·5

Fig 2. 3-D Discretization of End-Cover of pressure housing

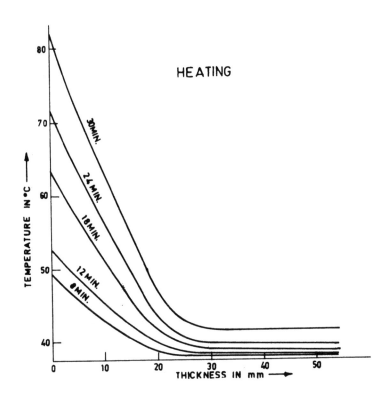

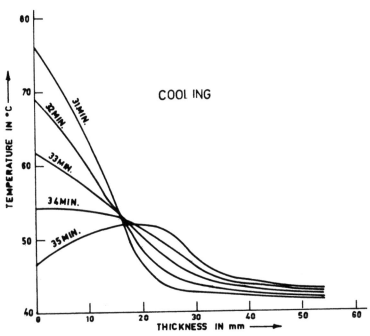

Fig 3. Temprature transient along the thickness for a
typical cross section of pressure housing.

TABLE 2. Usage factor correspond to different cyclic loading conditions.

Loading condition	Max.alternating stress intensity (Kg/cm^2)	Expected cycles	Allowable cycles	Usage factor
1 Bolting condition	861.0	50	1000000	.00005
2 Pressure Cycle of 0-122-0	1620.0	30000	100000	0.3000
3 Pressure Cycle of 0-175-0	2320.0	10	2000	0.0005
4 Thermal load	1530.0	200	180000	0.0012
		Total usage factor		0.30175

along with intersecting holes, is discretized by 3-D finite elements. There are 830, 8 noded brick elements connected by 1234 nodes, having three degrees of freedom per node, used in the analysis (fig.2). Same figure also shows different stress multipliers obtained for different holes.

4 TEMPERATURE TRANSIENTS ANALYSIS OF PRESSURE HOUSING

A possible cyclic thermal load is identified during refuelling operation. To obtain maximum alternating stress intensity due to this thermal cycle, a transient thermal analysis is done of the axisymmetric model. Computer code 'WELTEM' developed by Dutta (1981), is used for this purpose. The water temperature in the housing is assumed to be heated to 100°C in 30 min. from atmospheric temperature and then is cooled down to atmospheric temperature in 5 min time. A convective heat transfer coefficient of 0.000525 Kcal/hr-mm^2-°C is used assuming peripheral water velocity of 0.0701 m/sec due to 2 rpm rotation of magazine. Fig.3 shows temperature transients during heating and cooling at a typical cross section of the model. After computing stresses correspond to these temperature transients alternating stress intensities are calculated as per NB3216.2 of ASME Code.

5 COMPUTATION OF USAGE FACTOR

Various fatigue loadings identified (Sanat, 1985) for pressure housing are pressure cycle, testing cycle, mechanical assembly and disassembly cycles and thermal transient during refuelling. Different maximum stress intensities correspond to these cycles are calculated and their allowable cycles are compared with the expected cycles (Table 2). A total of 0.30175 usage factor is obtained for the pressure housing.

REFERENCES

Dutta, B.K.,Kushwaha,H.S. & Kakodkar,A. 1981. Computer program WELTEM (Analysis of two dimensional heat transfer problems by finite element technique)Theory and User's Manual. BARC/I-671
Sanat, A.1985. Letter No.005/35221/85/B/1135.

A simplified analysis of flange under thermal transient

A.B.Mukherjee, T.Narayanan & V.K.Mehra
Reactor Projects Division, Bhabha Atomic Research Centre, Trombay, Bombay, India

1 INTRODUCTION

Thermal transients such as startup, shutdown can cause temperature difference (thermal lag) between bolt and flange. The temperature difference results in differential thermal expansion of bolts and flange. This may lead to excessive overloading of bolts (or loss of pretension in the bolts) and excessive rotation of the flange. Discretely locoated bolts alongwith bolt holes in the flange call for 3 dimensional thermal analysis to get the temperature field in and around the bolt. Three dimensional thermal analysis not only becomes complicated and time consuming but structural analysis for the thermal lag become still more difficult. A simplified approach can replace the 3 dimensional thermal analysis modelling to a 2-dimensional axisymmetric simulation.

In this paper a method for 2-dimensional axisymmetric thermal analysis is prepsented. Effect of important parameters like contact resistance (between threads or washers), air conductivity, convective heat transfer on external surfaces etc. on the thermal lag is analysed by using the simplified 2-dimensional approach. Effects of these parameters on bolt stresses is also brought out.

2 AXISYMMETRIC SIMULATION

Figure-3 shows a plane thermal problem which represents a segment of a flange alongwith bolt and air gap surrounding it. Figure-1 shows the axisymmetric simulation of the same problem. The same problem has been solved earlier by 3-dimensional analysis (SPAAS, 1977). It may be noted that both the plane and axisymmetric problems have same boundary conditions as that of 3-dimensional problem. As a first step axisymmetric problem was restricted to radial heat flow only with a view to comparing results of axisymmetric problem with those of plane problem. Following dimensionless parameter is maintained same in both the problems for each of element like air gap, bolt etc.

$$\frac{\rho V C_p}{KA/\delta}$$

(where ρ is density, V is volume, is specific heat, K is thermal conductivity, A is area of heat transfer, δ is thickness of heat conducting layer)

For example, the V/A ratio of the simulated bolt (ring element) becomes twice to that of the original bolts, therefor density(ρ) of ring element is halved to keep the dimensionless parameter unchanged. Similarly axisymmetric modelling results in more heat flow through air gap, washer etc. due to their increased circumference therefore conductivity(k) of these materials are modified by the same geometrical factor (ratio betweend the actual air length to simulated air length. The correction is also done on density (ρ) to keep above dimensionless parameter unchanged.

Figure-4 confers results of plane problem and axisymmetric problem. Thermal lag on bolts is approximately same as calculated by the two methods, which shows the correctness of axisymmetric simulation.

3 PARAMETRIC STUDY

Refer to Fig.1 Axisymmetric simulation described in the previous paragraph is further modified to account for contact resistance between stud and flange, between stud and nut, between washers, between washers and flange and between washer and nut. Thermal resistance at each of these contact layers is based on results given in (G.E.C.). Figure-1 also shows the thermal transient used for the study.

Finite element program (Polirka, 1976) was used for heat transfer problem. Fig.5 shows the thermal lag between bolt and flange calculated by axisymmetric method described here. The same figure also shows the experimental as well as 3-D FEM results described in reference(SPAAS). Results of axisymmetric analysis fall between the experimental results and 3-dimensional FEM results. Difference in the results of the two analysis could be partly due to uncertainty in values of different contact resistance.

Influence of the following parameters are considered important for thermal lag.

1. Lower end of stud makes contact with flange.
This provides an additional heat flow path to stud and results in decrease in thermal lag by 9°C. Contact resistance between stud and flange was taken same as that of washer.

2. Contact resistance between threads.
If the thread contact resistance is made double then thermal lag increases both at bottom and top as shown in Fig.8.

3. Contact resistance between washer and nut.
If the value of contact resistance between washer and nut is doubled, thermal lag near top of stud increases by 5°C as shown in Fig.7.

4.Air conductivity.
If the air gap is reduced to half or air conductivity is increased to double, thermal lag decreases by 8°C.

5. Heat transfer on external surface.
Fig.8 shows the effect of convective heat transfer on thermal lag for a nominal value of H=20 x 10^{-6} kcal/mm^2 hr°C. This indicates that better insulation on flange can reduce the thermal lag.

Bolt stresses are shown in table as calculated by 2-dimensional axisymmetric analysis.

4. CONCLUSIONS

If conductivity at lower ends of stud is increased or heat transfer at external surface is reduced thermal lag reduces.

It is shown in this paper that a 2-dimensional axisymmetric analysis gives good results for thermal lag problem in the flange and it may not be necessary to carry out a 3-dimensional analysis. The axisymmetric thermal analaysis results can directly be used for 2-dimensional structural analysis of flange.

REFERENCES

SPAAS, HACM March 1977 - Some contribution to the structural design and analysis of pressure vessel and piping components. Department of Mech. Engg. Delft University of Technology, The Netherlands.

Heat transfer fluid flow data book: General Electric Company, N.Y.

Polirka, R.M. and Wilson, E.L. June 1976, DOT/DETECT FEM program Department of Civil Engg. University of California, Berkeley.

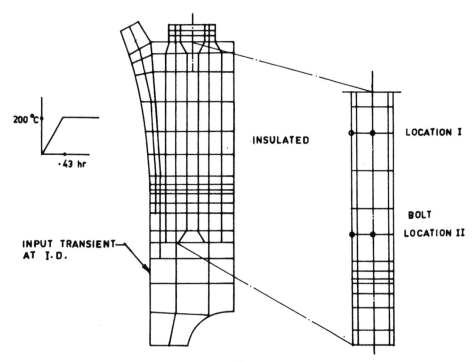

FIG. 1. F. E. MESH FOR AXI-SYMMETRIC MODEL

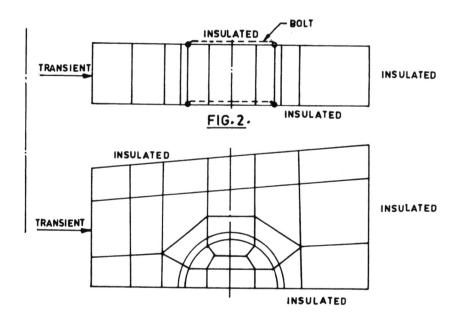

FIG. 2.

FIG. 3. F. E. MESH FOR PLANE MODEL

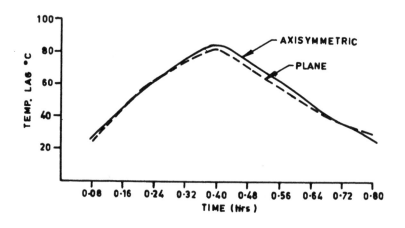

FIG. 4. THERMAL LAG FOR PLANE AND AXISYMMETRIC

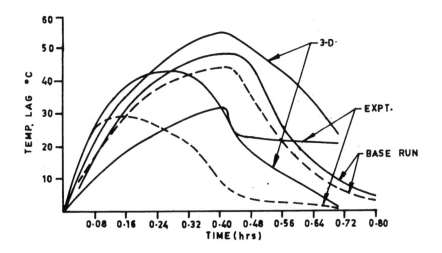

FIG. 5. BASE RUN, 3-D AND EXPERIMENTAL

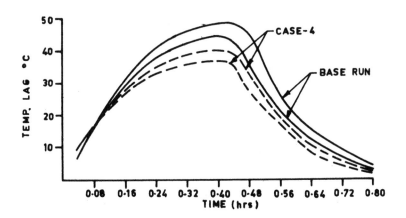

FIG. 6. THERMAL LAG FOR CASE-4

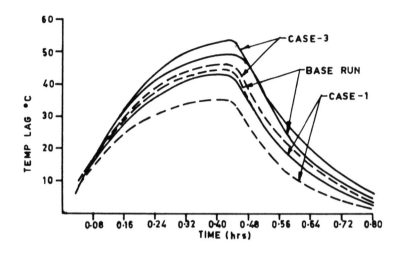

FIG. 7. THERMAL LAG FOR CASE 1 & 3

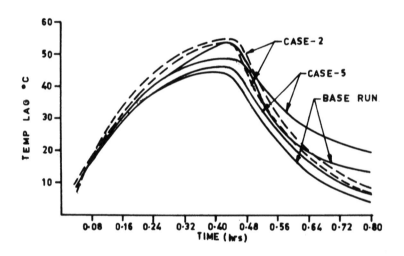

FIG. 8. THERMAL LAG FOR CASE 2 & 5

TABLE— BOLT STRESSES FOR THERMAL LAG

	σ_m	σ_b
BASE RUN	3·33	0·33
CASE-1	2·91	0·33
CASE-2	3·91	0·36
CASE-4	2·91	0·32

Stress and stability considerations for special multilayer type reactor vessel for advanced LWR

B.W.Wedellsborg

Containment Technology Consultants, Pleasant Hill, Calif., USA

1 INTRODUCTION

This paper presents a brief outline of the conceptual stage of the design of a special type multilayer reactor/pressure vessel (MLRPV) for advanced LWR's.

Basic design formulations are given, and the results of a preliminary stress evaluation of a model MLRPV are mentioned.

2 HISTORY

Construction of laminated or multilayer pressure vessels for non-nuclear applications is a well known, and well documented technology in Japan and in Europe.

The concept of using multilayer construction for nuclear reactor pressure vessels was considered in the U.S.A. and the U.K. in the early 1950's, when small scale test multilayer vessels were built.

These early multilayer vessels, however, were constructed by tightfitting or preshrunk shells, or by a single shell reinforced by circular or helical wrap-around posttensioned steelbands.

The multilayer reactor vessel (MLRPV) described in this paper differ from these early concepts in that the individual shells are sized to provide interspaces for the infilling of special metallic or other type filler materials.

Existing PWR's and BWR's have reactor vessels constructed of single shells or forgings. The stress levels reached locally in the walls, and nozzleregions during actual or postulated thermal and pressure transients may be high enough to pose special problems due to thermal-, and strain-ageing, low cycle fatigue, the cummulative effects of long gamma and neutron irradiation time histories, etc., as these reactors come of age.

The possibility of initiation, or propagation of existing crack in the nozzle or other regions of the singleshell vessel, which may lead to a LOCA, was one of the important safety reasons for contemplating the possibility of multilayer design for the primary reactor pressure vessel and portion of the primary coolant loop.

Other reasons were the eventual need for reactor vessels which can accommodate larger cores and/or carry larger pressure, or tolerate steeper thermal transients.

3 DESCRIPTION (SEE FIGURE 1)

The design of a Multilayer Reactor Pressure Vessel (MLRPV) involve the fabrication of two or more relatively thinwalled concentric steelshells, which are nested, one within the other, leaving narrow annular interspaces, which upon assembly of the shells, welding, weld stressrelieving and radiographing or other weld examination, are filled with special type materials.

Only the vessel head which contain most of the penetrations is built as a single shell.

The primary coolant nozzles and portions of the primary coolant loops may also be built multilayered.

The fillers are intended to function primarily as pressure transfer media. The filling operation may be preceded by differential heating of the component shells to obtain specified temperatures in order to create a prestressing effect upon the closure of the interspaces.

A special stiffener sleeve or interior concentric shell may be provided. This sleeve which serve as a combined stiffner and radiation and thermal shield is not welded to the MLRPV and may be removed or replaced if necessary.

The sleeve is built by special concrete or other material, encased in a 'hot liner.'

4 STRESS ANALYSES

Formulations used to perform the conceptual design of the MLRPV are given for the core midheight wall location.

4.1 Vessel Wall at Core Midheight (See Figure 2)

Assuming thin-shell axisymmetric geometry and linear elastic relations for each shell layer and disregarding the viscosity of the 'flowable' filler materials, the stress distribution in the wall of the MLRPV at a given time interval when the internal vessel pressure is (P_0) may be obtained from linear relations by first calculating the intershell filler pressure (P_j) from:

$$\bar{p} = A^{-1} B (\bar{T} - \bar{T}_0), \qquad\qquad \text{Eq. (1)}$$

Where ($\bar{p}$) is the intershell filler pressure vector, and $\bar{T}$ is the vector of average shell temperatures at the given time interval for the loadcase considered, and $\bar{T}_0$ is the vector of average shell temperatures at time of prestressing.

The hoop-, and vertical shell stress are obtained, subsequently from the following linear relations:

$$\bar{\sigma}_\theta = S_\theta \bar{p} + Q \ \bar{T}, \qquad\qquad \text{Eq. (2)}$$
$$\bar{\sigma}_z = S_z \bar{p} + Q_z \bar{T}, \qquad\qquad \text{Eq. (3)}$$

The v. Mises stress vector ($\bar{\sigma}_e$) is determined finally, from:

$$\bar{\sigma}_e = C_1 \ \bar{\sigma}_\theta = C_2 \ \bar{\sigma}_z, \qquad\qquad \text{Eq. (4)}$$

For the academic case of long cylindrical shells and zero vertical temperature gradients for each shell, then the non-zero coefficients of the matrices A, B, S, Q, and C are:

$$a_{jj-1} = -(R_{j-1})^2 \ (1 - \nu_{j-1}/2)/(H_{j-1}) \ (E_{j-1});$$
$$a_{jj+1} = -(R_j)^2 \ (1 - \nu_j/2)/(H_j) \ (E_j);$$

Assuming fillers are in the molten phase, denoting

$k_j \quad = (5-4\gamma_j)\ (R_j{}^o)^2(R_j{}^i)^2/(E_j)\left[(R_j{}^o)^2-(R_j{}^i)^2\right];$

$l_j \quad = \left[(R_j{}^o)^2+(R_j{}^i)^2\right];\ m_j = 6(\gamma_j)/(E_j);$

$a_{jj-1} = -2(k_{j-1});\ a_{jj+1} = -2(k_j);$

$a_{jj} \quad = (k_{j-1})(l_{j-1})/(R_j{}^i{}_{-1})^2+(k_j)(l_j)/(R_j{}^o)^2$
$\qquad\quad +(m_{j-1})(R_j{}^o{}_{-1})^2+(m_j)(R_j{}^i)^2;$

$b_{jj-1} = 6(\alpha_{j-1})(R_j{}^o{}_{-1})^2;\ b_{jj} = -6(\alpha_j)(R_j{}^i)^2;$

$b_j{}^F \quad = 6(\beta_j)\left[(R_j{}^i)^2-(R_j{}^o{}_{-1})^2\right];\ R_j{}^i = R_j-1/2\ H_j;\ R_j{}^o = R_j+1/2\ H_j;$

$S\theta_{jj} \quad = (R_j)/(H_j);\ S_{zjj} = (R_j)/2(H_j)$

$S^\theta{}_{jj+1} = -S^\theta{}_{jj};\ S_{zjj+1} = -S_{zjj};$

$q\theta_{jj-1} = (E_j)\ (\alpha_j)\ (H_j)/2(1-\nu_j)\ (R_{j+1}-R_{j-1});$

$q^\theta{}_{jj+1} = -q^\theta{}_{jj-1};\ q_{zjj-1} = q^\theta{}_{jj-1};\ q_{zjj+1} = q^\theta{}_{jj+1};$

$C_{1jj} \quad = \left[1+\zeta^2{}_j-\zeta_j\right]^{1/2};\ C_{2jj} = \zeta_j{}^{-1}\ C_{1jj}$

where ζ_j = the ratio of vertical to hoop stress in (j) shell α_j, β_j, and E_j, E_{Fj} are thermal expansion coefficients and moduli of elasticity respectively, for the (j) layer shell and filler.

4.2 Vessel Wall at Primary Coolant Nozzle Junction

The stress distribution which exist at a given time interval for a given loadcase may be estimated by 2 or 3-D finite element analysis, (Salim 1985) or approximately by using spherical shell geometry simulation for the vessel region around or close to the nozzle, deriving the thin shell vessel-nozzle discontinuity relations (Timoshenko 1959a) and then using procedure similar to the one described above.

4.3 Vessel Wall at Bolting Flange Junction

The stress distribution in this region may be estimated again from the similar procedure described earlier.

5 OPTIMIZATION

In order to reduce the operational and/or transient stress levels at critical highstress regions of the MLRPV for the controlling loadcase(s) an optimized design may be attempted by any of or a combination of the following two methods:

(a) Prestressing by preheating the shell layers to specified temperatures.

(b) Prestressing by pre-pressurization of the MLRPV prior to services, to specified pressures. Using the first method (a) and for example specifying an optimal v. Mises stress distribution at a critical location in the MLRPV wall for the controlling loadcase, then the prestressing temperature vector $\overline{T}_0$ may be estimated from relations (1) (2) (3) (4) so that a built-in prestressing effect leading to the desired optimal stress distribution is obtained.

The (linear) relation is:

$$\overline{T}_0 = U\ \overline{\dot{\sigma}}_e + V\overline{T}; \qquad \qquad \text{Eq (5)}$$

where the temperature vector $(\overline{T})$ for the loadcase considered is known, the V. Mises vector $(\overline{\sigma}e)$ is specified, and the matrices U and V are:

$U = C^{-1}\ B^{-1}\ AS_\theta{}^{-1};$
$V = I + B^{-1}\ AS_\theta{}^{-1}\ Q_\theta;$

where I is the unitary matrix. Similar procedure involving method (b) pre-pressurization of the MLRPV is derivable from eq's (1) (2) (3) and (4).

This method may result in an MLRPV with the outer layers pre-tensioned, and the inner layers pre-compressed.

6 STABILITY

The stability of the inner shell(s) is a consideration which is unique for this type of vessel.

A LOCA combined with maximum temperature levels in the inner shells of the MLRPV may cause at least one of the inner fillers to be in a molten-, or near molten state, where a hydrostatic pressure may exist or develop in addition to the regular uniform interlayer pressure acting on the inner shell and/or special stiffener sleeve.

The factor of safety (F.S.) against buckling for this loadcase must be specified and checked against the theoretical value, which may be estimated from:

$$F.S. = \frac{P_{icrit}}{P_i}; \qquad\qquad Eq. \ (6)$$

where (P_i) is the inner filler pressure including (average or maximum) hydrostatic, for the loadcase considered, and (P_{icrit}) is the elasto-plastic ovalling mode buckling pressure for the special stiffener sleeve, based on a non-perfect cylindrical shell geometry.

$$P_{icrit} = \frac{0.8 \ P_{ei}}{1+P_{ei} \ R_i \Delta R_i/M_{pi}}; \qquad\qquad Eq. \ (7)$$

(P_{ei}) is the buckling pressure for a perfect cylindrical shell (Timoshenko 1936b), (M_{pi}) is the plastic bending capacity, and (ΔR_i) the radial tolerance of the stiffener sleeve.

A similar stability evaluation must be performed for the construction loadcase.

7 RESULTS OF STRESS EVALUATION OF MODEL MLRPV

A model MLRPV was evaluated using the described procedure, and compared to a conventional single wall RPV of same size, dimensions, and shell materials.

The evaluation indicated substantial reduction in stresses caused by the postulated PTS-transient, and some reduction for the postulated LOCA-transient for the MLRPV as compared to the single shell RPV.

8. CONCLUSIONS

Some of the following advantages of the MLRPV design concept may be mentioned:

Ductile failure modes, crack propagation arrested at layer boundaries, use of small size offset welding, reduced need for weld stress relieving, less stringent tolerance requirements. Enhanced radiation-, and thermal shock protection for the loadcarrying outer shell layers. Large strain rates as may be caused by PTS, and steep blowdown transients are limited to the inside shell. Improved in-service leak monitoring, and higher vessel wall thermal conductivity than for conventional shrink-fit multilayer vessels.

These must be balanced, however, with the obvious disadvantages of more complicated nozzle-, and penetration design and the need for combined improved post construction 2-or 3D radiographic, and ultrasonic NDE of the MLRPV circumferential welds, and in general by the need for special, more costly construction.

REFERENCES

Salim, A., 1985, Two Dimensional Finite Element Evaluation of Stress Concentration Factors F18/12 Bruxelles: Transactions of 8th International Conference on SMiRT.
Timoshenko, S, 1959a, Theory of Plates and Shells, New York, McGraw-Hill.
Timoshenko, S., 1936b, Theory of Elastic Stability, New York, McGraw-Hill.

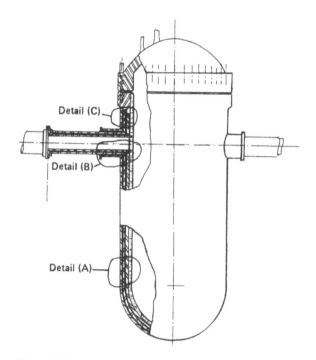

Figure 1 Multi-Layer Reactor Pressure Vessel (MLRPV)

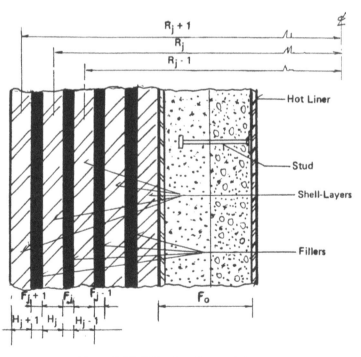

Figure 2 Wall Section, Detail (A)

404

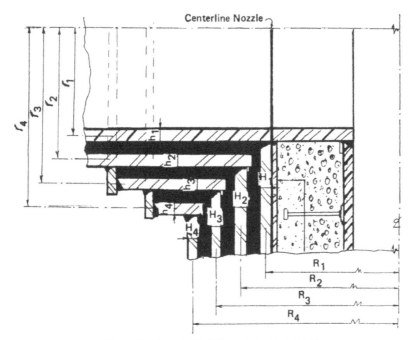

Figure 3 Nozzle-Wall Junction, Detail (B)

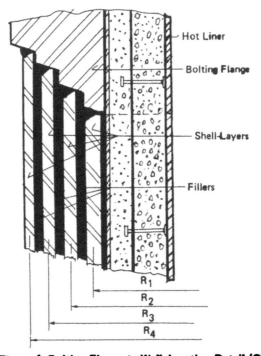

Hot Liner

Bolting Flange

Shell-Layers

Fillers

R_1

R_2

R_3

R_4

Figure 4 Bolting Flange to Wall Junction Detail (C)

405

Authors' index

For Product Safety Concerns and Information please contact our EU
representative GPSR@taylorandfrancis.com
Taylor & Francis Verlag GmbH, Kaufingerstraße 24, 80331 München, Germany